Lecture Notes in Electrical Engineering

Volume 1464

The book series *Lecture Notes in Electrical Engineering* (LNEE) publishes the latest developments in Electrical Engineering—quickly, informally and in high quality. While original research reported in proceedings and monographs has traditionally formed the core of LNEE, we also encourage authors to submit books devoted to supporting student education and professional training in the various fields and applications areas of electrical engineering. The series cover classical and emerging topics concerning:

- Communication Engineering, Information Theory and Networks
- Electronics Engineering and Microelectronics
- Signal, Image and Speech Processing
- Wireless and Mobile Communication
- Circuits and Systems
- Energy Systems, Power Electronics and Electrical Machines
- Electro-optical Engineering
- Instrumentation Engineering
- Avionics Engineering
- Control Systems
- Internet-of-Things and Cybersecurity
- Biomedical Devices, MEMS and NEMS

For general information about this book series, comments or suggestions, please contact leontina.dicecco@springer.com.

To submit a proposal or request further information, please contact the Publishing Editor in your country:

China

Jasmine Dou, Editor (jasmine.dou@springer.com)

India, Japan, Rest of Asia

Swati Meherishi, Editorial Director (Swati.Meherishi@springer.com)

Southeast Asia, Australia, New Zealand

Ramesh Nath Premnath, Editor (ramesh.premnath@springernature.com)

USA, Canada

Michael Luby, Senior Editor (michael.luby@springer.com)

All other Countries

Leontina Di Cecco, Senior Editor (leontina.dicecco@springer.com)

**** This series is indexed by EI Compendex and Scopus databases. ****

Arun Prakash Agrawal · Hari Mohan Pandey ·
Om Pal
Editors

Advances of Artificial Intelligence and Machine Learning in Image Processing, Security and Data Analytics

Select Proceedings of ICAAAIML 2023, Volume 2

Springer

Editors
Arun Prakash Agrawal
School of Computer Science Engineering
and Technology
Bennett University
Greater Noida, Uttar Pradesh, India

Hari Mohan Pandey
Bournemouth University
Poole, UK

Om Pal
Department of Computer Science
University of Delhi
North Campus
New Delhi, Delhi, India

ISSN 1876-1100 ISSN 1876-1119 (electronic)
Lecture Notes in Electrical Engineering
ISBN 978-981-95-0147-2 ISBN 978-981-95-0148-9 (eBook)
https://doi.org/10.1007/978-981-95-0148-9

This Springer imprint is published by the registered company Springer Nature Singapore Pte Ltd.
The registered company address is: 152 Beach Road, #21-01/04 Gateway East, Singapore 189721, Singapore

Preface

In an age defined by the exponential growth of data and the relentless pursuit of insight, the fusion of Artificial Intelligence and Machine Learning (AIML) with Image Processing, Security, and Data Analytics emerges as a beacon of innovation and opportunity. Volume 2 of this compendium embarks on a captivating exploration of this dynamic convergence, unraveling the transformative potential of AIML across diverse domains.

Within these pages, readers will embark on a captivating journey through the realms of image processing, where algorithms imbued with intelligence breathe life into pixels, unraveling the secrets hidden within visual data. From enhancing image quality to extracting meaningful features, the applications of AIML in image processing are as varied as they are profound.

Beyond the confines of pixels and colors, this volume delves into the critical domain of security, where AIML stands as a stalwart guardian against digital threats and vulnerabilities. Through advanced anomaly detection, behavioral analysis, and threat prediction, AIML fortifies the cyber landscape, safeguarding our digital infrastructure with unparalleled efficacy.

Furthermore, this volume illuminates the transformative power of AIML in the realm of data analytics, where algorithms sift through vast troves of information, distilling insights and patterns with unparalleled precision. From predictive modeling to prescriptive analytics, AIML empowers organizations to extract actionable intelligence from the data deluge, driving informed decision-making and unlocking untapped potential.

As we traverse the intricate landscapes of image processing, security, and data analytics, this volume serves as a beacon of knowledge and inspiration for scholars, practitioners, and enthusiasts alike. Each chapter is a testament to the boundless ingenuity of human endeavor, showcasing the myriad ways in which AIML is reshaping our world and catalyzing transformative change.

In closing, I extend my heartfelt gratitude to the contributors whose expertise and dedication have brought this volume to fruition. May their insights ignite new pathways of inquiry, inspire innovation, and foster collaboration in the ever-evolving realm of AIML.

Together, let us embark on this exhilarating odyssey through the realms of Image Processing, Security, and Data Analytics, as we chart a course toward a future illuminated by the transformative power of AIML.

Greater Noida, India — Arun Prakash Agrawal
Poole, UK — Hari Mohan Pandey
New Delhi, India — Om Pal

Contents

About the Editors

Arun Prakash Agrawal is a Professor at Bennett University, Greater Noida, with a Ph.D. in Computer Science and Engineering. He has over two decades of experience in teaching, research, and academic leadership. His research interests include software engineering, software testing, nature-inspired optimization, and machine learning. He has published extensively, mentored Ph.D. scholars, and received multiple awards, including Best Paper and Best Faculty honors. Dr. Agrawal has also taught internationally and organized several faculty development programs.

Hari Mohan Pandey is an Associate Professor at Bournemouth University, UK, specializing in artificial intelligence, soft computing, and natural language processing. He has authored several books and published over 100 research papers. Dr. Pandey serves on editorial boards of reputed journals and reviews for top international conferences. He has received multiple accolades, including the Global Award for Best Computer Science Faculty and recognition from the Prime Minister of India.

Om Pal is an Associate Professor at the University of Delhi and a former Scientist at MeitY, Government of India. With over 20 years of experience, his expertise spans cryptography, cybersecurity, quantum computing, and digital forensics. He holds multiple advanced degrees, two Indian patents, and has authored books with Elsevier and CRC Press. Dr. Pal has led national research initiatives and is a certified ethical hacker and AICTE-C-DAC Master Trainer in high-performance computing.

A Comparative Study of K-Means Clustering-Based Image Segmentation Methods

Simon Tongbram, Benjamin A Shimray, and Loitongbam Surajkumar Singh

Abstract In recent times, digital image processing has witnessed the emergence of several novel image segmentation methods. Image segmentation is an important part of image processing which involves the critical task of partitioning a digital image into discrete segments. This process plays a fundamental role in image analysis, facilitating essential functions such as feature extraction and image interpretation. Furthermore, it is widely used in medical science, like tissue classification, tumor identification and sizing, surgical planning, and more. Numerous image segmentation methods are available, and the k-means clustering approach is a commonly employed technique for this purpose. This research paper extensively reviews a variety of existing image segmentation techniques based on k-means clustering. However, it is essential to acknowledge that the k-means clustering approach does come with inherent limitations due to its dependence on priori values. These constraints have spurred the development of several modified models to mitigate these drawbacks. In this survey paper, our focus is on highlighting specific k-means clustering-based image segmentation techniques that have garnered recognition in the field. The paper also compares the segmentation techniques by considering evaluation metrics like MSE, Jaccard index, PSNR, and Correlation index to provide a deeper understanding of their segmentation capability and accuracy.

Keywords Image segmentation · Unsupervised learning · Clustering techniques · K-means · Hybrid segmentation

S. Tongbram (✉) · L. S. Singh
Department of Electronics and Communication Engineering, National Institute of Technology Manipur, Imphal, India
e-mail: simontongbram@gmail.com; simontongbram@nitmanipur.ac.in

L. S. Singh
e-mail: surajloi@nitmanipur.ac.in

B. A Shimray
Department of Electrical Engineering, National Institute of Technology Manipur, Imphal, India
e-mail: benjaminshimray@nitmanipur.ac.in

A. P. Agrawal et al. (eds.), *Advances of Artificial Intelligence and Machine Learning in Image Processing, Security and Data Analytics*, Lecture Notes in Electrical Engineering 1464, https://doi.org/10.1007/978-981-95-0148-9_1

1 Introduction

Segmentation is a process utilized in the partitioning of images into distinct regions, each sharing common characteristics like color, intensity, or texture. This facilitates a comprehensive understanding of the image for various analytical purposes [1]. The core objective of image segmentation is to divide an image into discrete segments or regions which are designed for specific applications. Over time, various methods and algorithms have emerged for image segmentation process [2]. Its applications consists of image enhancement, medical applications like tumor localization, tissue volume measurement, diagnosis, and computer-assisted surgery, among others. Moreover, image segmentation finds utility in advanced domains such as biometrics including fingerprint and facial recognition. Various techniques for image segmentation can be categorized into distinct classes, including edge-based, region-based, clustering-based, threshold-based, watershed-based segmentation, and artificial neural network-based segmentation. Each of these techniques addresses specific challenges and offers unique advantages (Fig. 1).

The segmentation process, i.e., edge-based, is considered as the most basic image segmentation process out of all the other processes. Detection of pixels between different segments of the image is performed with the edge-based segmentation process. The condition for different segments may be a rapid transition of intensity and, as a result, extraction of the pixels is performed and they are combined to form a closed boundary.

Segmentation based on Edge detection [3] is done on the assumption that the magnitude of the pixel connecting the foreground and the background is very specific. A number of different methods are introduced for the edge detection process and the most commonly used methods histogram based techniques. Gradient-based segmentation is mainly performed by varying the intensity of the two regions. The process

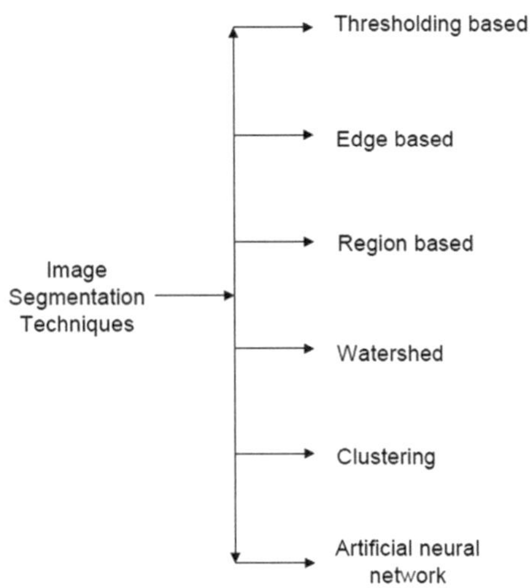

Fig. 1 Different image segmentation techniques

of histogram separation, depending on the grayscale, is performed by isolating the foreground from the background with the help of a selected value known as the threshold. Gradient-based segmentation is performed by changing the intensity of the two regions.

In the case of a threshold-based segmentation process, the image is divided into two or more image segments after it is compared to the threshold value. The primary method for isolating the foreground and background in the image segmentation process is the threshold method. Threshold methods, namely local, global, and adaptive are mainly implemented in the image segmentation process, depending on factors such as grayscale level and pixel size of the image. The global threshold method is used if the threshold operation depends on the gray level. If the threshold operation depends on the local properties and the gray level, the local threshold method is performed. The adaptive threshold method is achieved concerning gray intensity, adjacent pixels, as well as the coordinates of the image's pixel [4]. The disadvantage of this method is difficulty in finding a relevant threshold value for the separation of the image into segments and besides this, it also requires contrasting intensity values of an image's foreground and background.

The third category of image segmentation is region-based segmentation [5]. The algorithms used in this particular method include growing, splitting, and merging the regions. The region growing is defined as the process which includes extracting the connected regions consisting of a collection of pixels based on identical scales. For this process, a point known as the seed point is introduced and the points associated with the seed point, which are of the same intensity, are chosen and combine to form the growing regions. This procedure is performed until no pixel can be associated with the region. In contrast to this, in the case of the splitting and merging of the regions, the seed point is not selected; instead, the image is separated into independent regions and they are merged based on some specific condition. This method depends on the intensity along with its background and also produces a rough edge of the object which is being extracted.

The fourth category is watershed-based segmentation [6], which is performed based on the topological interpretation. So, to get the desired object from the image, the watershed transform algorithm is performed on the gradient image and during this, the image's body is similar to a catchment basin and the edge is similar to that of a watershed. In this particular method, the image's gradient is treated as the topographic surface and the pixels which have a lot of gradients are considered as the boundary. However, watershed's direct application results to over-segmentation issues because of the introduction of noise and other inconsistencies of the gradient, and this segmentation beyond the limit issues could be controlled with the help of marker-controlled watershed segmentation.

The fifth category of image segmentation is clustering-based segmentation [7, 8]. Clustering in image processing involves the classification of indistinguishable image primitives, which are then grouped together to create clusters. The criteria used to measure similarity between pixels are a crucial factor in the formation of these clusters. The process of gathering identical pixels into clusters operates on the principle of maximizing similarity within clusters while minimizing similarity between clusters. This survey aims to highlight commonly utilized methods for image segmentation in this category.

The sixth category based on artificial neural networks (ANNs). ANNs mimic the learning strategies of the human brain for making decisions [9]. For medical image segmentation, this method is mostly used. The image needed is separated from the background by it. The segmentation of the neural network is completely different from the traditional approach of segmentation. In this process, a neural network is used to map an image in which each neural represents one pixel. Then, by using dynamic equations, we extract image edges to guide each neuron's condition to the minimum energy specified by the neural net. Neural segmentation based on a network has three basic characteristics:

- Parallel fast computing capability, making it ideal for real-time applications.
- Unconstrained and nonlinear scale, and high interaction between processing units, enabling this method to set model for any operation.
- Satisfying reliability which makes it noise-insensitive.

2 Clustering

Clustering is the process of organizing an image's elements or objects into groups based on their features, and various clustering methods are available to improve segmentation effectiveness. K-means clustering, known for its simplicity and robustness, is widely used. However, its performance can be influenced by the initial selection of centroids. To address this, several techniques have been introduced to reduce k-means clustering limitations. This paper presents an in-depth investigation into image segmentation methods based on k-means clustering.

2.1 K-Means Clustering Algorithm

The algorithm [10, 11] is extensively utilized in data partitioning, where data is grouped into "k" distinct clusters based on their resemblance to centroids. Each cluster is defined by its centroid, which represents the point where the total distances from all objects within the cluster are minimized. This algorithm is applied to determine centroids for all clusters.

Let's take an example of an image with dimensions $g * h$, which needs to be clustered into "k" data groups. We consider an input pixel, denoted as $p(g, h)$, with "c" representing the center. The algorithm follows these steps:

i. Initialize the center and "k".
ii. Calculate the Euclidean distance denoted as "d" for each pixel in the image using the following relationship:

$$d = \|p(g, h) - c_k\|$$

iii. Utilizing the acquired "d" value, allocate the pixels to their nearest center.
iv. Compute the updated center position using the subsequent formula.
v. The process is repeated until the value of tolerance is satisfied.

The k-means algorithm is renowned for its simplicity and benefits, but it does have certain limitations. Its precision is significantly impacted by the initial centroid choice. To address these drawbacks, several algorithms have been created. The next section offers a review of the literature concerning image segmentation techniques using k-means clustering.

2.2 *Related Works*

The paper [12] proposes a new algorithm for uniform distribution of data points by decreasing the computational complexity of the k-means method. The proposed method is compared with most delegate approaches of k-means and it prove that it has better cluster result and more efficient.

In [13], an improved variant of the k-means algorithm, referred to as k-means++, is introduced. It aims to identify initial cluster centers by assigning probabilities based on their proximity to the nearest center. The underlying idea is to determine the initial seed selection by considering the squared differences between cluster centers and the outcomes.

The paper [14] proposes an improved method to boost the precision and effectiveness of the clustering algorithm for k-means. They introduced an enhanced k-means method that incorporates an orderly two-approach method. The first is to locate the initial center, and the other is to give the clusters the data point. They took different initial centroid and measured the time and precision of execution. Thus, they can conclude from the outcome that the proposed algorithm lowers the complexity of the time giving the same precision of the clusters.

The method describe in [15] introduce an efficient approach based on k-means clustering with better initial center. In the algorithm, a new approach is implemented to find the initial center and it offers an efficient path to allocate the points of datasets to appropriate clusters giving lesser time complexity. They have proven that their proposed method give more precise output when comparing with the original k-means clustering algorithm having less computational time. This algorithm requires no additional input, such as threshold value, however, the proposed algorithm still has a drawback to initialize k number of cluster and determining the value of k will be the future work to come.

In [16], the paper discusses brain tumor segmentation through a combination of the k-means clustering and FCM algorithms, focusing on region calculation. The paper is structured into three main sections: image pre-processing, enhanced k-means, and fuzzy c-means, and finally, feature extraction. Initially, image quality is improved through a filtering process during image pre-processing. Subsequently, the proposed k-means technique is applied to cluster the image, followed by the utilization of fuzzy c-means. The resulting segmented image is then employed for feature extraction within the region of interest. For their analysis, an MRI image was used, and the size of the tumor region extracted from the image was calculated.

The paper [17] introduce an enhanced version of k-means algorithm. They optimized the running time of the k-means algorithm by reducing the number of dataset for every iteration. They noticed that a very limited number of the data changes its cluster after certain iterations and therefore, they put a boundary between those data which are not going to change and those who might change it. In other words, the borders between clusters are made of those points which are close to the edge of their cluster such that they could switch clusters when the next centroid shift.

The paper [18] introduces a new approach of k-means algorithm known as DK-means algorithm. It defines a deterministic initialization of initial centroids of k-means algorithm by finding a group of possible centroids using a constrained bi-partitioning strategy. When the suggested approach is compared to enhanced k-means algorithms such as k-means++ and MinMax, as well as other deterministic initialization algorithms, it is observed that the new method produces better results in respect that is has fast and stable convergence and it gives improve quality when comparing with other methods.

In [19], the paper introduces an enhanced version of the k-means algorithm to address its limitations. To improve the initialization process, an automatic parameter determination algorithm is suggested, utilizing image histogram information for more accurate parameter determination. The modified k-means method is then assessed both objectively and subjectively, considering output quality and time complexity. Additionally, when compared to other soft clustering algorithms, objective evaluations demonstrate the superior performance of the proposed enhanced k-means algorithm.

In [20], a novel approach is introduced, known as dynamic k-means clustering, which starts by determining the cluster centroids and the number of clusters through the calculation of a threshold value. If the Euclidean distance between data points is less than or equal to this threshold value, they are grouped together; otherwise, a new cluster is created with the dissimilar data point. Evaluation results demonstrate that this dynamic k-means clustering approach outperforms other methods.

In [21], a novel method designed for high-dimensional data is introduced. This method utilizes hybrid k-means clustering to identify the optimal number of clusters. Through experimentation, it is observed that this approach is more accurate in determining the exact number of clusters without relying on validation indices. Additionally, given the limitations of validation indices in assessing the quality of clusters in large datasets, this proposed method proves to be valuable in precisely determining the number of clusters.

The paper [22] introduces a new method for segmentation and classification of microCT images used in dentistry. In the proposed method, cluster centroids are determine more accurately by using improve k-means++ and also verifies its superiority over other clustering methods. The approach improves k-means clustering based on its histogram to enhance the precision of classification.

In [23], a k-means-based image segmentation method is presented, which is improved by integrating Ant Colony Optimization (ACO). ACO is utilized during the k-means clustering process for pixel assignment to different clusters. This method is compared with c-means and the magic wand algorithm, and its time complexity is assessed through drift analysis. Experimental findings illustrate that the proposed approach produces more precise segmentation results.

In [24], a technique for subspace segmentation that categorizes high-dimensional data into their respective subspaces is presented. They introduce a novel low-rank representation method called the self-regularized latent k-means method. Extensive experiments on various datasets were conducted to validate the efficacy of this proposed algorithm.

In [25], a method is introduced that combines k-means clustering, morphological filtering, and Chan-Vese active contour segmentation. The detection of dendritic cells is initially performed through k-means clustering, further refined using morphological filters. For more precise cell segmentation, the Chan-Vese active contour method is applied. The proposed approach is evaluated using real dendritic cell images and compared against ground truth images provided by biological experts. Experimental results indicate that the proposed method achieves higher accuracy.

3 Simulation and Discussion

In this section, we present the experimental results of our comprehensive study to compare the segmentation efficiency and effectiveness of various recently developed K-Means-based clustering techniques. The experimentation was performed by using the images from Berkeley database [26]. We have performed a rigorous evaluation on various k-means-based clustering method describe in [13, 15–17, 20] and compared with traditional k-means technique. To evaluate the efficiency of these techniques, we have computed certain validity indices including Mean square error (MSE), Jaccard Index, Peak signal to noise ratio (PSNR), and the correlation coefficient (ρ).

MSE serves as a performance measure comparing the input and output images, quantifying the disparity between these images. A low MSE value signifies high-quality output image results.

$$\mathrm{MSE} = \frac{1}{g * h}\sum\nolimits_{i=1}^{h}\sum\nolimits_{j=1}^{g}\left[x_{ij} - y_{ij}\right]^2 \tag{1}$$

Here, "g" represents the number of rows, "h" denotes the number of columns, "x_{ij}" corresponds to the input image, and "y_{ij}" represents the segmented image.

PSNR serves as a measure of image quality by comparing the highest possible signal power to the power of noise affecting the image. A low PSNR value signifies lower image quality, suggesting that more errors have been introduced to the image. Conversely, a high PSNR indicates good image quality, with minimal introduced error.

$$PSNR = 10log\frac{[x_{max}]^2 \cdot g \cdot h}{\sum_{i=1}^{h}\sum_{j=1}^{g}\left[x_{ij} - y_{ij}\right]^2} \tag{2}$$

The Jaccard index is employed to measure the likeness between two images. Its values fall within the 0–1 range, with higher values signifying a stronger resemblance between the images.

$$J = \frac{x_{ij} \cup y_{ij}}{x_{ij} \cap y_{ij}} \tag{3}$$

The correlation coefficient quantifies the degree of similarity between the input and output images. It is defined by the following equation:

$$\rho = \frac{\frac{1}{g.h}\sum_{i=1}^{h}\sum_{j=1}^{g}\left(x_{ij} - \underline{x}\right)\left(y_{ij} - \underline{y}\right)}{\sqrt{\frac{1}{g.h}\sum_{i=1}^{h}\sum_{j=1}^{g}\left(x_{ij} - \underline{x}\right)}\sqrt{\frac{1}{g.h}\sum_{i=1}^{h}\sum_{j=1}^{g}\left(y_{ij} - \underline{y}\right)}} \tag{4}$$

Here, $\underline{x}$ represents the average value of the input image and $\underline{y}$ represents the average value of the output image, a higher value signifies greater segmentation efficiency.

These evaluation indices presents a quantitative basis for comparing the efficiency and segmentation accuracy of the techniques. Our comparative analysis on the various methods revealed valuable insights into the strengths and weaknesses of each k-means-based technique. Figure 2 shows the original images as well as segmented images by using various hybrid k-means methods. Tables 1, 2, 3, 4, 5 and 6 gives the evaluation indices of the images. While each method exhibited unique characteristics, we observed that these hybrid method are more superior compared to traditional k-means method. The approach outlined in [16, 20] demonstrates consistent effectiveness and precision across a wide range of datasets when compared to other methods. However, it's important to note that the choice of the most suitable technique may depend on the specific characteristics of the dataset and on the type of application.

4 Conclusion

A comprehensive review was carried out, providing insights into various hybrid methodologies that employ k-means clustering for image segmentation. Image segmentation, an essential step in image processing, entails partitioning an image into distinct regions, each of which represents a significant component of the image.

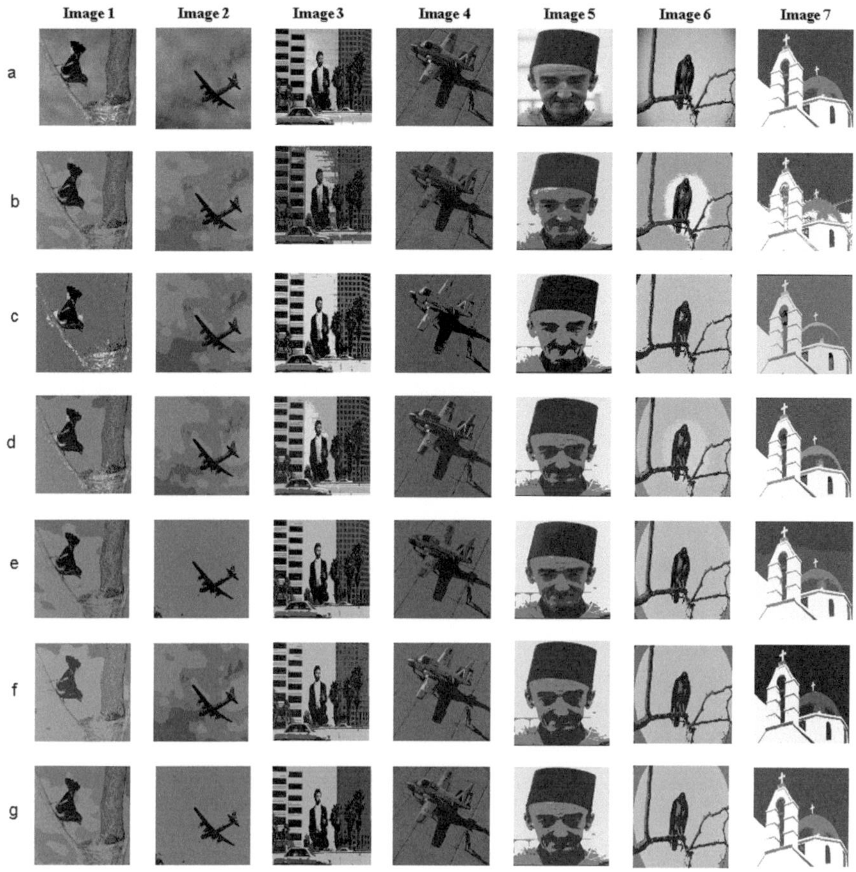

Fig. 2 **a** Original images, segmentation results by **b** k-means clustering, **c** [13], **d** [15], **e** [16], **f** [17], and **g** [20]

Table 1 Evaluation indices on image 1

Method	MSE	PSNR	Jaccard index	Correlation
K-means	46.45	30.57	0.95	0.93
[13]	39.36	32.26	0.97	0.95
[15]	37.25	35.25	0.97	0.96
[16]	38.78	34.62	0.96	0.95
[17]	36.32	35.87	0.98	0.97
[20]	36.89	36.11	0.98	0.97

Table 2 Evaluation indices on image 2

Method	MSE	PSNR	Jaccard index	Correlation
K-means	27.90	30.45	0.90	0.75
[13]	24.06	32.24	0.96	0.93
[15]	27.38	31.62	0.97	0.94
[16]	21.51	34.80	0.97	0.92
[17]	22.94	33.71	0.95	0.94
[20]	25.63	35.03	0.96	0.90

Table 3 Evaluation indices on image 3

Method	MSE	PSNR	Jaccard index	Correlation
K-means	65.75	29.98	0.91	0.81
[13]	63.23	31.93	0.92	0.88
[15]	52.59	32.83	0.94	0.92
[16]	45.57	35.73	0.94	0.92
[17]	56.72	33.17	0.95	0.91
[20]	47.62	35.14	0.95	0.93

Table 4 Evaluation indices on image 4

Method	MSE	PSNR	Jaccard index	Correlation
K-means	27.38	30.68	0.93	0.92
[13]	26.48	32.58	0.94	0.95
[15]	25.62	32.22	0.93	0.96
[16]	22.59	34.52	0.95	0.95
[17]	24.48	32.48	0.96	0.96
[20]	21.51	34.80	0.95	0.96

Table 5 Evaluation indices on image 5

Method	MSE	PSNR	Jaccard index	Correlation
K-means	53.89	25.26	0.89	0.91
[13]	45.47	28.41	0.92	0.92
[15]	40.21	30.74	0.93	0.91
[16]	38.82	32.38	0.91	0.92
[17]	34.28	34.27	0.94	0.95
[20]	39.37	32.86	0.93	0.92

Table 6 Evaluation indices on image 6

Method	MSE	PSNR	Jaccard index	Correlation
K-means	28.92	30.45	0.92	0.85
[13]	24.16	32.34	0.94	0.92
[15]	26.58	31.52	0.94	0.93
[16]	22.51	34.87	0.93	0.92
[17]	23.84	33.51	0.93	0.93
[20]	24.43	34.23	0.93	0.90

Among the various techniques utilized for this purpose, k-means clustering is notably prominent. Despite the development of a number of image segmentation methods, there is still no general method suitable for segmenting any form of image, and therefore research into the image segmentation method remains a major concern. Indeed, more research is required in order to establish an efficient segmentation technique for partitioning and defining objects or boundaries in an image in a simple and relevant way to meet advanced application requirements. And, in this paper, we have also highlighted some recent hybrid k-means clustering-based image segmentation techniques and compared their segmentation accuracy and concluded that these hybrid models are more superior to the conventional methods and delivers better segmentation results. Further investigations could explore hybrid approaches that combine the strengths of multiple k-means-based techniques for enhanced clustering performance. Additionally, research on optimizing these techniques for specific domains or data types may yield further improvements in efficiency and accuracy.

References

1. Raut SA, Raghuwanshi M, Dharaskar R, Raut A (2009) Image segmentation - a state-of-art survey for prediction. In: 2009 international conference on advanced computer control. IEEE
2. Umaa Mageswari S, Sridevi M, Mala C (2013) An experimental study and analysis of different image segmentation techniques. Procedia Eng 64:36–45
3. Bhardwaj S, Mittal A (2012) A survey on various edge detector techniques. Procedia Technol 4:220–226
4. Sivakumar V, Murugesh V (2014) A brief study of image segmentation using thresholding technique on a noisy image. In: International conference on information communication and embedded systems (ICICES2014). IEEE
5. Kaganami HG, Beiji Z (2009) Region-based segmentation versus edge detection. In: 2009 fifth international conference on intelligent information hiding and multimedia signal processing. IEEE
6. Vincent L, Soille P (1991) Watersheds in digital spaces: an efficient algorithm based on immersion simulations. IEEE Trans Pattern Anal Mach Intell 13:583–598
7. Sharma P, Suji J (2013) A review on image segmentation with its clustering techniques. IJSIP 9:209–218
8. Tongbram S (2020) Clustering-based image segmentation techniques: a review. JARDCS 12:701–707

9. Hosseini M, Zekri M (2012) Review of medical image classification using the adaptive neuro-fuzzy inference system. J Med Signals Sens 2:49
10. Celebi ME, Kingravi HA, Vela PA (2013) A comparative study of efficient initialization methods for the k-means clustering algorithm. Expert Syst Appl 40:200–210
11. Pena JM, Lozano JA, Larrañaga P (1999) An empirical comparison of four initialization methods for the K-means algorithm. Pattern Recogn Lett 20:1027–1040
12. Napolean D, Lakshmi PG (2010) An enhanced k-means algorithm to improve the efficiency using normal distribution data points. Int J Comput Sci Eng 2:2409–2413
13. Arthur D, Vassilvitskii S (2007) k-means++: the advantages of careful seeding. In: Proceedings of the 18th annual ACM-SIAM symposium on discrete algorithm. Society for Industrial and Applied Mathematics, Philadelphi, PA, pp 1027–1035
14. Nazeer AKA, Sebastian MP (2009) Improving the accuracy and efficiency of the k-means clustering algorithm. In: Proceedings of the world congress on engineering, London, UK, p 1
15. Yedla M, Pathakota SR, Srinivasa TM (2010) Enhancing K-means clustering algorithm with improved initial center. Int J Comput Sci Inf Technol 1(2):121–125
16. Jose A, Ravi S, Sambath M (2014) Brain tumor segmentation using K-means clustering and fuzzy C-means algorithm and its area calculation. Int J Innov Res Comput Commun Eng 2(2)
17. Poteraş CM, Mihăescu C, Mocanu M (2014) An optimized version of the K-means clustering algorithm. In: Proceedings of the 2014 federated conference on computer science and information systems. IEEE
18. Jothi R, Mohanty SK, Ojha A (2017) DK-means: a deterministic K-means clustering algorithm for gene expression analysis. Pattern Anal Appl 22:649–667
19. Khan Z, Ni J, Fan X, Shi P (2017) An improved k-means clustering algorithm based on an adaptive initial parameter estimation procedure for image segmentation. Int J Innov Comput Inf Control. 13(5)
20. Hossain MdZ, Akhtar MdN, Ahmad RB, Rahman M (2019) A dynamic K-means clustering for data mining. IJEECS 13:521
21. Alibuhtto MC, Mahat NI (2020) Distance based k-means clustering algorithm for determining number of clusters for high dimensional data. https://doi.org/10.5267/j.dsl.51-58
22. Wang L, Li S, Chen R, Liu S-Y, Chen J-C (2017) A segmentation and classification scheme for single tooth in MicroCT images based on 3D level set and k-means++. Comput Med Imaging Graph 57:19–28
23. El-Khatib SA, Skobtsov YA, Rodzin SI (2019) Theoretical and experimental evaluation of hybrid ACO-k-means image segmentation algorithm for MRI images using drift-analysis. Procedia Comput Sci 150:324–332
24. Wei L, Zhou R, Wang X, Zhu C, Yin J, Zhang X (2019) Subspace segmentation via self-regularized latent K-means. Expert Syst Appl 136:316–326
25. Braiki M, Benzinou A, Nasreddine K, Hymery N (2020) Automatic human dendritic cells segmentation using K-means clustering and Chan-Vese active contour model. Comput Methods Programs Biomed 195:105520
26. Martin D, Fowlkes C (2001) The Berkeley segmentation database and benchmark. Berkeley University, Computer Science Department

Convolutional Neural Network-Based Deep Learning Method for Fruit Classification

Mrinal Paliwal, Sunil Kumar Chawla, Punit Soni, Anurag Jain, Tanupriya Choudhury, and Ketan Kotecha

Abstract Fruits are the most popular demanding product in the market, so the quality of those products should be very good. Automation in the field of fruit classification can bring great continence to fruit mongers. Creating large data banks to store information about diseases and use them with techniques such as deep learning will be helpful for economic development. This paper will be helpful to the agricultural industry. However, some pears and apple varieties have many similarities, and these sorts of fruit are typically popular with consumers. These things make the task of identifying rotten fruits more difficult. To solve this issue, this research proposes a convolutional neural network-based deep learning method for fruit classification. The dataset used contains images of different fruits. Applying a convolutional neural

M. Paliwal · S. K. Chawla · P. Soni
Chitkara University Institute of Engineering and Technology, Chitkara University, Chandigarh, Punjab, India
e-mail: mrinal.paliwal@chitkara.edu.in

S. K. Chawla
e-mail: sunil.chawla@chitkara.edu.in

P. Soni
e-mail: punit.soni@chitkara.edu.in

A. Jain (✉)
School of Computer Science, IILM University, Gurugram, India
e-mail: anurag.jain@iilm.edu

T. Choudhury
Adjunct Professor, CSE Department, Symbiosis Institute of Technology, Symbiosis International (Deemed University), Lavale Campus, Pune 412115, Maharashtra, India
e-mail: tanupriya@ddn.upes.ac.in; tanupriya.choudhury@sitpune.edu.in

K. Kotecha
Symbiosis Centre for Applied Artificial Intelligence, Symbiosis Institute of Technology, Symbiosis International (Deemed University), Pune, Maharashtra, India412115,
e-mail: director@sitpune.edu.in

T. Choudhury
Professor,SoCS, University of Petroleum and Energy Studies (UPES), Dehradun 248007, Uttrakhand, India

A. P. Agrawal et al. (eds.), *Advances of Artificial Intelligence and Machine Learning in Image Processing, Security and Data Analytics*, Lecture Notes in Electrical Engineering 1464, https://doi.org/10.1007/978-981-95-0148-9_2

network, we will categorize the images of fruit into fresh and rotten categories. Several tests have been conducted on those images to get an efficiency of 94.92%.

Keywords Convolutional neural network · Transfer learning · Deep learning · Fruit classification

1 Introduction

Fruit classification is a fundamental task with regard to computer vision, with applications ranging from agricultural automation and food quality assessment to retail inventory management and health monitoring. After grains and vegetables, fruits are the majorly grown product and its industry is the third largest industry worldwide. Fruit classification is a major challenge faced by the fruit industry nowadays, which has brought about significant economic benefits. It is obvious from existing literature that Convolutional Neural Networks (CNNs) is well-suited to computer vision [1, 2]. It is also well-suited for image classification tasks like fruit and vegetable recognition [3–7]. Deep learning techniques, especially CNNs, have shown remarkable success in image recognition due to the ability to learn automatically the relevant features from raw pixel data [8]. CNNs can effectively describe spatial hierarchies and patterns in a given image, making them an ideal choice for complex tasks like fruit classification, where identifying subtle differences in shape, color, and texture is crucial. Fruit classification using CNNs has been actively used in areas of research due to its numerous applications in agriculture, the food industry, and automated fruit sorting systems. A summary of some key papers that were influential in this field is given below:

Bargoti and Underwood [9]: In this paper, an object recognition framework is discussed. Faster R-CNN was utilized, and the database comprised mango, apple, and almond detection.

Sahu et al. [10]: The researchers applied CNNs to classify banana diseases using image data. Their work demonstrated the effectiveness of deep learning in distinguishing various diseases affecting banana crops.

Abade et al. [11]: This study discusses various methods for handling features of disease detection, dataset properties, studied pathogens and crops. The results of the paper can be utilized to have an idea of the most recent developments in the field of CNNs for the diagnosis of plant diseases as well as to identify any research gaps that need to be addressed by the researchers.

Jiménez et al. [3]: This paper explored fruit recognition using range/attenuation images and investigated various CNN architectures for the task. It provided insights into the performance of different models and image modalities.

Mureşan and Oltean [4]: This research focused on fruit recognition using deep learning approaches, including CNNs and transfer learning techniques. It addressed the challenges of limited training data and explored methods for improving classification accuracy.

Xiao et al. [5]: Various deep learning models and methods for fruit detection and recognition tasks were covered in this review study. It compared different convolutional architectures and highlighted the strengths and limitations of each approaches individually.

Sa et al. [6]: This paper proposed a deep fruit recognition system using CNN (Faster R-CNN) for fruit classification. They evaluated different CNN architectures and presented their results on a large-scale fruit dataset.

Ahmed et al. [7]: This paper provided a comparative analysis of various CNN architectures for fruit classification tasks. It compared the effectiveness of various models and highlighted the best strategy for fruit recognition.

Dhiman et al. [12]: The deep neural network (DNN) model proposed in this study uses citrus fruits as training data to identify the disease and its severity degree. That has been categorized into high, medium, low, and healthy categories. In this paper transfer learning is used with VGGNet is utilized to build a multi-classification framework for each severity class.

Trivedi et al. [13]: In this paper, the concept of convolutional neural network and image segmentation is used. CNN uses the features like edges, texture, and picture color.

2 Methodology

The research methodology used in the current work is demonstrated in Fig. 1. In this proposed technique, the authors developed a deep learning based fruit classification system using CNN using Kaggle dataset [14] which is used for both training and testing purpose. Once the preprocessing was completed, we applied a random fruit image from gallery for experimental validation. This system was able to classify whether the input random image is of a rotten fruit or a fresh fruit based on the proposed CNN model.

2.1 Algorithm and Dataset

Step 1: Input the images

Step 2: Split Dataset: Validation data and Train

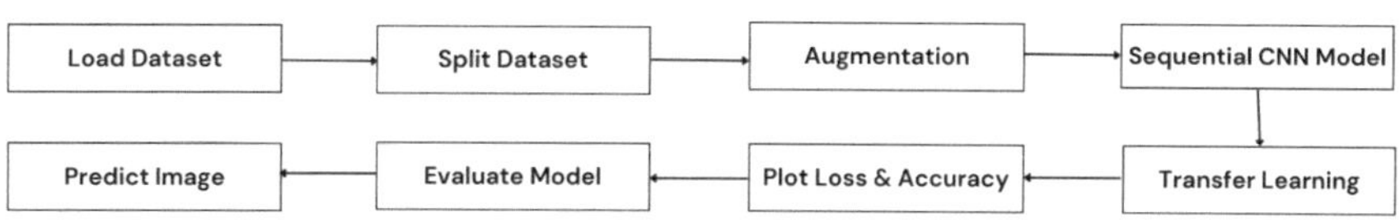

Fig. 1 Implementation phases of the proposed model

Step 3: Classification them using CNN go to Step 4

Step 4: Augmentation

Step 5: Build CNN model:

CNN has three main layers:

1. Convolutional layer
2. Pooling layer
3. Fully connected layer

Step 6: Train the Model

Step 7: Plot the result

Step 8: Plot Confusion Matrix

The first step was to gather a diverse and extensive dataset of fruit images. The authors have used the open-source and widely used Kaggle dataset [15], which included 395 photographs of fresh apples, 381 images of fresh bananas, 388 images of fresh oranges, 601 images of rotting apples, 503 images of rotten bananas, and 403 images of rotten oranges. With this dataset will serve as the foundation for training, validating, and testing the CNN model. The sample and quality of the fruit are shown in Fig. 2.

As it is visible in Fig. 2, the dataset contains images of fruits with complex backgrounds, different angles, and different light conditions. The entire dataset has 2,671 images in total, which are split across various categories. The concept of data augmentation, a method that increases the count of images in the database, was applied in this study.

2.2 *Preprocessing*

The authors have preprocessed the dataset before training. The database was separated into three parts, namely training, testing, and validation data. Data preprocessing is essential to ensure that CNN can learn meaningful representations from the images [15]. To increase the diversity of the training dataset, common preprocessing techniques include resizing images to a predetermined size, normalization, and data augmentation were used.

2.3 *Augmentation*

Data augmentation is an effective technique often used in convolutional neural network-based models to artificially increase the amount of the training dataset by performing various alterations on the existing images. The total number of images

Fig. 2 Images of dataset used for training and testing

will increase to 7,308 for fresh fruit and 3,593 for rotten fruit images. It helps in improving the model's generalization and robustness, especially when the original dataset is limited or imbalanced. The main idea behind data augmentation is to introduce variations in the training data while preserving the original label information. This process enables the CNN to learn more diverse features and reduces the risk of overfitting by presenting the model with a wider range of examples.

Table 1 shows the dataset used for the training and testing purpose, whereas Fig. 3 shows how an image of orange is augmented and small images are made from the big image. These images were used for training purposes. This augmentation helped us to increase the count of cropped images used in the dataset.

Table 1 Type of datasets used

Fruits images training	Number	Fruits images testing	Number
Fresh apple images	1695	Fresh apple images	395
Fresh banana images	1583	Fresh banana images	381
Fresh oranges images	1466	Fresh oranges images	388
Rotten apple images	2342	Rotten apple images	601
Rotten banana images	2224	Rotten banana images	530
Rotten oranges images	1595	Rotten orange images	403

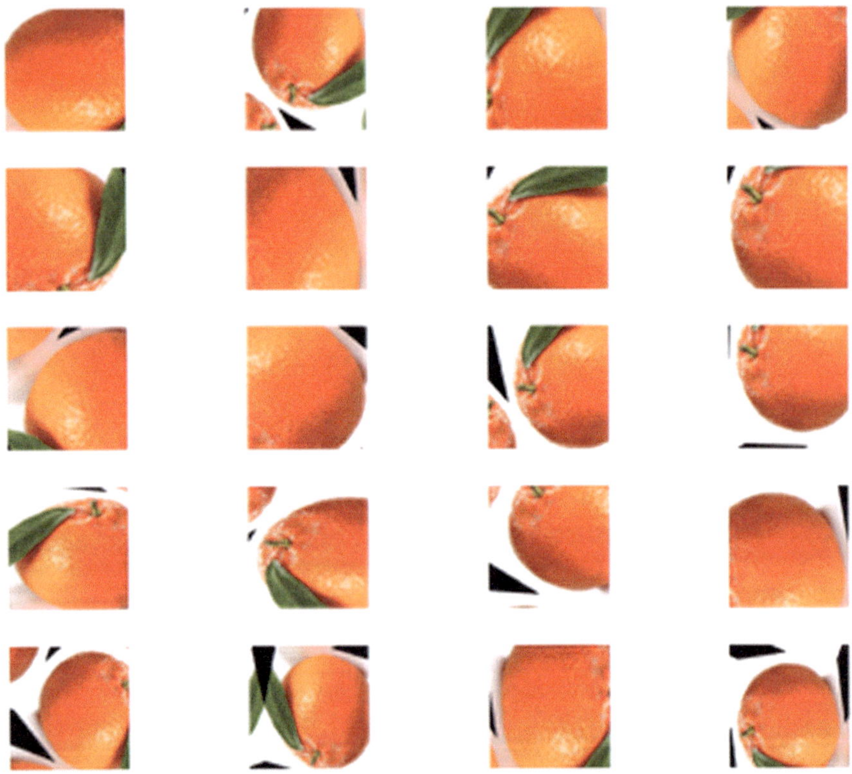

Fig. 3 Image of orange after data augmentation

2.4 CNN Architecture

In present study, convolutional neural network-based architecture is designed and tailored for the fruit classification task. This architecture is used as a sequential model with many convolutional layers (conv2d), pooling layers (max_pooling2d), dropouts, and fully connected layers to recognize and categorize images of fruits.

Table 2 CNN-based fruit recognition method's architecture (sequential model)

S. No.	Layer	Output shape	Parameters
1	Convolutional layer 1 (conv2d)	(None, 148, 148, 32)	896
2	Pooling layer 1(max_pooling 2d)	(None, 74, 74, 32)	0
3	Dropout 1 (dropout)	(None, 74, 74, 32)	0
4	Convolutional layer 2(conv 2d_1)	(None, 72, 72, 64)	18,496
5	Pooling layer 2(max_pooling 2d_1)	(None, 36, 36, 64)	0
6	Dropout 2 (dropout_1)	(None, 36, 36, 64)	0
7	Convolutional layer 3(conv 2d_2)	(None, 34, 34, 128)	73,856
8	Pooling layer 3 (max_pooling 2d_2)	(None, 17, 17, 128)	0
9	Dropout 3 (dropout_2)	(None, 17, 17, 128)	0
10	Flatten (flatten)	(None, 36,992)	0
11	Dense 1 (dense)	(None, 512)	18,940,416
12	Dropout 4 (Droupout_3)	(None, 512)	0
13	Dense 2 (Dense_1)	(None, 1)	513

Total number of parameters: 19,034,177
Total trainable parameters: 19,034,177
Total on-trainable parameters: 0

Table 2 defines the detail about the different layers used in convolutional neural network architecture. The model was trained with the collected and preprocessed dataset. During the training phase, the model was iteratively adjusted for its internal parameters to minimize a chosen loss function and improve classification accuracy. The parameter details of the model used in current research work are given in Table 2.

2.5 *Transfer Learning*

Transfer learning (TL) is a machine learning technique where information learned from the solution of one problem is used to solve another problem that is unrelated but nonetheless important. Table 3 shows the details of various layers of transfer learning used in implementation. The different layers are carefully selected, through which the data will pass and train the model. It involves utilizing neural network models that have already been trained on a sizable dataset to enhance the performance of a novel task for dataset with insufficient labeled data.

Table 3 Transfer learning implementation through various layers

S. No.	Layer name	S. No.	Layer name
1	Input_2	10	Block3_pool
2	Block1_conv1	11	Block4_conv1
3	Block1_conv2	12	Block4_conv2
4	Block1_pool	13	Block4_conv3
5	Block2_conv2	14	Block4_pool
6	Block2_pool	15	Block5_conv1
7	Block3_conv1	16	Block5_conv2
8	Block3_conv2	17	Block5_conv3
9	Block3_conv3	18	Block5_pool

Table 4 Evaluate each epoch's loss and accuracy

Epoch	Val_loss	Step—loss	Val_accuracy	Accuracy
1	0.2312	0.3502	0.8987	0.8606
2	0.2049	0.2551	0.9146	0.8946
3	0.1838	0.2062	0.9210	0.9112

Accuracy test: 0.949221670627594
Loss test: 0.12792862951755524

2.6 *Evaluation*

The performance of the model under investigation was evaluated using a separate validation dataset and various performance metrics such as accuracy, val_loss, val_accuracy, and step-loss were computed. Table 4 shows the performance of the proposed system under the above-mentioned parameters for each epoch. The accuracy that we achieved in this work is 94.92% and the loss test is 12.79%.

2.7 *Testing*

The final step involves testing the trained model on a separate test dataset to assess its generalization and real-world performance. Figure 4 shows that as the number of epochs increases, the loss decreases in both training and validation. Figure 5 demonstrates that accuracy increases with the number of epochs increases; this shows that the neural network is well-trained for the given set of data.

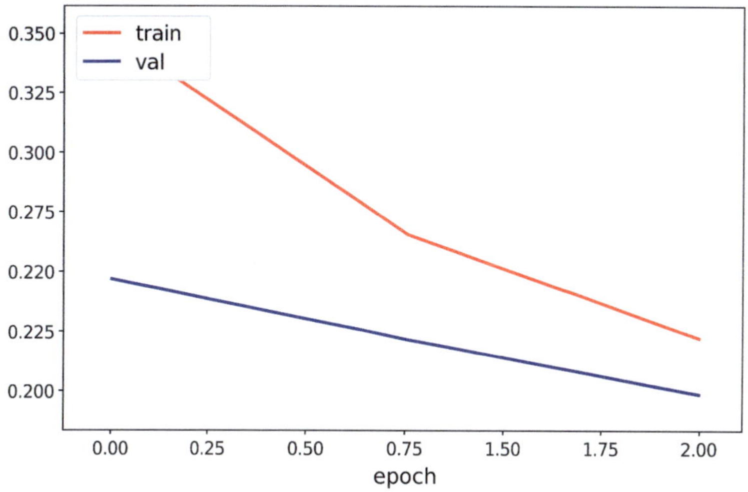

Fig. 4 Training and validation loss

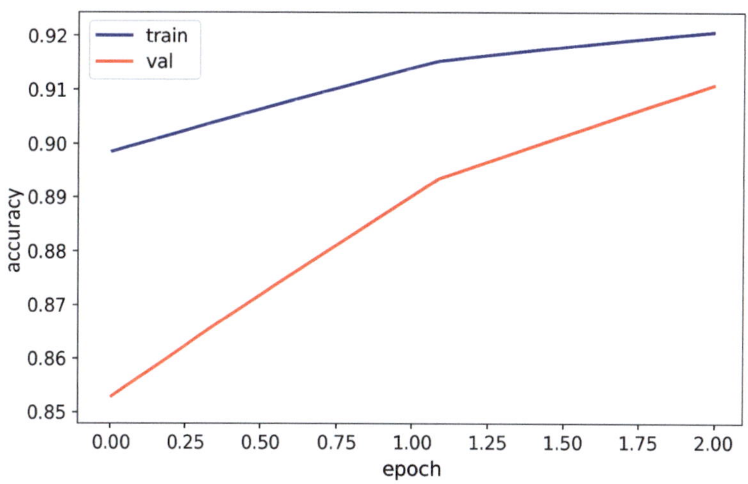

Fig. 5 Training and validation accuracy

3 Discussion

The model has been tested with random images as shown in Figs. 6 and 7. The results were calculated with 53 different images of different fruits, some of them are shown in Fig. 6, where the prediction percentage is given for the fresh fruit image. Figure 7 shows the images of rotten fruit prediction percentage for different fruit images tested on our system. The proposed system can identify if the given input image of a fruit is a rotten fruit or a fresh fruit based on the trained neural network. Table 5 shows the comparison of other techniques which can be used for fruit classification. The commonly used techniques are extracted wavelet entropy, k-means clustering, and CNN. The neural network techniques work well in the case of classification as shown in Table 5. The different CNN technologies such as six-layer CNN and LeNet-5 CNN have also been used and the results are mentioned in Table 5. Our proposed work also used CNN and gives an accuracy of 94.92%.

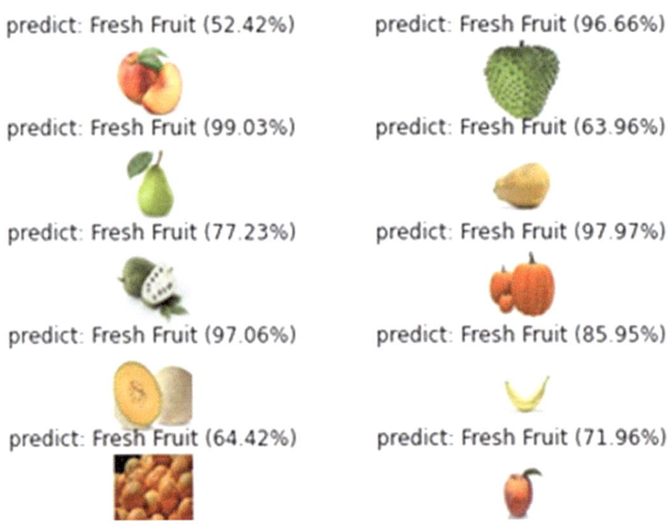

Fig. 6 Results of prediction of fresh fruits

Fig. 7 Results of prediction of rotten fruits

Table 5 Comparison of different methods used with the proposed technique

S. No.	Reference	Method used	Accuracy achieved (%)
1	[16]	Extracted wavelet entropy	89.47
2	[17]	K-means clustering	90
3	[18]	Six-layer CNN	91.44
4	[19]	LeNet-5 CNN	95.67
5	[20]	Max-wins-voting SVM with Gaussian radial basis kernel	88.2
5	Proposed	Convolutional neural network	94.92

4 Conclusion

This research paper demonstrates how deep learning and CNNs can be used to tackle real-world issues like fruit classification and to offer a reliable and effective method for correctly identifying and classifying various fruits from digital images captured from different angles. Our goal is to design a useful and expandable fruit classification model that can be improved upon and used for automated fruit recognition across a variety of areas.

References

1. Rani P, Kumar R, Jain A, Lamba R, Sachdeva RK, Choudhury T (2023) PCA-DNN: a novel deep neural network oriented system for breast cancer classification. EAI Endorsed Trans Perv Health Technol 9:1–18
2. Sachdeva RK, Singh KD, Bathla P, Jain A, Choudhury T, Kotecha K (2023) Empowering hepatitis diagnosis using RFE feature selection. In: 2023 7th international symposium on multidisciplinary studies and innovative technologies (ISMSIT), Ankara, Turkiye, pp 1–5. https://doi.org/10.1109/ISMSIT58785.2023.10304999
3. Jiménez AR, Jain AK, Ruíz RC, Rovira JLP (1999) Automatic fruit recognition: a survey and new results using range/attenuation images. Pattern Recogn 32(10):1719–1736
4. Mureşan H, Oltean M (2018) Fruit recognition from images using deep learning. Acta Univ Sapientiae Inform 10(1):26–42
5. Xiao F, Wang H, Xu Y, Zhang R (2023) Fruit detection and recognition based on deep learning for automatic harvesting: an overview and review. Agronomy 13(6):1625
6. Sa I, Ge Z, Dayoub F, Upcroft B, Perez T, McCool C (2016) DeepFruits: a fruit detection system using deep neural networks. Sensors 16(8):1222
7. Ahmed MI, Mamun SM, Asif AUZ (2021) DCNN-based vegetable image classification using transfer learning: a comparative study. In: 2021 5th international conference on computer, communication and signal processing (ICCCSP). IEEE, pp 235–243
8. Munir K, Umar AI, Yousaf W (2020) Automatic fruits classification system based on deep neural network. NUST J Eng Sci 13:37–44
9. Bargoti S, Underwood J (2017) Deep fruit detection in orchards. In: 2017 IEEE international conference on robotics and automation (ICRA). IEEE, pp 3626–3633. https://doi.org/10.1109/ICRA.2017.7989417
10. Sahu P, Chug A, Singh AP, Singh D, Singh RP (2022) Classification and activation map visualization of banana diseases using deep learning models. In: International conference on innovative computing and communications: proceedings of ICICC 2021, vol 3. Springer, Singapore, pp 751–767
11. Abade A, Ferreira PA, de Barros Vidal F (2021) Plant diseases recognition on images using convolutional neural networks: a systematic review. Comput Electron Agric 185:06–125
12. Dhiman P, Kukreja V, Manoharan P, Kaur A, Kamruzzaman MM, Dhaou IB, Iwendi C (2022) A novel deep learning model for detection of severity level of the disease in citrus fruits. Electronics 11(3):495
13. Trivedi NK, Gautam V, Anand A, Aljahdali HM, Villar SG, Anand D, Goyal N, Kadry S (2021) Early detection and classification of tomato leaf disease using high-performance deep neural network. Sensors 21(23):7987
14. Fruits fresh and rotten for classification. https://www.kaggle.com/datasets/sriramr/fruits-fresh-and-rotten-for-classification. Accessed 10 May 2023
15. Zeng PP, Li LS (2019) Research on fruit image classification and recognition based on convolutional neural network. Mech Des Res 35(1):23–26
16. Wang S, Zhang Y, Ji G, Yang J, Wu J, Wei L (2015) Fruit classification by wavelet-entropy and feedforward neural network trained by fitness-scaled chaotic ABC and biogeography-based optimization. Entropy 17(8):5711–5728
17. Sidehabi SW, Suyuti A, Areni IS, Nurtanio I (2018) Classification on passion fruit's ripeness using K-means clustering and artificial neural network. In: 2018 international conference on information and communications technology (ICOIACT). IEEE, pp 304–309

18. Lu S, Lu Z, Aok S, Graham L (2018) Fruit classification based on six layer convolutional neural network. In: 2018 IEEE 23rd international conference on digital signal processing (DSP). IEEE, pp 1–5
19. Wang SH, Chen Y (2020) Fruit category classification via an eight-layer convolutional neural network with parametric rectified linear unit and dropout technique. Multimed Tools Appl 79:15117–15133
20. Zhang Y, Wu L (2012) Classification of fruits using computer vision and a multiclass support vector machine. Sensors 12:12489–12505

Exploring the Role of Deep Learning and Explainable AI in Android Malware Detection: A Systematic Review

Mandeep Kumar, Abhishek Kajal, and Kulbir Singh

Abstract This article examines the use of Explainable Artificial Intelligence (XAI) in conjunction with Deep Learning models to detect Android malware. Grad-CAM, LIME, SHAP, IG, and LRP are XAI techniques that improve model interpretability and provide deeper insights into decision-making processes. Deep Learning models, specifically CNNs and Graph Convolutional Networks (GCNs), outperform traditional methods in identifying malware families from datasets like as VirusShare, Android Malware Genome Project, Drebin, CICMalDroid2020, and AMD. Actual malware samples improve the models' resilience and effectiveness in real-world circumstances. However, issues like as vulnerability to adversarial attacks, scalability, and generalization across different malware types remain. Researchers and practitioners must work together to improve XAI techniques and strengthen cybersecurity defenses. The combination of XAI and deep learning offers exciting prospects to improve Android malware detection while also providing transparent and dependable solutions for a safer digital world.

Keywords Explainable AI · Deep learning · Android malware detection · Convolutional neural networks · Cyber security

M. Kumar (✉) · A. Kajal
Department of CSE, GJUS&T, Hisar, India
e-mail: mkpachar11@gmail.com

K. Singh
IEEE, Chicago, IL, USA

A. P. Agrawal et al. (eds.), *Advances of Artificial Intelligence and Machine Learning in Image Processing, Security and Data Analytics*, Lecture Notes in Electrical Engineering 1464, https://doi.org/10.1007/978-981-95-0148-9_3

1 Introduction

1.1 Cybersecurity

In today's linked digital landscape, comprehensive cybersecurity measures are more important than ever, especially in the realm of Android smartphones. As mobile technology, particularly Android smartphones and tablets, becomes more popular, the convergence of cybersecurity with the Android ecosystem poses serious concerns. This convergence brings with it both creative prospects and major concerns in terms of protecting our digital domain. Cybersecurity, the art of protecting digital assets from malevolent actors, is a never-ending battle against ever-changing dangers. With millions of apps accessible for Android, the platform has sadly become a breeding ground for harmful software ranging from spyware to ransomware, posing substantial hazards to individuals, enterprises, and governmental authorities. Improving cybersecurity entails hardening the environment against illegal access and strengthening information security for legitimate users. Utilizing cutting-edge technology, such as deep learning inside Artificial Intelligence (AI), has the potential to alter cybersecurity processes by allowing faster and more precise identification and mitigation of cyber threats utilizing sophisticated neural networks. In this dynamic environment, researchers and cybersecurity specialists go further into comprehending the complexities of deep learning algorithms, using Explainable Artificial Intelligence (XAI) methodologies as a critical component. XAI seeks to decipher the complexities of these models, offering information on their decision-making processes in order to improve malware detection. The objective is to create clear and dependable solutions that enable security experts to understand, trust, and fine-tune Android malware defenses. This combination of deep learning, XAI, and Android malware research is a once-in-a-lifetime potential to take cybersecurity to new heights. Envisioning a future in which Android users comfortably explore the digital environment, secure in the knowledge that AI-backed security is there at every step, represents the possibilities of this groundbreaking journey. The dramatic increase in malware infection rates from 2009 to 2018, from 12.4 million to a startling 812.67 million impacted identities, highlights the dynamic nature of cyber threats and highlights the critical significance of effective cybersecurity measures in combatting such attacks [1].

The AV-TEST Institute records approximately 450,000 newly found dangerous programs (malware) and potentially unwanted applications (PUA). The data in these files is stored after they have been properly inspected and categorized based on their unique properties. The information is subsequently converted into updatable graphs by visualization programs, which generate current and relevant malware statistics represented in Fig. 1 [2].

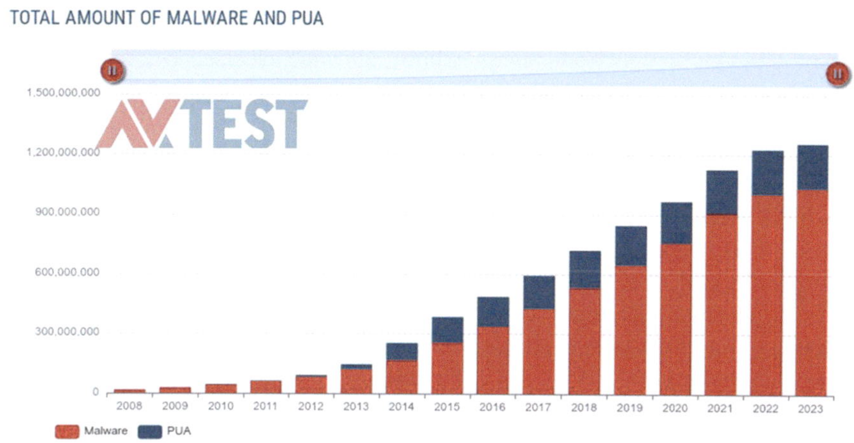

Fig. 1 Malware infection rate [2]

1.2 *Explainable Artificial Intelligence*

Explainable AI, also known as XAI, is a subfield of machine learning (ML) that aims to provide a better understanding of model behavior by explaining how a model generated a prediction. Understanding how a model acts, and how its training dataset influences it, everyone who produces or uses ML has tremendous new capabilities for improving models and increasing confidence in their predictions, and recognize when things go wrong. AI approaches that can be explained are very beneficial because they do not rely on a model when you understand an Explainable AI technique has a wide range of applications [20].

1.3 *Android Malware Detection*

The fast expansion of mobile devices, as well as the widespread acceptance of Android operating systems, has transformed our connection with technology. While this progress offers many advantages, it has also exposed users to the growing threat of Android malware. These harmful programs infiltrate networks, compromising crucial data, interrupting operations, and posing serious security risks to both persons and companies. Although traditional malware detection approaches have shown to be effective, they are unable to keep up with the rising sophistication of cyber threats. To address this significant issue, a combination of deep learning and Explainable Artificial Intelligence (XAI) will be used to improve Android malware detection systems. Deep learning, an advanced subset of machine learning, has achieved great success in fields such as image and language processing. Its promise for enhancing malware detection from its capacity to interpret patterns from large datasets, for improving accuracy and efficiency. Integrating XAI methods into malware detection systems also provides transparency and interpretability, allowing cybersecurity specialists to understand the inner workings of deep learning models. XAI enables these models to formulate human-comprehensible explanations for their conclusions, revealing light on why a specific program is classified as malware. This transparency not only builds trust in the detection system but also facilitates in the discovery of possible flaws. The major goal of this analysis is to thoroughly investigate current developments in Android malware detection using deep learning and XAI. We intend to critically analyze existing research, methodology, and datasets in this domain by analyzing them. Through synthesizing community knowledge, we aim to illuminate potential hurdles and future prospects in this rapidly evolving field.

Key highlights of this review paper include:

Deep learning architectures and methods used in Android malware detection, including Convolutional Neural Networks (CNNs), Recurrent Neural Networks (RNNs), and Transformer models, are thoroughly reviewed.

Exploration of feature extraction approaches and their use in improving DL model performance for malware detection.

Examining the representativeness of common datasets used for training and comparing deep learning-based malware detection algorithms to real-world malware samples.

A critical examination of XAI methodologies used to give insights into deep learning models' decision-making processes.

A thorough examination of the performance measurements, assessment procedures, and benchmarks employed in the literature.

Identification of open issues and research gaps, as well as prospective future research initiatives in the field of Android malware detection using deep learning and XAI.

Finally, the purpose of this study is to provide cybersecurity researchers, practitioners, and policymakers with an up-to-date and comprehensive knowledge of the advances achieved in Android malware detection using deep learning with XAI. As the security environment evolves, fully utilizing the potential of these novel techniques is critical in protecting Android users from more complex and harmful malware threats.

2 XAI Background

Explainability of AI is not much concerned with the interest in the AI domain, but how explaining affects the brain and psychology. Various psychologists make their psychology more productive with the help of neurobiology for relating their mindset image with brain working. In the scenario of Sigmund Freud's "Project of Scientific Psychology," compiled in 1895 which obtains knowledge is viewed as network complexity, and learning viewed as the successful transfer relationships in network, and neurologist John Hulingson Jackson viewed the brain as a multilayer distributed system in 1869. These techniques are useful, Friedrich Hayek creates a model for learning Hebb's synapses, assuming that the brain can represent Hebb's network. After this theory, computers are available for scientists to create a neural network which is a perceptron. Joseph Licklider gives the concept of adaptable networks and improved human–computer interaction through mental models. And described features of the Internet in 1962 and which helped create the ARPANET prototype. This gives the long-term goal of enhancing interaction between humans and programs. Licklider realized that neural networks can't replace a few human functions, which decreases AI investments. After that researchers explored Aristotle's ideas and built AI systems based on inference and symbolic manipulation rules. These are more understandable but demonstrate the fragility and incapability of evolution. Neural networks have resurged and automated processes with their strong computing power. The combined use of rule-based approaches and neural networks by DeepMind's AlphaGo series shows the best results in the higher use of AI which impacts critical areas.

DARPA started the "Explainable Artificial Intelligence" program in 2015 with scientists from various disciplines to make AI that gives a better human–computer interface with the help of explanation specialists.

As a result, their research is a great method for understanding challenges in Explainable AI [3].

2.1 Categorizations of XAI

Several categorizations have been proposed in the literature to classify various explainability methodologies. However, these classification systems are not always exact and might vary greatly depending on methodological factors. As a result, they may produce overlapping or non-overlapping groupings. This article briefly addresses many categorization approaches and taxonomies associated with explainability strategies. Figure 2: Flow diagram of XAI Methods categorization shows the flow diagram demonstrating various methods' categorization [19].

Global Methods Versus Local Methods. Local explainable techniques are primarily concerned with explaining the results of a single model. They accomplish this by creating ways for deciphering particular predictions or conclusions produced by the model. Global approaches, on the other hand, dive into the inner workings of a prototype by exploiting its knowledge base, mentoring, and related data. These global approaches try to give a more thorough knowledge of the model's behavior [19].

Surrogate Methods Versus Visualization Methods. Surrogate techniques are collections of models that are used to test other black-box models. By analyzing the surrogate model's judgements and comparing the black-box model's and surrogate

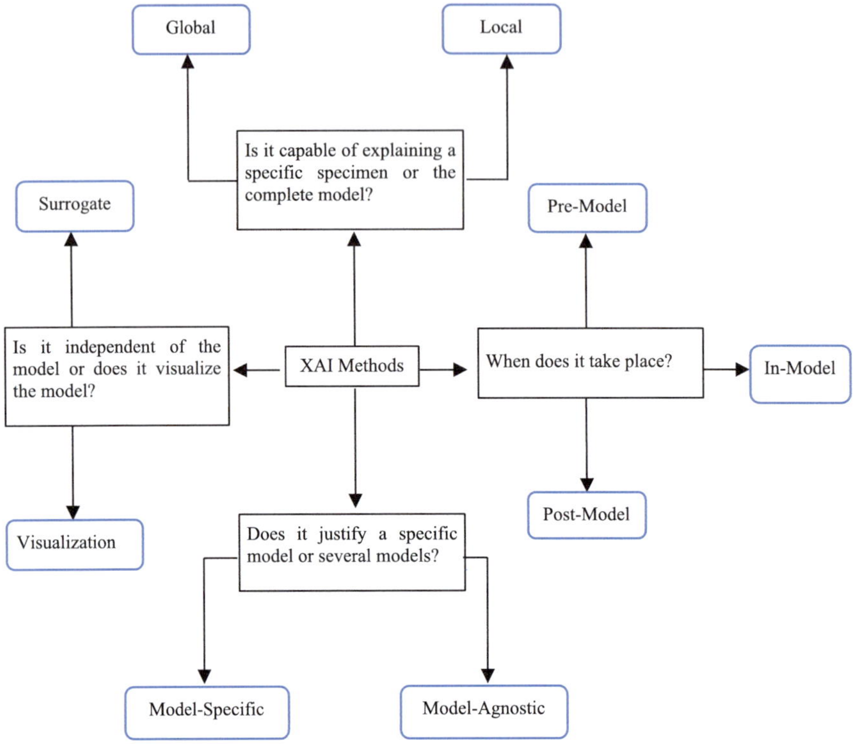

Fig. 2 Flow diagram of XAI methods classification

model's judgements, the black-box models may be successfully grasped. The decision tree illustrates surrogate approaches. The visualization approaches, such as activation maps, are not independent models, but they assist clarify some characteristics of the models through the imaging process [19].

Model Specific Versus Model-Agnostic. Model-specific analysis processes are adjusted to each unique model's metrics. The GNNExplainer (Graph Neural Network Explainer) is one such solution, which is especially developed for explainability in circumstances when the data representation complexity necessitates the explicit usage of graph neural networks. Model-agnostic techniques, on the other hand, are used for post-processing evaluation and are not connected to a certain model architecture. These strategies may be applied to a variety of models without requiring any alterations. They do not, however, allow direct access to the underlying model weights or architectural features [19].

Pre-model Versus In-Model Versus Post-model. Pre-model procedures are stand-alone methods that do not necessitate the usage of a particular model architecture. Principal Component Analysis (PCA) and t-Distributed Stochastic Neighbor Embedding (t-SNE) are two such methodologies that may be used to data prior to modeling.

In-model techniques, on the other hand, are approaches to interpretability that are built directly into the model itself. During the model's training phase, several strategies give information. After the model has been developed and trained, post-model approaches are applied. Through tactics such as mentorship and additional analysis, they can offer useful insights into what the model has learnt [19].

3 XAI Framework and Datasets for Malware Detection

3.1 XAI Framework for Malware Detection

On the basis of the papers reviewed for malware detection, we demonstrated a generic XAI framework diagram for malware detection in this section. Figure 3 depicts the framework diagram for Explainable AI applications in Malware detection. This graphic is regarded as general as it can be to demonstrate the process of implementing XAI in the malware field. This work is divided into different phases, illustrated with the help of simple examples appearing at each stage. The cycle of framework begins with the data processing using the feature extraction technique, which is subsequently implemented in the malware detection model. With a decent interface design, several explainability approaches are employed to describe the model. The entire process of explaining a model is further validated using some matrices: correctness, interpretability, and user satisfaction.

3.2 Malware Datasets

Nowadays, one of the biggest threats to the entire cyberspace around the globe is malware attacks, and "Android Malware" is the most vulnerable classification among various types of malware attacks. Big reason behind this is the exponential growth of android smartphones by each passing day, and the most vulnerable as well. There is availability of the different types of datasets for different malware types. Some publicly available datasets in the context of Android malware categories are listed below in Table 1.

4 XAI Applications for Defending Against Malware

A DNN works like a black-box model which doesn't disclose anything about their internal working structure. To overcome this Yan et al. [14] proposed a method for extracting rules from DNN to detect malicious network traffic. He used Droid-collector as a traffic dataset and split it before extracting features. Malicious apps

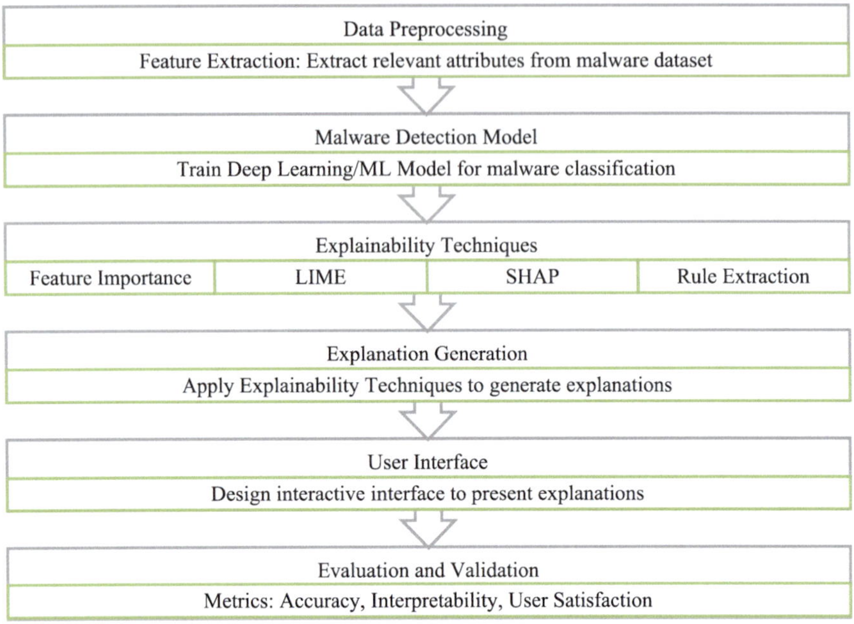

Fig. 3 The XAI framework for malware detection

datasets downloaded from VirusShare.com and different app markets. Later, set the parameters affecting the experiment by checking the impact of packets for feature extraction and configuring the neurons with the best configuration. Finally, compared with other methods to validate its results comparatively better. This method uses the extracted rules from a deep neural network to detect malicious traffic.

Grad-CAM is used to visualize the area of images that helps to interpret and understand the decision-making process of DL models. Iadarola et al. [15] proposed a method that uses images as input to DL models for malware detection and family identification. The method used two parts: standard model training and explainability steps. Grad-CAM is used as a debugging model in image detection problems but here specific shapes to recognize in the images are not available. Model ID 2 has the best result with an accuracy of 0.977 on a dataset containing 8446 samples from seven different malware families. A method that combines human interaction with automated techniques to enhance the explainability of models is called semi-automated explainability-driven. Iadarola et al. [17] SCIBA (Smali Code to Images and BAck) was introduced as a tool for analyzing Android malware. Collected 8,439 real-world Android malware samples from six different malware families. The tool consists of two modules, TAMi and CATI, where CATI converts.apk files to.png image files while TAMI is responsible for training and testing CNN models. Grad-CAM is used to visualize the area of image decisions which are also made by the CNN model.

Table 1 Available datasets in context of Android Malware category

Sr. No.	References	Dataset Name	Year	Dataset details
1	[5]	VirusShare	Updating	System currently contains 65,424,639 malware samples
2	[6]	Android Malware Genome Project	2012	Most of the existing android malware families are covered in this dataset by over 1200 malware samples during the period Aug 2010–Oct 2011
3	[7]	Drebin	2014	The research intends to collect a wide range of application features by conducting an exhaustive static analysis of the Android platform. It evaluates a large dataset that includes 123,453 apps from various markets, as well as 5,560 current malware samples from 179 different malware families
4	[8]	CICMalDroid2020	2020	This dataset contains a collection of 11,598 instances. Out of a total 11,598 instances, 1795 are benign and 9,803 of are malignant
5	[9]	AMD	2021	AMD (Advanced Malware Dataset) contains 24,553 samples divided into 135 categories and covers 71 unique malware families from 2010 to 2016
6	[10]	ImageNet	2020	14,197,122 images, 21,841 synsets indexed
7	[11]	CIFAR-10	2019	The CIFAR-10 dataset consists of 60,000 color pictures, each measuring 32 × 32 pixels and divided into ten classes. There are 6,000 photos in each of these categories. For evaluation and model assessment, the dataset is further divided into 50,000 training photos and 10,000 test images
8	[11]	MNIST	2019	This collection contains 60,000 grayscale pictures, each 28 × 28 pixels in size, of handwritten single digits spanning from 0 to 9

(continued)

Table 1 (continued)

Sr. No.	References	Dataset Name	Year	Dataset details
9	[12]	National Internet Emergency Center	2021	The National Internet Emergency Centre provided 10,010 samples, with an emphasis on recent harmful incidents rather than older datasets like Genome from 2011
10	[13]	AndroidZoo	2022	AndroZoo is a growing archive of Android apps gathered from many sources, including the official Google Play app store. There are now 22,969,574 different APKs in the collection. Numerous antivirus programs analyze each APK to identify whether applications are tagged as malware (or will soon do so)

Convolutional neural networks are used for image classification and recognition tasks. There are also some difficulties interpreting black-box model predictions. Kinkead et al. [16] presented a method for CNN model explainability on Android malware detection using LIME. CNN achieved good detection performance of the Drebin dataset, which consists of 5560 malicious and benign apps from different malware families. They compare important locations identified by CNN and LIME, where CNN aligns closely with those identified by LIME. Manthena et al. [18] analyze malware behavior in the cloud environment with the help of ML and DL models using performance metrics for SVM, RF, FFNN, and CNN. In these matrices, CNN performs better than all others with 97.01% accuracy. The study discusses the effectiveness of DL models, especially in detecting and classifying malware with the use of preprocessing, feature extraction, model training, and evaluation. For different colors of images, different methods are used. Marais et al. (2022) three models are trained on the Ember dataset using their respective feature extraction approach. With the aid of three hybrid models, propose CNN to identify malware using grayscale photographs as input. Then, using the HIT approach, train another CNN on an RGB picture. In the grayscale model, CNN + RF archives higher accuracy of 0.8804. On RGB, CNN model accuracy is 0.94. Static analysis tests three algorithms XGBoost, LightGBM, and (DNN) where DNN has a better accuracy of 0.9071.

In the evaluation of explainable artificial intelligence interpretability is possible by different methods. Marais et al. (2022) introduce three models that are trained on the Ember dataset using their respective feature extraction approaches. With the aid of three hybrid models, propose CNN to identify malware using grayscale photographs as input. Then, using the HIT approach, train another CNN on an RGB picture. In the grayscale model, CNN + RF archives higher accuracy of 0.8804. On RGB, CNN model accuracy is 0.94. Static analysis tests three algorithms XGBoost, LightGBM, and (DNN) where DNN has a better accuracy of 0.9071 [21].

A form of neural network architecture developed exclusively to function on graph-structured data is called Graph Convolutional Network (GCN). GCNs have received a lot of attention in the subject of graph representation learning and have proved to be useful in tasks including node classification, link prediction, and graph classification. Gao et al. (2021) present GDroid which is a GCN-based approach to Android malware detection and family classification. GDroid improves classification and API usage accuracy which is good when compared with the help of real-world data. It is best compared to methods that enhance malware classifiers as opposed to model-extracting APIs with similar Android API documentation. It properly represents real-world events [22].

There is also an AdMat method which is a research technique for detecting and classifying Android malware. It makes use of a Convolutional Neural Network (CNN) architecture on matrices produced from Android application (APK) files. The goal of AdMat is to identify and categorize Android malware based on structural factors. Vu and Jung [9] introduce AdMat which solves the preprocessing of APK files for NN by transferring raw opcodes into images type objects while keeping the API call relationships. The workflow of AdMat contains the conversion of call graphs in GML, feature engineering, and CNN with two dense layers. By using these it makes APKs that fit into deep learning-based malware detection and classification.

Evaluating deep learning explanation techniques in security entails examining their efficacy, interpretability, and resilience in the context of security applications. Warnecke et al. [4] investigate six popular explanation methods and their contributions to security systems for malware detection, and vulnerability checkers. In comparing white-box, and black-box explanation methods the white-box method achieves the best. IG and LRP give the best performance in all methods. If there is no availability of white-box there, black-box method LIME is recommended (Table 2).

5 Analysis and Discussion

The data presented here provides a complete review of research publications focusing on the detection of Android malware utilizing Explainable Artificial Intelligence (XAI) in conjunction with deep learning models. To address the difficult issues of malware detection, the research applies several deep learning architectures such as convolutional neural networks and GCNs. Furthermore, the use of XAI techniques such as Grad-CAM, LIME, SHAP, IG, and LRP enables the interpretation and visualization of decision-making processes inside models.

Diverse Datasets: The evaluated publications train and evaluate their models using a variety of datasets. VirusShare, Android Malware Genome Project, Drebin, CICMalDroid2020, AMD, and others are among the datasets used. The usage of real-world malware samples and malware families assures the resilience and efficacy of the models in real-world circumstances.

Deep Learning Models Outperform Traditional techniques in Android Malware Detection: The results show that deep learning models, notably CNNs and GCNs,

Table 2 XAI applications in defending mechanism against malware

Ref	Models	Year	XAI techniques										
			Local	Global	Model-specific	Model-agnostic	Post hoc	Intrinsic	Text	Visual	Arguments	Models	XAI methods
[1]	DNN	2020	✔		✔		✔					✔	Generated trees
[7]	DNN	2014	✔			✔	✔		✔		✔	✔	SHAP
[9]	CNN	2021	✔				✔		✔	✔			Grad-CAM, Heatmap
[10]	CNN	2021	✔			✔	✔			✔	✔		LIME
[16]	CNN	2021	✔	✔	✔			✔		✔			Grad-CAM
[12]	RF,SVM,DNN	2023	✔	✔	✔	✔	✔		✔	✔	✔	✔	SHAP
[14]	DNN	2020	✔		✔		✔			✔	✔	✔	BP, GBP, GGCAM, OCC, FA, LIME
[17]	DNN	2020	✔	✔	✔	✔	✔		✔				LRP, Gradients, IG, LEMNA, LIME, SHAP

outperform traditional techniques in Android malware detection tasks. These models are highly accurate and efficient in detecting malware samples from various malware families.

Enhancements to Interpretability: XAI approaches play a critical role in making black-box deep learning models more transparent and interpretable. Grad-CAM, for example, visualizes critical areas in pictures to help in comprehension of the models' judgements. When we look closely at how XAI techniques such as Grad-CAM, LIME, SHAP, IG, and LRP work in real-life situations, we can see how they affect how understandable models are. Figure 4 shows how much better we can understand these models after using these techniques. It highlights how they help us grasp the complex decisions models make.

Comparisons of performance: Comparative evaluations of various XAI approaches and model designs provide important information. Several research show that approaches like SHAP and LIME are useful for elucidating deep learning model predictions. These research make use of Explainable AI (XAI) tactics that are classified into two types: model-specific and model-agnostic approaches. Model-specific approaches are designed to understand a single model, but model-agnostic approaches are adaptable and may be used to any model, increasing their versatility. Furthermore, these research papers mainly depend on post hoc explanations inside XAI techniques, allowing for insights to be gained after a model's training period. This feature makes it easier to apply them to existing models, enhancing their interpretability and usefulness.

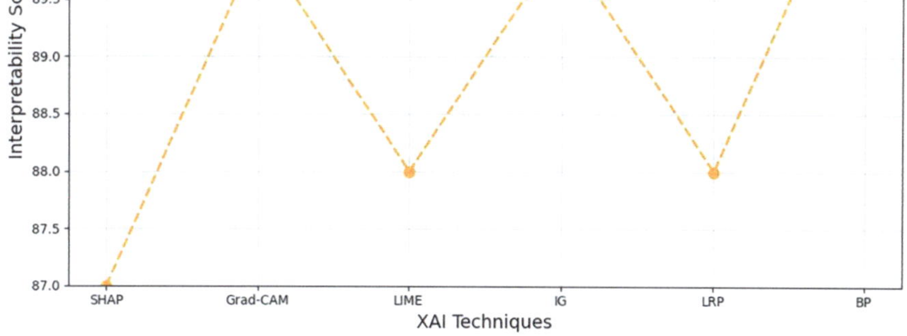

Fig. 4 Impact of XAI techniques

6 Conclusion

A robust solution for identifying Android malware is Explainable Artificial Intelligence (XAI) utilizing deep learning models. An examination of multiple studies reveals that these combined methodologies have a high success rate in precisely identifying varied malware families and harmful network activity. This work emphasizes the importance of combining deep learning's ability to grasp complicated data patterns with XAI's ability to explain and show these patterns, putting light on the decision-making processes within opaque algorithms. This emphasis on transparency and interpretability is critical in creating trust and assurance in detection systems, making them more user-friendly and accessible to cybersecurity specialists. The XAI approach BP earned the highest score among single-target trojaned AlexNet models in four of six trojaned AlexNet models. However, the effects for many targets are less certain. In this regard, GGCAM and OCC offer considerable possibilities. LIME, SHAP, and LEMNA are more efficient in managing huge amounts of malware samples for malware analysis. When considering assessment criteria, IG and LRP stand out as the top methodologies overall. Explainability techniques must be enhanced and polished further to address difficulties such as robustness to adversarial attacks, scalability to handle large-scale datasets, and generalizability across different malware types.

Finally, we can conclude that the combination of XAI with deep learning for Android malware detection has enormous potential for strengthening cybersecurity defenses. As this dynamic subject evolves, academics and practitioners must collaborate and push the frontiers of knowledge to design more secure and resilient systems to protect our digital environment in future.

References

1. Kajal A, Nandal SK (2020) Cyber security against DDoS, Malware, spoofing attacks using machine learning with genetic algorithm. Int J Adv Sci Technol 29(5):5388–5400
2. Malware Statistics & Trends Report|AV-TEST. https://www.av-test.org/en/statistics/malware/. Accessed 20 Dec 2023
3. O. Chergykalo D, Klyushin DA (2023) Fundamental fallacies in definitions of explainable AI: explainable to whom and why? Intell Syst Ref Lib 232:25–42
4. Warnecke A, Arp D, Wressnegger C, Rieck K (2020) Evaluating explanation methods for deep learning in security. In: Proceedings - 5th IEEE European symposium on security and privacy, euro S and P 2020, pp 158–174
5. VirusShare.com. https://virusshare.com/. Accessed 29 July 2023
6. Zhou Y, Jiang X (2012) Dissecting android malware: characterization and evolution. In: Proceedings - IEEE symposium on security and privacy, pp 95–109
7. Arp D, Spreitzenbarth M, Hübner M, Gascon H, Rieck K (2014) Drebin: effective and explainable detection of android malware in your pocket
8. Alani MM, Awad AI (2022) PAIRED: an explainable lightweight android malware detection system. IEEE Access 10:73214–73228
9. Vu LN, Jung S (2021) AdMat: a CNN-on-matrix approach to android malware detection and classification. IEEE Access 9:39680–39694

10. Lin YS, Lee WC, Celik ZB (2020) What do you see? Evaluation of explainable artificial intelligence (XAI) interpretability through neural backdoors. In: Proceedings of the ACM SIGKDD international conference on knowledge discovery and data mining, pp 1027–1035
11. Fidel G, Bitton R, Shabtai A (2019) When explainability meets adversarial learning: detecting adversarial examples using SHAP signatures. In: Proceedings of the international joint conference on neural networks
12. Wu B, Chen S, Gao C, Fan L, Liu Y, Wen W, Lyu MR (2020) Why an android app is classified as malware? Towards malware classification interpretation. ACM Trans Softw Eng Methodol 30(2)
13. Liu Y, Tantithamthavorn C, Li L, Liu Y (2022) Explainable AI for android malware detection: towards understanding why the models perform so well? In: Proceedings - international symposium on software reliability engineering, ISSRE, October 2022, pp 169–180
14. Yan A, Chen Z, Zhang H, Peng L, Yan Q, Hassan MU, Zhao C, Yang B (2021) Effective detection of mobile malware behavior based on explainable deep neural network. Neurocomputing 453:482–492
15. Iadarola G, Martinelli F, Mercaldo F, Santone A (2021) Towards an interpretable deep learning model for mobile malware detection and family identification. Comput Secur 105
16. Kinkead M, Millar S, McLaughlin N, O'Kane P (2021) Towards explainable CNNs for android malware detection. Procedia Comput Sci 184:959–965
17. Iadarola G, Casolare R, Martinelli F, Mercaldo F, Peluso C, Santone A (2021) A semi-automated explainability-driven approach for malware analysis through deep learning. In: Proceedings of the international joint conference on neural networks, July 2021
18. Manthena H, Kimmel JC, Abdelsalam M, Gupta M (2023) Analyzing and explaining black-box models for online malware detection. IEEE Access 11:25237–25252
19. Chopra M, Kumar A (2023) Evaluation measures and applications for explainable AI. Intell Syst Ref Lib 232:89–110
20. Gianfagna L, Di Cecco A (2021) The landscape. In: Explainable AI with python, pp 1–25
21. Marais B, Quertier T, Morucci, S (2022) AI-based malware and ransomware detection models.
22. Gao H, Cheng S, Zhang W (2021) GDroid: Android malware detection and classification with graph convolutional network. Computers & Security, 106, 102264.

Network Intrusion Detection Using LSTM-Based Models

Aditya Bhushan, Ashutosh Kumar Singh, and Vijay Kumar Dwivedi

Abstract A major challenge in detecting anomalies in dataset is the need for labeled data to train and validate models. In this paper, we analyze a novel approach that use the Long Short-Term Memory (LSTM) to identify attacks in an imbalanced dataset. Importantly, our models are trained exclusively on examples from the normal class. The LSTM is specifically trained to recognize the typical traffic patterns in normal data and to create a compressed representation of the input data. With a focus on real-time detection, this paper illuminates the path to a new era of proactive network security, where dynamic, context-aware anomaly detection stands as the vanguard against the ever-evolving landscape of cyberattacks. With a commitment to achieving more than 99% accuracy, we meticulously outline our methodology for preprocessing the CICIDS2018 dataset, encompassing data cleaning and feature engineering, particularly in the presence of imbalanced datasets where normal instances significantly outnumber anomalous ones. We detailed the methodology for preprocessing the CICIDS2018 dataset.

Keywords CICIDS2018 dataset · Network intrusion detection · Machine learning · LSTM · Security

1 Introduction

In today's world, where networks form the foundation of modern communications and commerce, the security of these networks is critical. The growth of connected devices and changing cyber threats have made traditional network access systems ineffective at protecting sensitive data and critical systems. To counter this evolving threat, this work has delved into deep learning, specifically short-term temporal

Ashutosh Kumar Singh and Vijay Kumar Dwivedi—These authors contributed equally to this work.

A. Bhushan (✉) · A. K. Singh · V. K. Dwivedi
Department of Computer Science and Engineering, United College of Engineering and Research, Prayagraj, Uttar Pradesh, India
e-mail: adityabhushan2017@gmail.com

A. P. Agrawal et al. (eds.), *Advances of Artificial Intelligence and Machine Learning in Image Processing, Security and Data Analytics*, Lecture Notes in Electrical Engineering 1464, https://doi.org/10.1007/978-981-95-0148-9_4

Long Short-Term Memory (LSTM)-based error detection models, on a new method for detecting network access. Network Intrusion Detection Systems (NIDS) have long been a cornerstone of defense against such threats. These systems serve as sharped-eyed gatekeepers, monitoring network traffic in real time, and indicate any suspicious or malicious activities [1]. Traditional NIDS, often rule-based, have played a crucial role in identifying known threats based on predefined patterns. However, as cyber threats continue to evolve and grow in complexity, these rule-based systems face limitations in adaptability and the ability to detect novel attacks effectively. By learning from historical data, anomaly detection models develop a baseline of normal behavior, enabling them to discern when new data points deviate significantly from this baseline. These deviations, or anomalies, can manifest as outliers, unusual patterns, or unexpected behaviors, which may indicate critical events or anomalies that demand attention and action [2].

The importance of anomaly detection extends far beyond identifying security threats. In healthcare, it can help detect rare diseases or medical conditions early, potentially saving lives. In manufacturing, it aids in predictive maintenance by flagging equipment abnormalities before they lead to costly breakdowns. In finance, it safeguards against fraudulent activities by identifying suspicious transactions. In environmental monitoring, it can signal unusual pollution levels or natural disasters. An essential component of network security is the early detection and reduction of these intrusions [3]. Traditional security measures, such as firewalls and signature-based intrusion detection systems (IDS), play a vital role in defending against known threats. These systems excel at recognizing predefined patterns of malicious activity.

However, they often struggle when confronted with novel, previously unseen attack vectors or sophisticated, evasive tactics employed by malicious actors. Anomaly detection, a sub-field of machine learning and data analytics, offers a complementary approach to network security. Instead of relying on predefined patterns of malicious behavior, anomaly detection seeks to identify deviations from normal network activity, it becomes possible to flag any activity that falls outside this baseline as potentially suspicious or malicious. The model architecture is shown in Fig. 1. In the network security, anomaly detection can be a game changer. It has the potential to identify and detect insider threats, where legal users may engage in malicious activities, often flying under the radar of traditional security measures.

Researchers in this field have explored various methodologies and approaches to enhance the detection of abnormal activities and potential threats within network traffic. Different approaches leverage the power of feature extraction and pattern recognition to distinguish normal network behavior from potentially malicious activities. Moreover, the research community has delved into the evaluation and bench marking of intrusion detection systems using standardized datasets, to ensure fair and objective comparisons among different techniques [4]. Overall the field anomaly detection continues to evolve as researchers explore innovative methods, adapt to changing network environments, and collaborate to enhance the security of digital networks against increasingly sophisticated threats. Table 1 shows the percentage distribution for the network traffic represented by CSE-CIC-IDS2018 [5]. Most of the

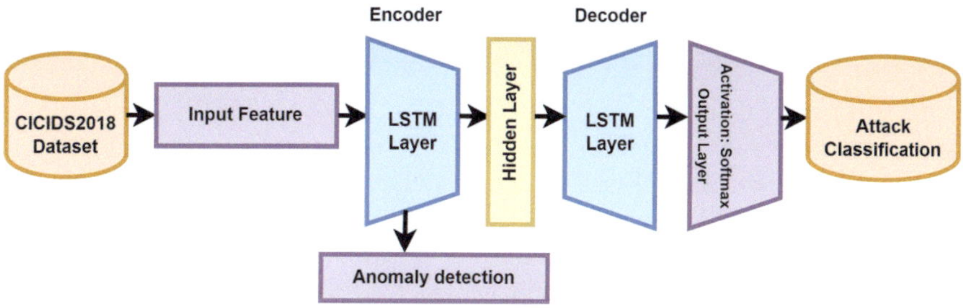

Fig. 1 Anomaly classification by used method

Table 1 CICIDS2018: Network traffic classifications

Traffic types	Distributions(%)
Benign	63.67
SSH-brute force	17.89
FTP-brute force	18.44

attacks made by intruders are malignant (almost 700k) FTP-Brute-force and SSH-Brute-force type attacks are less in numbers (less than 200k). Most of the intruders try to make a malignant attack on network systems.

We seek to uncover the strengths and limitations of LSTM networks in handling a wide spectrum of network intrusion outline, including both known and unknown threats. By leveraging LSTM-based anomaly detection models, we aim to enhance our network security posture, enabling the proactive identification of suspicious activities that deviate from established network norms [6]. These systems primarily rely on predefined signatures and rules to identify known threats, leaving them ill-equipped to detect novel or previously unseen attack patterns. In this paper, the real dataset from the CSE-CIC-IDS2018 represents dynamic, modifiable, repeatable, scalable data, using LSTM model which is one of the acceptable learning models in dealing with prediction and obtaining an accuracy rate high in detection up to 99%.

The major contributions of this paper as follows:

- Explored LSTM-based models for anomaly detection in addition to traditional intrusion detection.
- Investigated techniques to make LSTM-based models more interpretable.
- Investigated techniques to create lightweight LSTM models suitable for deployment in resource constrained environments.

The remaining work in this paper is organized into the sections. Section 2 review and summarize existing literature relevant to the research topic. Section 3 provide a detailed description of the model design, data collection methods, and analytical techniques used. Section 4 presents the findings of the research and interprets the results, and paper is concluded in Sect. 5.

2 Related Work

In this section, we discusses some of public IDS databases since 1998, highlighting their shortcomings and problems, thus highlighting the real need for detailed and reliable information. Several datasets are available for network intrusion detection research and practice. When using these datasets, it's essential to review the documentation and understand their specific characteristics, including the types of attacks included, data preprocessing steps, and any known limitations. Particular often use these datasets to train and evaluate intrusion detection models, helping to improve the security of networks and systems.

Lin et al. [7] utilizes the deep learning enables the system to uncover intricate patterns and anomalies within network traffic, even in large-scale and complex environments. The CIC-IDS 2018 dataset was employed to train the proposed form, and the results analysis indicates the accuracy of 96.22%, a detection rate of 15%, and a recall rate of 96%.

Zhou and Pezaros [8] understand the strengths and limitations of various machine learning classifiers, and formulate the six different deep learning methods were applied to the CIC-AWS-2018 dataset for the purpose of detecting attacks, classify zero-day attacks and recorded the 100% of intrusion rate, a zero-day intrusion accuracy rate of 96%, and a 5% false-positive rate.

Traditional signature-based intrusion detection methods often prove inadequate in defending against these sophisticated threats. To find the gap, one can turned to machine learning as a powerful tool for identifying novel and emerging threats. To do so effectively, we leverage a carefully curated dataset that reflects real-world network traffic scenarios and includes instances of zero-day attacks [9].

2.1 KDD Cup 1999 Data

The KDD Cup 1999 dataset is one of the most widely used datasets for intrusion detection research. It contains network traffic data collected over a week, with various types of attacks injected into the traffic. This dataset is often used for bench marking intrusion detection systems. While working with the KDD Cup 1999 dataset, you'll likely need to perform data preprocessing, handle missing values, and deal with class imbalance issues having the statistics, number of records is 4,898,431, 41 features, 23 classes(including various types of network attacks), data type be both continuous and categorical [10]. KDD Cup 1999 Dataset pre-processed the features are extracted to create a more structured dataset for data mining tasks.

NSL-KDD is an improved version of the KDD Cup 1999 dataset. It addresses some of the limitations of the original dataset, such as redundancy and the presence of duplicate records. NSL-KDD is more suitable for modern intrusion detection research. This dataset has the 41 features, 2 labels classes (binary), and 22 different attack types.

2.2 *CICIDS2017*

The Canadian Institute for Cybersecurity Intrusion Detection Evaluation Dataset (CICIDS2017) is a recent dataset that includes both benign and malicious network traffic. It includes a diverse range of network traffic data captured in a controlled environment [11]. The dataset is designed to resemble real-world network traffic conditions and cyber threats, providing a more accurate representation of the challenges faced by intrusion detection systems. One can use this dataset as a standard to compare the performance of different IDS. The dataset supports various research areas, including anomaly detection, signature-based detection [12], and network security analysis.

2.3 *UNSW-NB15*

The University of New South Wales Network Intrusion Detection Evaluation Dataset (UNSW-NB15) is another contemporary dataset designed for intrusion detection research[13]. It includes different types of attacks in network traffic, and it has been pre-processed for ease of use. This dataset is valuable for training and evaluating intrusion detection systems and other cybersecurity related machine learning models. In September 2021, it consists of approximately 2.8 million records. UNSW-NB15 includes various types of attacks, categorized into different classes. Examples of attack categories include denial-of-service (DoS), probing, intrusion detection evasion (IDE), and more.

2.4 *DARPA Intrusion Detection Data*

DARPA 1999 Dataset was collected in 1999 during the DARPA Intrusion Detection Evaluation Program (IDEP) [14]. It contains network traffic data captured from a simulated military network environment. DARPA 2000 dataset like the 1999 dataset, it includes network traffic data with both normal and attack scenarios. The dataset is designed to support intrusion detection research and evaluates the performance of intrusion detection systems in a more complex environment.

3 Method

This section outlines the development of a network intrusion detection model employing deep learning techniques and LSTM. The model is engineered to effectively handle the intricate and multidimensional characteristics of network traffic data. To enhance the detection capabilities, a system that combines both anomaly detection and the classification of diverse types of attacks is employed. This system is implemented and evaluated using a real-world dataset called CSE-CICIDS2018, which encompasses a comprehensive array of attack scenarios.

3.1 Real Dataset

In this part, the utilized data for implementing IDS is described. The dataset comprises with 80 features. The dataset contains a large number of records, specifically around 16 million instances. CICIDS2018 is a multi-class dataset, typically represent different types of network activities, including normal traffic and various attack types. Common attack categories found in the dataset include denial-of-service (DoS) attacks, intrusion attempts, and more. When working with the CICIDS2018 dataset, researchers typically split it into training and testing subsets to train and evaluate their intrusion detection models. Similarly, some other dataset, describe in Table 2 , provide the information like the CICIDS2018 dataset which provide a framework to work upon the IDS.

3.2 Long Short-Term Memory

Deep learning is a powerful method for accurate detection and prediction in complex data like videos, images, and text. It uses multiple layers to process data, improving accuracy and reducing costs in attack and malware detection. Compared to traditional machine learning, deep learning methods, such as LSTM, automatically extract features, making them superior for intrusion detection in cybersecurity. LSTM, specifically, is valuable for analyzing sequential network traffic data, aiding in tracking current and past network activity to address long-term dependencies and enhance detection.

In anomaly detection, LSTM model can be trained on historical data to learn the regular patterns and relationships within the data describe in Algorithm 1. Once trained, it can continuously analyze incoming data and flag instances that deviate significantly from the learned patterns as potential anomalies. This capacity to model complex temporal dependencies makes LSTM a valuable tool for identifying irregularities and potential threats in various domains, including cybersecurity, finance, and industrial monitoring.

Table 2 Comparison between different public available datasets

	Network	Traffic	Label	Interact	HTTP	Email	SSH	FTP	Bforce	DoS	Scan	DNS	Features
KDD'99 [5]	Yes	NO	Yes	Yes	Yes	Yes	Yes	Yes	Yes	Yes	Yes	NO	Yes
DARPA [13]	Yes	Yes	Yes	Yes	Yes	Yes	Yes	Yes	Yes	Yes	Yes	NO	NO
CDX [11]	NO	NO	NO	Yes	NO	NO	NO	NO	NO	Yes	NO	Yes	NO
KYOTO [7]	Yes	NO	Yes	Yes	Yes	Yes	Yes	Yes	Yes	Yes	Yes	Yes	Yes
ISCX2012[15]	Yes	NO	Yes	Yes	Yes	Yes	Yes	Yes	NO	Yes	Yes	NO	NO
CICIDS2018 [10]	Yes	Yes	Yes	Yes	Yes	Yes	Yes	Yes	Yes	Yes	Yes	Yes	Yes

Algorithm 1 Long short-term memory

```
Input: CICIDS2018 Dataset (features and labels)
Output: Trained LSTM model
Data Preprocessing:
Load and preprocess the CICIDS2018 dataset (e.g., normalization, one-hot encoding).
Split Dataset:
Split the dataset into training, validation, and test sets.
Initialize LSTM Model:
Create an LSTM-based model for anomaly detection.
Training:
for each epoch do
    for each batch in training set do
        Forward pass: Compute predictions.
        Calculate loss (e.g., using Mean Squared Error).
        Backpropagation: Update model parameters.
    end for
    Validate the model on the validation set.
    Early stopping if validation loss does not improve.
end for
Testing:
Test the trained model on the test set.
Evaluation:
Evaluate model performance (e.g., accuracy, precision, recall, F1-score).
Save Model:
Save the trained LSTM model for future use.
Return Trained LSTM model.
```

Table 3 Implemented experiment description

Unit	Description
Processor	3.4 GHz Intel Core i3
RAM	8 GB
OS	Windows 10
Package used	TensorFlow, sklearn, numpy

3.3 Experimental Environment

The model was implemented using Visual Studio Code 2019 with Python 3.9 version. Our experiments were carried out using the following system setup shown in Table 3 .

4 Results and Discussion

In this section, the experimental results are presented. After evaluating the performance of the results using the main measures, we then proceed to compare the models.

Table 4 Parameter of proposed model

Parameter name	Value
Hidden layer	2
First hidden layer	128 node
Second hidden layer	64 node
Epoch	10
Dropout rate	0.5
Activation function	ReLU
Learning rate	0.001
Batch size	64

4.1 Results

In this section, we analyze the hyperparameters that must be set to avoid the overfitting, as shown in Table 4 . The first hidden layer has 128 nodes with ReLU (Rectified Linear Unit) activation. This means that there are 128 hidden nodes in the first hidden layer. The "ReLU" activation function is applied to the outputs of these nodes. After the first hidden layer, there is a dropout layer with a dropout rate of 0.5. Dropout is a regularization technique that randomly drops a fraction of the nodes during training to prevent overfitting shown in Fig. 2.

The second hidden layer follows the dropout layer and has 64 nodes with ReLU activation. There is an output layer with the number of nodes equal to the number

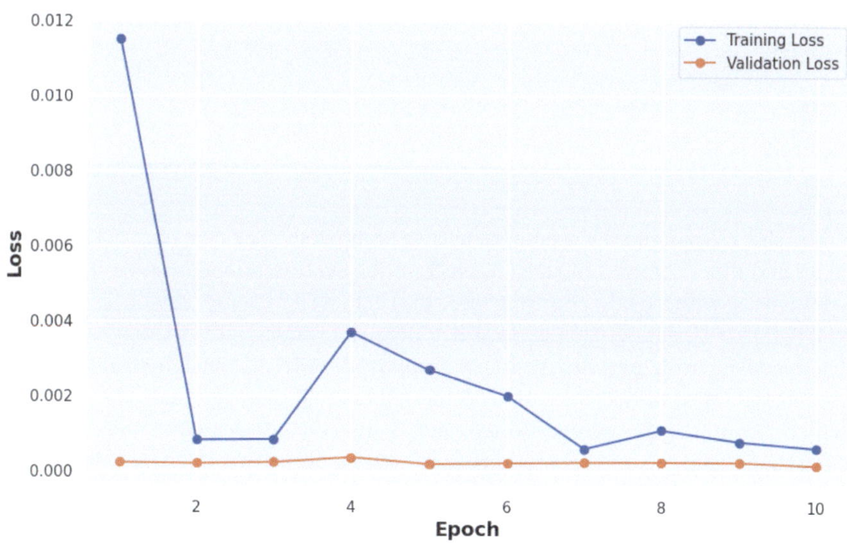

Fig. 2 Training/Validation loss

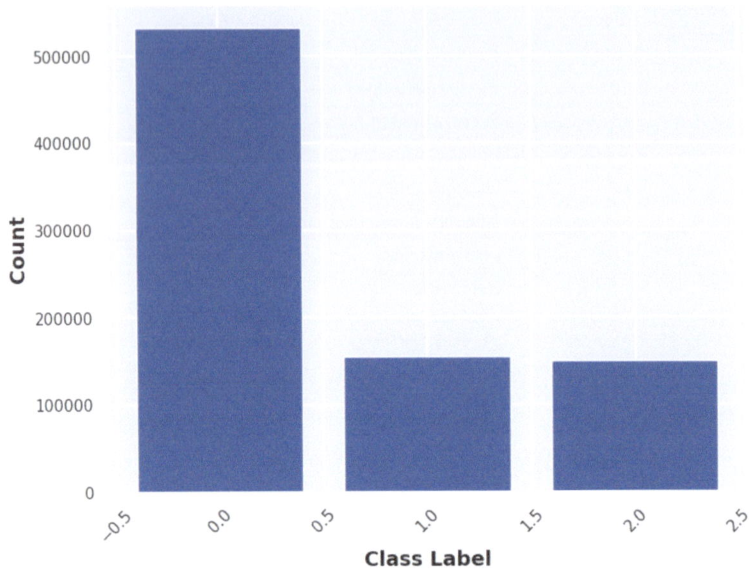

Fig. 3 Epoch class distribution probabilities

of classes in your classification problem. The output layer uses softmax activation, which is commonly used in multi-class classification problems to obtain class probabilities shown in Fig. 3, and the code defines two hidden layers with 128 and 64 nodes, respectively, in the neural network model. These are the hidden nodes responsible for performing computations on the input data to learn representations and make predictions.

4.2 Discussion

In this part, we will provide a comprehensive comparison between our LSTM-based model and other selected models. Table 5, in the section provides an listing of the papers, along with the best performance scores achieved by some previously work for the CICIDS2018 dataset, comparison will involve evaluating the models' accuracy. The objective is to determine whether our proposed model demonstrates superior performance in terms of detecting network intrusions or if there are areas for improvement. By analyzing this comparison, we aim to light on the strengths and weaknesses of our model and provide perception into its potential real-world applicability.

The proposed LSTM-based model, implemented using TensorFlow and Scikit-learn, stands out with a perfect accuracy of 100.00%. While other models show strong performance, the choice between them may depend on factors like computational complexity, ease of implementation, and interpretability. Consider the computational

Table 5 Comparative analysis between proposed and other model

References	Method	Environment	Accuracy (%)
Lin et al. [7]	LSTM	TensorFlow	96.20
Li et al. [16]	Deep autoencoder	Python	–
Zhao et al. [17]	Deep autoencoder	Python, TensorFlow	97.90
Kanimozhi et al. [18]	MLP	Python, Scilit-learn	99.97
Chadza et al. [19]	Baum–Welch	MATLAB	97.00
Atefnia et al. [20]	Modular DNN	Python, Scilit-learn	100.00
Filho et al. [21]	Random forest	Python, Scilit-learn	–
The proposed model	LSTM	TensorFlow, sklearn	100.00

Table 6 Test accuracy for model

Total samples	Accuracy
6540	0.999

resources required for each model, as this could impact practical applicability. By examining this comparison, the proposed model appears promising, but a deeper analysis of various metrics and practical considerations will provide a more nuanced understanding of its strengths and areas for potential improvement. The data in Table 6 indicates that there are 6540 total samples in the test set, and the corresponding accuracy achieved by the model is 99.9%. An accuracy of 99.9% suggests that the model made correct predictions for almost all of the 6540 samples. This high accuracy level indicates a strong performance of the model in distinguishing between normal and intrusive network activities.

5 Conclusion

We have presented an in-depth analysis of LSTM-based detection models for network. We have demonstrated that LSTM-based models exhibit promising performance in detecting network intrusions and anomalies. These models can effectively capture sequential dependencies within network traffic data, enabling the identification of subtle and previously unseen threats. We have explored the application of LSTM neural networks in the context of identifying anomalous network behaviors and potential security threats. In conclusion, our study underscores the promise of LSTM-based anomaly detection models in network intrusion detection. Our experiments have shown that LSTM-based models are adaptable to a wide range of network datasets, making them versatile tools for network intrusion detection across different environments and network architectures. It represent a valuable addition to the arsenal of cybersecurity tools, offering enhanced capabilities for identifying emerging threats and safeguarding critical network resources for future purpose.

References

1. Verkerken M, D'hooge L, Wauters T, Volckaert B, De Turck F (2022) Towards model generalization for intrusion detection: unsupervised machine learning techniques. J Netw Syst Manag 30:1–25
2. Yuan S, Wu X (2022) Trustworthy anomaly detection: a survey. arXiv:2202.07787
3. Gupta S, Aga D, Pruden A, Zhang L, Vikesland P (2021) Data analytics for environmental science and engineering research. Environ Sci Technol 55(16):10895–10907
4. Hossain MD, Ochiai H, Doudou F, Kadobayashi Y (2020) Ssh and ftp brute-force attacks detection in computer networks: Lstm and machine learning approaches. In: 2020 5th international conference on computer and communication systems (ICCCS). IEEE, pp 491–497
5. Leevy JL, Khoshgoftaar TM (2020) A survey and analysis of intrusion detection models based on cse-cic-ids2018 big data. J Big Data 7(1):1–19
6. Althubiti SA, Jones EM, Roy K (2018) Lstm for anomaly-based network intrusion detection. In: 2018 28th international telecommunication networks and applications conference (ITNAC). IEEE, pp 1–3
7. Lin P, Ye K, Xu C-Z (2019) Dynamic network anomaly detection system by using deep learning techniques. In: Cloud computing–CLOUD 2019: 12th international conference, held as part of the services conference federation, SCF 2019, San Diego, CA, USA, June 25–30, 2019, Proceedings 12. Springer, pp 161–176
8. Zhou Q, Pezaros D (2019) Evaluation of machine learning classifiers for zero-day intrusion detection–an analysis on cic-aws-2018 dataset. arXiv:1905.03685
9. Moustafa N, Koroniotis N, Keshk M, Zomaya AY, Tari Z (2023) Explainable intrusion detection for cyber defences in the internet of things: opportunities and solutions. IEEE Commun Surv Tutor
10. Protić DD (2018) Review of kdd cup '99, nsl-kdd and kyoto 2006+ datasets. Vojnotehnički glasnik/Military Technical Courier 66(3):580–596
11. Panigrahi R, Borah S (2018) A detailed analysis of cicids2017 dataset for designing intrusion detection systems. Int J Eng Technol 7(3.24):479–482
12. Singh A, Prakash J, Kumar G (2022) Intrusion detection system: a comparative study of machine learning-based ids
13. Ahmad M, Riaz Q, Zeeshan M, Tahir H, Haider SA, Khan MS (2021) Intrusion detection in internet of things using supervised machine learning based on application and transport layer features using unsw-nb15 data-set. EURASIP J Wirel Commun Netw 2021(1):1–23
14. Özgür A, Erdem H (2016) A review of kdd99 dataset usage in intrusion detection and machine learning between 2010 and 2015
15. Wei G, Wang Z (2021) Adoption and realization of deep learning in network traffic anomaly detection device design. Soft Comput 25(2):1147–1158
16. Li X, Chen W, Zhang Q, Wu L (2020) Building auto-encoder intrusion detection system based on random forest feature selection. Comput Sec 95:101851
17. Zhao F, Zhang H, Peng J, Zhuang X, Na S-G (2020) A semi-self-taught network intrusion detection system. Neural Comput Appl 32:17169–17179
18. Kanimozhi V, Jacob TP (2019) Artificial intelligence based network intrusion detection with hyper-parameter optimization tuning on the realistic cyber dataset cse-cic-ids2018 using cloud computing. In: 2019 international conference on communication and signal processing (ICCSP). IEEE, pp 0033–0036
19. Chadza T, Kyriakopoulos KG, Lambotharan S (2019) Contemporary sequential network attacks prediction using hidden markov model. In: 2019 17th international conference on privacy, security and trust (PST). IEEE, pp 1–3
20. Atefinia R, Ahmadi M (2021) Network intrusion detection using multi-architectural modular deep neural network. J Supercomput 77:3571–3593
21. Lima Filho FSd, Silveira FA, Medeiros Brito Junior A, Vargas-Solar G, Silveira LF (2019) Smart detection: an online approach for dos/ddos attack detection using machine learning. Sec Commun Netw 2019:1–15

Advantages of Generative AI in Image Processing

Pan Singh Dhoni and Anshul Jain

Abstract The inexorable advancement of artificial intelligence (AI) in image processing emerged as a topic of discussion during an informal gathering among peers, and friends. During a discussion on image enhancement techniques, the potential of AI tools in photo editing, specifically in areas such as color improvement, noise reduction, sharpening, and enhancing overall photo quality, was brought to the forefront. This led to an introspective inquiry into the realm of generative AI and its multifaceted applications in image processing. This research was anchored on a thorough understanding of the foundational principles of AI, the innovative strides of generative AI, and the intricacies of the neural and diffusion/Generative Adversarial Network (GAN) techniques. Through a combination of diverse methodological approaches and practical exposure to advanced image processing tools such as OpenAI's DALL-E [1, 2] and Midjourney [3], our study unearthed the transformative potential of text-to-image conversion. Such advancements are not just technological marvels; they have profound implications on industries reliant on image processing. Our analysis further delved into the multifarious advantages generative AI offers across several sectors, with specific focus on its applications in retail, healthcare, banking, forensics, art, automotive, defense, and agriculture. The overarching conclusion of our study postulates that generative AI is not merely a disruptive force but a transformative agent. Its capacity to enhance performance metrics, catalyze innovation, and solidify competitive positioning across industries is unparalleled. Its tangible impacts are manifold: from elevating image quality standards and saving lives in healthcare settings, to catalyzing revenue growth in sectors like retail and automotive, facilitating prompt and efficient criminal investigations, boosting agricultural productivity, and more.

Keywords AI · Art · Auto industry · Fashion · Generative AI · GAN · Image · ML · Medical · Retail · Photo · Forensics

P. S. Dhoni (✉)
Five Below Inc, Philadelphia, PA, USA
e-mail: ps.dhoni@gmail.com

A. Jain
EPAM Systems, Inc, Newtown, PA, USA

A. P. Agrawal et al. (eds.), *Advances of Artificial Intelligence and Machine Learning in Image Processing, Security and Data Analytics*, Lecture Notes in Electrical Engineering 1464, https://doi.org/10.1007/978-981-95-0148-9_5

1 Introduction

In today's rapidly advancing digital landscape, images have transcended their traditional roles and now profoundly shape our cognitive processes and memories. Their ubiquity, coupled with their innate capacity to convey complex ideas and feelings. The timeless adage, "A picture is worth a thousand words," resonates more than ever in our image-centric world.

An extensive review of existing literature on AI for image processing, including [4, 5], revealed a persistent need for improvements in image quality, turnaround time, and cost-effectiveness across various sectors, ranging from retail [6, 7], medical [4], art, auto [8], to forensic imaging [9]. This paper addresses this need by highlighting the potential of generative AI in image processing to address these long-standing challenges. We also study numerous papers on image processing using AI [10, 11], these papers also, highlighting continue improvement required on image processing.

In this research paper, we delve into the realm of AI, providing a brief discussion on generative AI, followed by an in-depth examination of our chosen methodology. We have implemented a codebase to generate images using generative AI models. In the subsequent steps, we conducted a comparative analysis between non-generative AI and generative AI, representing our findings using tables.

Finally, research paper delves into the multifaceted potential of generative AI in the realm of image processing. Our exploration spans an array of sectors where this burgeoning technology offers transformative advantages, from the vibrant world of art and meticulous realm of photo editing to dynamic industries like retail, medical imaging, automotive design, and beyond.

At the conclusion of this paper, we assert that there is significant potential for the application of generative AI in image processing, with the expectation of continued enhancements in image quality. As technological frontiers expand, we anticipate a broader accessibility to tailored generative AI tools for image processing.

2 What Is Artificial Intelligence

Machines exhibit Artificial Intelligence (AI) through the abilities of perception, synthesis, and inference, distinguishing it from the cognitive capabilities inherent in humans and other animals [12, 13]. Within the domain of AI, machine learning emerges as a specialized branch, further divided into deep learning subsets like Natural Language Processing (NLP), Convolutional Neural Networks (CNNs), and Recurrent Neural Networks (RNNs). Central to these fields are two pivotal learning methodologies: supervised learning, which relies on labeled data, and unsupervised learning, which does not (Fig. 1).

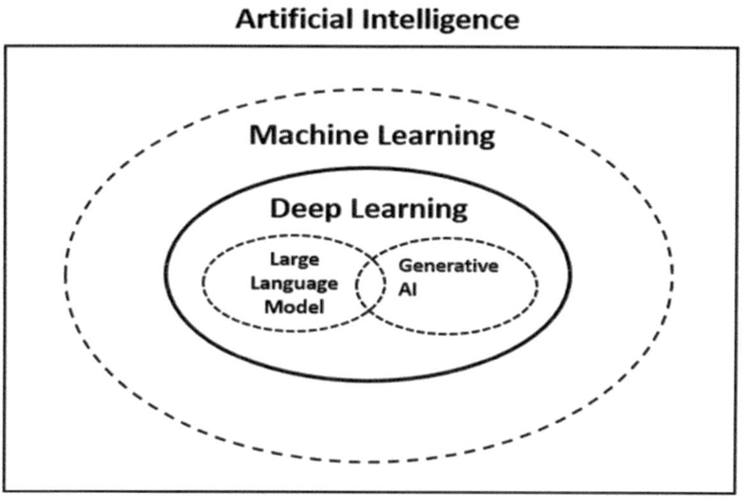

Fig. 1 Artificial intelligence [6]

3 Generative AI

As generate word is self-explanatory, it means it is generating new and original content such as text, image, audio, and videos. As per Wikipedia [14] and research article [15], Generative artificial intelligence generated text, audio, video, and images using generative models. Generative AI models are powered by a variety of techniques, including Generative adversarial networks (GANs), Variational autoencoders (VAEs), and Transformers.

Generative Adversarial Networks (GANs)

GANs [16, 17] are another type of generative AI model that pits two models against each other: a generator and a discriminator. The generator is responsible for generating new data, while the discriminator is responsible for distinguishing between real and generated data. The two models are trained in an adversarial process, where the generator tries to fool the discriminator into thinking that its generated data is real, while the discriminator tries to get better at distinguishing between real and generated data.

4 Methodology

4.1 Generating New Images Using OpenAI and Python

In our pursuit to comprehensively understand and harness the capabilities of OpenAI, we undertook a systematic experimentation process. Utilizing the OpenAI DALL-E API, which stands at the forefront of generative image techniques, we interfaced it with the versatility of the Python programming language. Our central aim was to devise a prototype application that dynamically interprets textual prompts and, in response, conjures up illustrative images that align with the essence of the input provided [18]. This approach not only allowed us to gauge the responsiveness and accuracy of the DALL-E model but also provided insights into its potential applications in real-world scenarios.

For clarity and transparency, we have detailed the underlying code utilized in this experimentation. Additionally, a curated set of sample results offers a glimpse into the visual outputs derived from our methodological approach (Fig. 2).

This code demonstrates how powerful generative AI can be in generating new and previously non-existent images quickly.

4.2 Literature Review and Exploring MONAI Framework

Begin by conducting a thorough literature review to establish a solid foundation in the realm of Generative AI, particularly its application in image processing. Utilize various online resources for a comprehensive exploration. This includes investigating websites such as "midjourney.com" and examining the functionalities of AI tools like "DALL-E." Analyze current blogs, research papers, and articles to discern emerging trends and innovations in Generative AI, with a special focus on image processing applications.

Framework Exploration and Implementation. Employ the MONAI framework [19], specifically designed for medical image processing, and provide detailed sourcing. Implement and evaluate the MONAI framework using code from GitHub [20, 21] repositories, focusing on its application in Generative AI. Leverage Amazon Web Services (AWS) [22] for an efficient and scalable cloud environment to test the MONAI framework. Highlight AWS's features like AI-assisted annotation and their integration with MONAI.

Test and apply generative AI models within the MONAI framework for image synthesis and related tasks. Assess the performance of these models in terms of efficiency, accuracy, and speed, particularly in medical image processing. Record and discuss observations, challenges, and advantages of employing generative AI models in the MONAI framework.

```
import openai
from time import sleep

# read the openai API key
OPENAI_API_KEY = "sk-[redacted]
openai.api_key = OPENAI_API_KEY

# method to create an image from a text prompt provided as input
# uses the openai DALL-E API

def generate_image(prompt_text, number_of_images):
    result = openai.Image.create(
        prompt = prompt_text,
        n=number_of_images, # number of images to generate. Count of images (urls) depends on this value
        size="256x256" # size of the image
    )

    # the result is a JSON with "data" as the key containing 'url's as values (depending on the value of n)
    # extracting just the url
    # Example JSON strcuture:
    #   {
    #           "created": xxxxxxxxxxx,
    #           "data": [
    #                   {
    #                       "url": "https://<image_url>"
    #                   },
    #                   {
    #                       "url": "https://<image_url>"
    #                   },
    #                   {
    #                       "url": "https://<image_url>"
    #                   },
    #               ]
    #   }

    # collect all image urls (based on the value of n)

    image_url_list = []

    for img_url in result["data"]:
        image_url_list.append(fr'{img_url["url"]}')


    return image_url_list

prompt_text = "old man with a dog"
number_of_images = 3
image_url_list = generate_image(prompt_text, number_of_images)

for img_url in image_url_list:
    print(img_url)
✓ 7.1s

https://oaidalleapiprodscus.blob.core.windows.net/private/org-jJqvnKdBvNyiJlcXWAMVdFfx/user-1Ca5M9GmNeqxJpWl6RzkFZSD/img-xdpbKc
https://oaidalleapiprodscus.blob.core.windows.net/private/org-jJqvnKdBvNyiJlcXWAMVdFfx/user-1Ca5M9GmNeqxJpWl6RzkFZSD/img-CGq3Bw
https://oaidalleapiprodscus.blob.core.windows.net/private/org-jJqvnKdBvNyiJlcXWAMVdFfx/user-1Ca5M9GmNeqxJpWl6RzkFZSD/img-7gH59x
```

Fig. 2 Gen AI output

Outcome Analysis. Analyze the results produced by the generative AI models in the MONAI framework. Focus on benefits such as reduced processing times and improved image quality. Discuss the potential implications of advanced image processing in the medical sector, highlighting the life-saving applications made possible by generative AI technology.

Results:

Fig. 2 (continued)

5 Case Study: Image Processing Using Non-generative AI Versus Generative AI

We performed a comparison of the traditional AI-based image processing methods versus generative AI-based image processing methods [23–25]. These were some of the metrics we considered, Accuracy, Computational Efficiency (Processing Time), Generalization, and Interpretability. For the sake of brevity, we are sharing a quantitative and statistical analysis of one sample input image processed via a traditional AI algorithm, namely, "Non-local Means Denoising and then again via a generative AI-based algorithm, namely, "Convolutional Neural Network (CNN) trained for image denoising" (Fig. 3).

In summary, the generative approach (CNN) achieves higher denoising accuracy and better generalization but at the cost of increased training time and potentially lower interpretability compared to the traditional approach (NLM) (Tables 1 and 2).

Graphical representation for PSNR (Fig. 4).

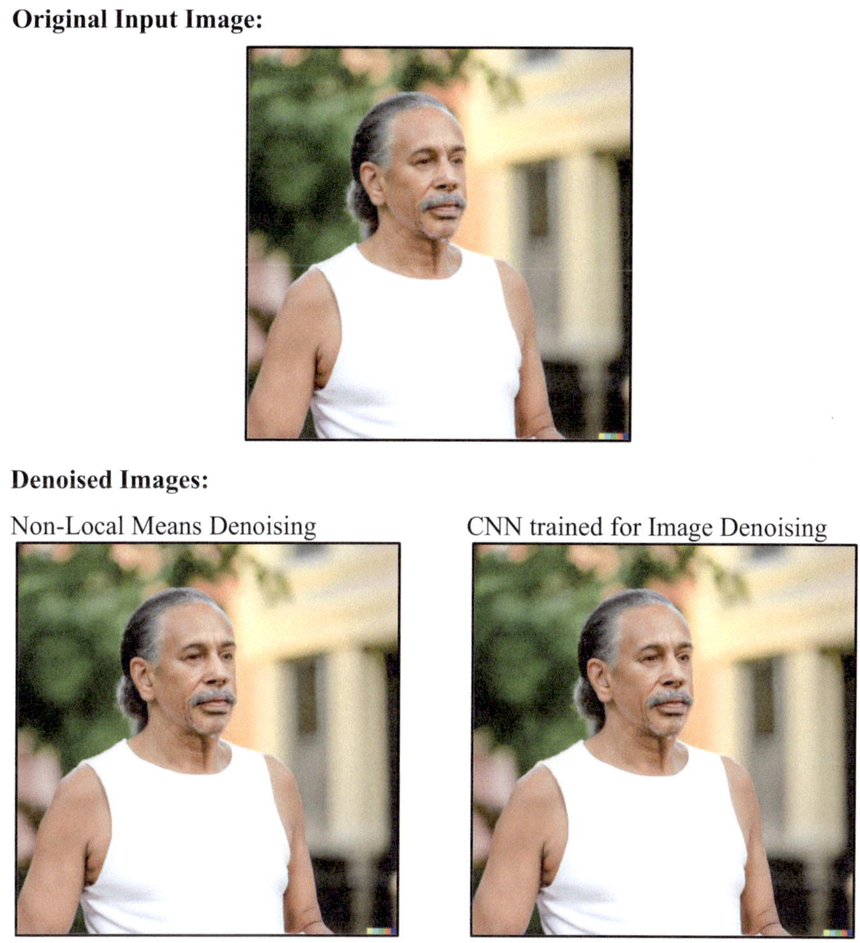

Fig. 3 Non-generative AI versus generative AI images

Table 1 Image processing using non-generative AI versus generative AI

Parameter/Metric	Non-local means denoising	CNN trained for image denoising (GenAI)
Processing time	25 s	First time Processing time is more, subsequent run, it will reduce
Generalization	Moderate	High
Interpretability	High	Not as high due to the inherent complexity of CNNs
Interpretability	High	Not as high due to the inherent complexity of CNNs
Accuracy (MSE)	15	PSNR = 18 dB (inversely related to MSE)

Table 2 NLM versus CNN PSNR value for multiple images

Image	NLM PSNR	CNN PSNR
Image 1	15	18
Image 2	14	16
Image 3	18	20
Image 4	12	16

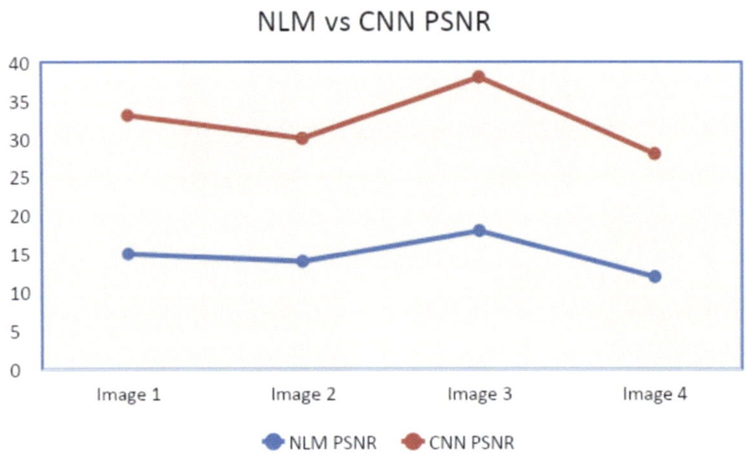

Fig. 4 Graph Comparison between NLM versus CNN

6 Results: Generative AI Applications on Image Processing

Based on above methodology (4.1) and original research done in the form of case study, we have concluded benefits in different business domains (couple of domains highlighted at Fig. 4), more details will be given below (Fig. 5).

Fig. 5 Domain for generative AI image processing

6.1 *Image Synthesis*

Art Generation. Generative Adversarial Networks (GANs) networks are increasingly employed in the digital arts domain for the automated creation of visual content, ranging from illustrations to paintings, across a multitude of styles. This burgeoning application of GANs has given rise to a new genre commonly referred to as "AI art."

Photorealistic Rendering. GANs are capable of generating synthetic images that closely mimic reality, and these images are employed in various fields such as video game development, architectural visualization, and product design.

Image Compression. Generative Adversarial Networks (GANs) can be leveraged to develop enhanced image compression algorithms, which reduce file sizes while maintaining high levels of image quality.

Anomaly Detection and Super-Resolution. Generative Adversarial Networks (GANs) have the capability to model the underlying normal distribution of an image dataset, thereby serving as an effective tool for the identification of anomalies or outliers within image sets.

Generative models possess the ability to perform super-resolution on low-resolution images, thereby improving both image quality and the level of detail.

6.2 *Image Editing and Enhancement*

Object Removal. Generative Adversarial Networks (GANs) have demonstrated efficacy in image inpainting tasks, where they can interpolate and reconstruct missing regions within an image after the removal of objects.

Colorization and Background Removal. In the realm of colorization, generative models have shown proficiency in converting grayscale images to full-color versions.

In the context of background removal, Generative Adversarial Networks (GANs) demonstrate the ability to segment and eliminate background elements from images.

6.3 *Image Processing Benefit at Retail Industry*

In the retail industry, generative AI promises substantial benefits, as highlighted in several studies. These benefits extend across customer segmentation, marketing strategies, customer experience enhancement, and inventory management.

Shelves monitoring, Foot traffic Analysis. We can use generative AI-based image processing to quantify store shelf and compare with inventory. This will help to reduce out of stock products. Similarly, we can process camera captured images to major or foot traffic at store and analyze this for revenue growth. We can also use generative AI image processing to reduce theft activity.

Benefits in Product Design. By analyzing historical sales data, customer preferences, behaviors, and other related metrics, generative AI can significantly aid in product design. This design will be more attractive and some of never seen before.

Improvements in Product Image Quality. Historical data underscores the direct correlation between high-quality product images and sales; products with clearer images have witnessed a sales surge of up to 250%. Generative AI's image processing capabilities stand poised to revolutionize this sector. With the assistance of generative AI tools, e-commerce platforms can feature high-resolution, enhanced images, ensuring better product visibility and potentially driving sales.

6.4 Medical Imaging and Impact at Healthcare Industry

Generative AI, especially, is manifesting its potential in the realm of medical imaging [26]. Clearer and more precise image outputs are becoming the norm, enabling medical professionals to gain insights into underlying diseases and facilitating more accurate treatments. For a deeper understanding of the advancements in medical image processing, we referred to NVIDIA's blog post [27].

Synthetic Data Generation and Image Reconstruction. Generative models have demonstrated their ability to create synthetic medical images. These not only aid in training diagnostic AI models but also bolster data privacy and supplement limited datasets.

By combining low-resolution images with a few high-quality ones, generative AI can reconstruct superior quality images, enhancing the diagnostic process.

Image Segmentation. According to an article on [28], generative AI models can segment medical images based on the distinct structures and tissues present. Such segmentation proves invaluable when focusing on specific anatomical regions like the brain or lungs (Fig. 6).

Frameworks like MONAI [29]. These provide specialized tools and methods for refining medical images, showcasing how technology intersects with healthcare to produce more accurate outcomes.

Image interpretation and workflow Improvement at Radiology: Generative AI will help at Radiology for interpreting images and workflow to get patient old history and information by providing the key parameters.

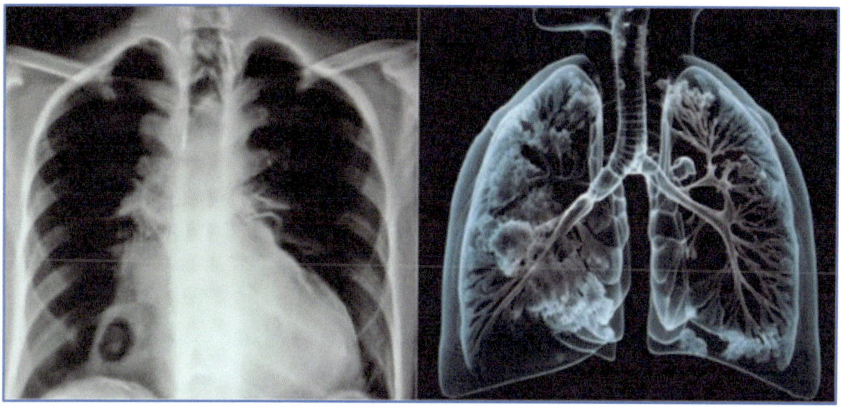

Fig. 6 Generative AI images based on prompt chest X-ray of pneumonia and given dataset (figure taken from [28])

6.5 Face Generation and Modification

Deepfake Technology. In the sphere of deepfake technology, Generative Adversarial Networks (GANs) are utilized to generate deepfake videos in which facial features can be seamlessly replaced within existing video footage.

Age Progression/Regression. In the area of age progression and regression, generative models are capable of simulating temporal facial transformations, either by projecting the potential aging of an individual's face over a span of time or by recreating a more youthful appearance.

6.6 Data Augmentation

Generative Adversarial Networks (GANs) exhibit several significant applications in the domain of synthetic data generation. Predominantly, they are employed to produce data to train auxiliary networks. Such applications span diverse fields, encompassing facial synthesis, medical imaging, and sports camera calibration [30].

Generating Synthetic Data. In the realm of data augmentation, generative artificial intelligence (AI) models demonstrate the capacity to generate synthetic data, thereby enriching small or limited datasets. This capability is especially valuable in the context of training machine learning algorithms where labeled data are scarce.

6.7 Forensics and Criminal Investigations

Image Reconstruction. Generative AI algorithms can improve the quality of grainy or low-resolution images, which may be crucial for identifying suspects or understanding crime scenes.

Authentication of Media. With the rise of deepfakes, it's becoming increasingly important to verify the authenticity of digital media. Generative AI can be trained to identify manipulated images or videos, providing an additional layer of verification.

Generating Hypothetical Situations: Generative AI could be used to simulate how a crime could have taken place, based on the available evidence. This could help investigators test various hypotheses more quickly.

Automated Sorting and Tagging of Data. Generative AI could assist in automatically categorizing and tagging vast amounts of data, which could be particularly useful in large-scale investigations where thousands of images and videos need to be analyzed.

Automated Report Writing and Predictive Policing. Generative AI can be programmed to write preliminary reports based on the data it analyzed, freeing up investigators to perform more nuanced tasks that require human judgment.

Generative AI can analyze historical crime data to predict where and when crimes are likely to occur, helping law enforcement agencies to allocate resources more effectively.

6.8 Auto Industry

GAN AI capabilities for images will help auto industry, it will help to design concept vehicle for future. In the realm of automotive design, artists initiate their process by sourcing visual references, often termed "scraps", which are aligned with contemporary vehicular styling trends. These references are typically extracted from image repositories, filtered through keyword searches. The methodology encompasses a thorough examination of vehicles throughout the industry's history and present landscape. Subsequent to meticulous human curation, a synthesis of prevailing designs combined with novel inspirations rooted in a company's signature style materializes. This convergence subsequently informs the artists' preliminary 2D sketches, which later evolve into 3D digital models and tangible clay prototypes [31].

7 Conclusion

Our research has illuminated the pervasive and expanding role of generative AI in the field of image processing. The utilization of AI-based image generation is experiencing significant growth, with a rising number of users relying on text-to-image generation algorithms. Importantly, the applications extend far beyond rudimentary uses and have shown immense potential in a variety of domains, including but not limited to, artistic creation, retail, forensics, medical imaging, and criminal investigations. For future research, it is anticipated that more advanced models will be developed, characterized by reduced CPU usage and lower memory requirements, thereby enhancing computational efficiency and performance.

Key insights from our research:

- Generative AI, when applied to image processing, holds a promise that is vast and largely untapped. Our current understanding offers but a preliminary glimpse into its potential, reminiscent of surveying merely the tip of an immense iceberg.
- The ethical implications of integrating generative AI into image processing, especially as it pertains to content creators and artists, demand rigorous scrutiny and dialogue. However, prevailing trends indicate that these technologies are on course to secure a dominant presence in the domain of content production.
- As technological frontiers expand, we foresee a broadened accessibility to generative AI tools tailored for image processing. The emergence and adoption of no-code and low-code platforms are set to act as significant accelerants, propelling this field to new heights.

References

1. Hierarchical text-conditional image generation with CLIP latents. https://cdn.openai.com/papers/dall-e-2.pdf. Accessed 13 Apr 2022
2. Bubeck S et al (2023) Sparks of artificial general intelligence: early experiments with GPT-4.
3. Midjourney. https://en.wikipedia.org/w/index.php?title=Midjourney&oldid=1182467614. Accessed 29 Oct 2023
4. Kaul V et al (2020) History of artificial intelligence in medicine. Elsevier, pp 807–812. https://doi.org/10.1016/j.gie.2020.06.040
5. He K, Gan C, Li Z, Rekik I, Yin Z, Ji W, Gao Y, Wang Q, Zhang J, Shen D (2022) Transformers in medical image analysis. Elsevier, pp 59–78. https://doi.org/10.1016/j.imed.2022.07.002
6. Marder M, Harary S, Ribak A, Tzur Y, Alpert S, Tzadok A et al (2015) Using image analytics to monitor retail store shelves. IEEE Xplore IBM 3:1–3:11. https://doi.org/10.1147/JRD.2015.2394513
7. Gangopadhyay A et al (2001) An image-based system for electronic retailing. Decis Supp Syst 107–116. https://doi.org/10.1016/S0167-9236(01)00105-1

8. Michalopoulos PG et al (1991) Vehicle detection video through image processing: the autoscope system, pp 21–29. https://doi.org/10.1109/25.69968
9. Carew RM et al (2019) Imaging in forensic science: five years on. Elsevier, pp 24–33. https://doi.org/10.1016/j.jofri.2019.01.002
10. Stoitsis J et al (2006) Computer aided diagnosis based on medical image processing and artificial intelligence methods. Elsevier, pp 591–595. https://doi.org/10.1016/j.nima.2006.08.134
11. Mesejo P et al (2020) A survey on artificial intelligence techniques for biomedical image analysis in skeleton-based forensic human identification. MDPI, Multidisciplinary Digital Publishing Institute. https://doi.org/10.3390/app10144703. Accessed 08 Sep 2020
12. Koul A (2019) Exploring the landscape of artificial intelligence, what is AI (Chap. 1). In: Practical deep learning for cloud, mobile, and edge: real-world AI, computer-vision projects using python, keras, tensorflow. O'Reilly Media, pp 3–
13. Artificial intelligence. https://en.wikipedia.org/w/index.php?title=Artificial_intelligence&oldid=1161337952. Accessed 22 June 2023
14. Generative artificial intelligence. https://en.wikipedia.org/w/index.php?title=Generative_artificial_intelligence&oldid=1174554739. Accessed 09 Sep 2023
15. Dhoni P (2023) Unleashing the potential: overcoming hurdles and embracing generative AI in it workplaces: advantages, guidelines, and policies. https://doi.org/10.36227/techrxiv.23696709.v1
16. Wang K et al (2017) Generative adversarial networks: introduction and outlook. IEEE/CAA, pp 588–598. https://doi.org/10.1109/jas.2017.7510583
17. Goodfellow I, Pouget-Abadie J, Mirza M, Xu B, Warde-Farley D, Ozair S, Cour-ville A, Bengio Y (2014) Generative adversarial nets. In: Advances in neural information processing systems, vol 27, pp 2672−2680
18. Generate images with OpenAI in python. https://www.geeksforgeeks.org/generate-images-with-openai-in-python/. Accessed 18 May 2023
19. Generative AI for medical imaging: extending the MONAI framework. https://arxiv.org/abs/2307.15208. Accessed 27 July 2023
20. GenerativeModels. https://github.com/Project-MONAI/GenerativeModels. Accessed 19 May 2023
21. Generative AI for Medical Imaging: extending the MONAI framework. https://github.com/Warvito/generative_monai. Accessed 27 July 2023
22. AI-assisted annotation of medical images using MONAI label on AWS, https://aws.amazon.com/blogs/industries/ai-assisted-annotation-of-medical-images-using-monai-label-on-aws/. Accessed 04 May 2023
23. Image Enhancer. https://imgupscaler.media.io/app/sharpener
24. VANCEAI. https://vanceai.com/denoise-ai/
25. OpenCV. https://docs.opencv.org/3.4/d5/d69/tutorial_py_non_local_means.html
26. Li X et al (2021) When medical images meet generative adversarial network: recent development and research opportunities - discover artificial intelligence. SpringerLink, Springer International Publishing
27. Accelerating medical image processing with NVIDIA DALI (2022). https://developer.nvidia.com/blog/accelerating-medical-image-processing-with-dali/. Accessed 18 Jan 2022
28. How imaging generative AI will transform the medical radiological practice. Health Manag 23(2) (2023). https://healthmanagement.org/c/healthmanagement/issuearticle/how-imaging-generative-ai-will-transform-the-medical-radiological-practice
29. Medical open network for artificial intelligence (2023). https://monai.io/. Accessed 15 Sep 2023
30. A review of synthetic image data and its use in computer vision (2022). https://www.ncbi.nlm.nih.gov/pmc/articles/PMC9698631/. Accessed 21 Nov 2022
31. Generative AI revs up new age in auto industry, from design and engineering to production and sales (2023). https://blogs.nvidia.com/blog/2023/08/09/generative-ai-auto-industry/. Accessed 09 Aug 2023

A Framework for Intrusion Detection System Using Deep Learning

Gauri Gera and Supriya Raheja

Abstract With the exponential increase in transmission of data daily, data has become highly susceptible to security breaches and intrusions. Although a set of rules can help a software navigate through intrusions and alert the networks users, they still struggle with zero-day attacks. This study aims at drawing comparisons between different novel frameworks utilizing various machine learning algorithms as well as contrasting the performance of deep learning algorithms with that of ensemble learning technique. Performance of algorithms such as Long Short-Term Memory (LSTM), Feed-Forward Neural Network (FFNN), and Extreme Gradient Boosting (XGBoost) is explored. Noteworthy results are obtained with LSTM demonstrating 99.9976%, FFNN achieving 99.9731%, and XGBoost showcasing 99.7358% accuracy.

Keywords Neural networks · XGBoost · Long short-term memory networks · Gated recurrent units · Zero-day attacks

1 Introduction

The inception of Intrusion Detection Systems (IDS) can be traced to the emergence of intrusion attacks witnessed in the 1980s. The Morris worm is regarded to be one of the first worms to spread through the internet on a substantial scale. Spread on 2 November [1], it not only caused a widespread disruption of the Unix services but also acted as a catalyst for sparking interest in enforcing network security. Morris worm and subsequent attacks played a major role in raising awareness about the need for stricter security measures. With the evolution of security measures, the intrusion attacks continue to evolve as well. It is a difficult task to counter them with existing intrusion detection systems. Novel and zero-day attacks have become prevalent, giving hackers a chance to exploit the vulnerability in a software or a code

---

G. Gera · S. Raheja (✉)
Amity Institute of Engineering and Technology, Amity University, Noida, Uttar Pradesh, India
e-mail: supriya.raheja@gmail.com

A. P. Agrawal et al. (eds.), *Advances of Artificial Intelligence and Machine Learning in Image Processing, Security and Data Analytics*, Lecture Notes in Electrical Engineering 1464, https://doi.org/10.1007/978-981-95-0148-9_6

before the developers have the chance to address it [2]. These attacks have been particularly challenging because the traditional IDS tend to be ineffective toward them. Traditional IDS mostly rely on known patterns for detecting malicious behaviors [3]. However, zero-day attacks leave no scope for familiar patterns. Zero-day attacks act as distraction and a means for stealing sensitive data, disrupt operations, and compromise systems. As the vulnerabilities in the software are already being exploited, the developers need to mitigate the threat in short amounts of time. The impact of a zero-day attack can be devastating, leading to large-scale damage to the victims of these attacks.

To tackle the problems associated with security breaches, this paper proposes a simple intrusion detection framework to classify normal and abnormal behaviors. An intrusion detection framework can be regarded as a framework which monitors networks and checks for abnormal activities. This paper also aims at drawing comparison between various algorithms and their performance on unseen datasets. This study primarily focuses on building a framework for classifying normal and abnormal behaviors using deep learning, which is a sub-domain of machine learning. The major challenge of this study is to build deep learning models that can give high accuracy with low false alarm rate [4].

The paper is further divided into four major components. Section 2 provides a comprehensive Birdseye view of the work done by other researchers in the domain of intrusion detection systems. Section 3 discusses the procedure followed in training our models for reaching results demonstrated in Section 4. Section 4 also shows a detailed comparison of all the algorithms used in our models. Lastly, Sect. 5 summarizes our results, discusses drawbacks, and provides future scope.

2 Literature Review

This section of the paper deals with the attempts made by various researchers in the field of developing an IDS using machine learning [5]. Many researchers have made use of datasets like the KDDCup99 dataset for analysis of their models, but this dataset is 20 years old and is not updated to precisely depict the current scenarios [6].

Wang et al. [7] proposed an approach called FC-ANN which made use of artificial neural networks and incorporated fuzzy clustering technique. This technique is good for datasets where a lot of overlapping occurs and there exists no proper distinction between datapoints.

Jabez et al. [8] made use of outlier detection approach as a method for detection of unknown attacks. This approach is based on detecting outlier points, i.e., points isolated from the general cluster. The advantage of this approach is that the execution time is significantly lower than other machine learning approaches [9].

Hesham et al. [9] made use of a Bayesian statistical model to detect intrusions. Bayesian models have a wide range of use in computation due to their ability to make conclusions based on probabilities.

Singh and Jang-Jaccard proposed the approach of MSCNN-LSTM-AE [10]. This approach was particularly addressing the problems with consideration of seven signals or processes with respect to time and space. Their approach made use of Long Short-term memory (LSTM) which was based on autoencoder network for processing of temporal features. Mbow et al. improvised handling of intrusion detection by handling class imbalances [11]. Class imbalances occur when there are not enough points present in some classes, which results in models being trained unevenly. The classes with less datapoints may go undetected upon intrusion. Zhen Lie et al. implemented a fast Fourier transform method on neural networks which helped in addressing slow and tedious detection of Distributed Denial of Service attacks (DDoS) [12].

All the techniques proposed by various researchers make use of statistical calculations to come to a result. There can also be a rule-based IDS that can be used to detect intrusions. However, a rule-based approach does not work for modified attacks [13]. Moreover, it is a tedious task to manually create and rectify rules as the attacks increase.

The learning of deep learning models can be improved by using dropout techniques. Introduced by the paper titled "Dropout: A Simple Way to Prevent Neural Networks from Overfitting" by Nitish Srivastava et al. [13], dropout technique efficiently suggests dropping random neurons to reduce the model's probability to overfit.

3 Methodology

This section of the paper deals with all the steps followed to train the various models and reach their respective conclusions.

3.1 Understanding the Dataset

This section focuses on understanding the dataset and correlation of datapoints with the important categories. This study makes use of the UNSW-NB15 dataset as it depicts an accurate situation of the modern scenarios of IoT security [6]. The UNSW-NB15 dataset does a great job at capturing the kind of frequency that can be expected in real-world scenario of different kinds of attacks such as DoS, D-Dos, backdoors, and worms [14]. This dataset is far more complex than other benchmark datasets which increases the reliability of this dataset [14]. The UNSW-NB15 is divided into two parts, i.e., the training dataset and the testing dataset. Both the datasets are used to draw final conclusions.

The UNSW-NB15 consists of 49 features [6]. Some of the most remarkable features are as follows:

1. srcip, dstip

indicate the source IP address and the destination IP address, respectively.

2. sport, dport

indicate the source port and the destination port, respectively.

3. sbytes, dbytes

indicate the number of bytes sent by the source and the destination in the network, respectively. A sudden increase in the number of bytes sent on the network can usually be an indication of malicious activity on the network.

4. durr

indicates the duration of transmission of bytes in the network. It can be crucial in detecting intrusions occurring over a network.

5. proto

indicates the protocol used in data transfer. The protocol used is an important factor when considering the identification and determination of network intrusions.

6. service

indicates the service in use. The kind of service in use is also a key factor in determining the type of intrusion.

Here are various protocols and services used in transaction of data. Figure 1 represents the frequencies of attacks on the services used. It is noted that DNS services observed the most abnormal behavior (abnormal behavior is represented by blue).

The dataset represents 10 kinds of situations of attacks, namely: Analysis, Backdoors, Dos, Exploits, Fuzzers, Generic, Normal, Reconnaissance, Shellcode, and worms. Figure 2 shows the distribution of the datapoints based on the attacks. We observe that the dataset contains imbalanced classes, with not enough datapoints for the model to learn about some attack categories.

Use of color sequence points in Fig. 3 is made to understand the correlation of datapoints present in the dataset with respect to the attack categories.

3.2 Data Preprocessing and Feature Selection

To obtain optimal results from our models, we make use of data preprocessing techniques by understanding the correlation of various features and their importance in the training of various models. This study makes use of correlation matrix (as shown in Fig. 4) to understand the range of correlation between the features of the dataset.

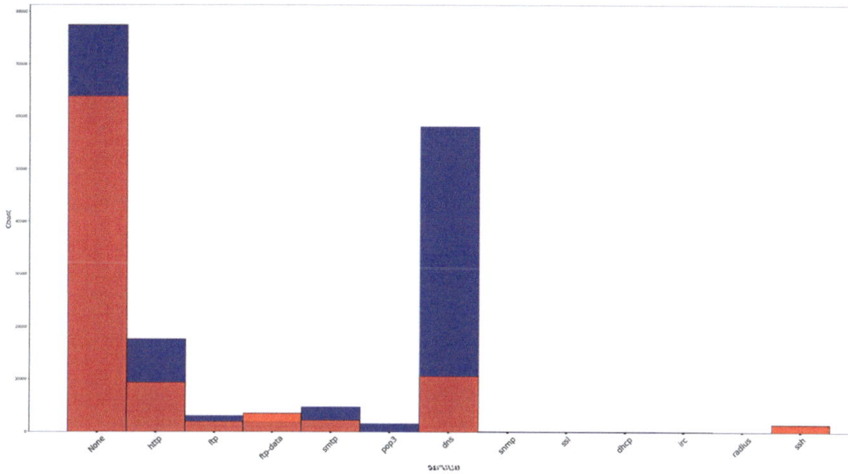

Fig. 1 Distribution of malicious and benign behavior with different services

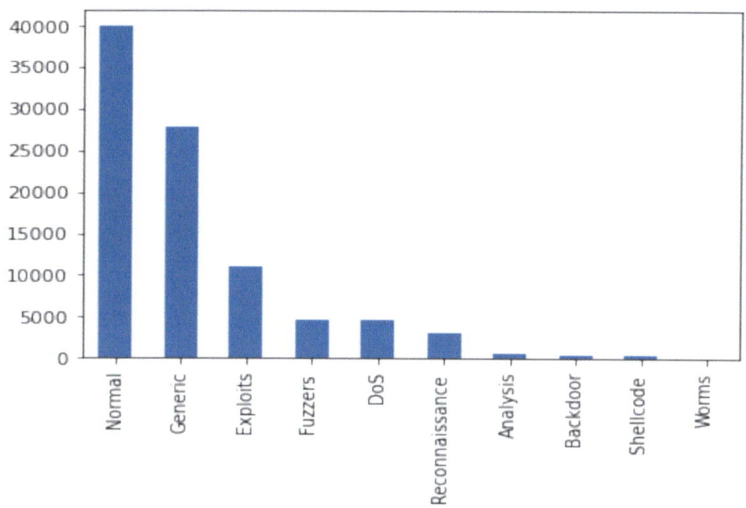

Fig. 2 Frequencies of different behaviors over the networks

A correlation matrix helps in analyzing correlation of each feature with each of the other features in the dataset. This step is used to eliminate features that have a high correlation in the dataset [15, 16]. Removal of highly correlated features is important as it introduces multicollinearity and higher dimensionality [17], which can make it challenging for the models to distinguish between the effects each of the features have on the outcome(s) individually. This step is crucial to prevent overfitting and yield accurate results. Figure 5 shows features which have correlation greater than 0.9, which are removed in the preprocessing of the dataset.

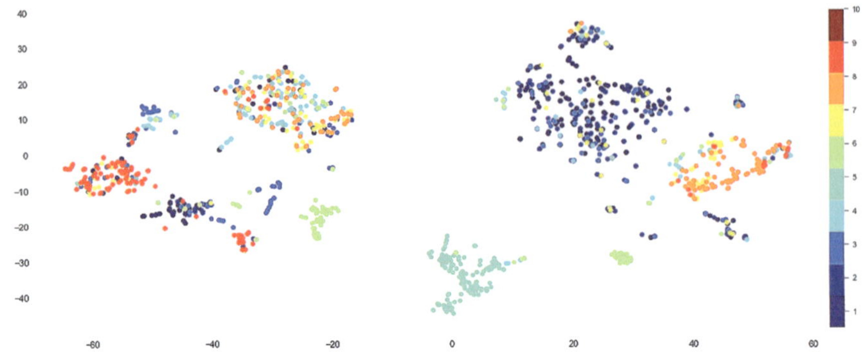

Fig. 3 Datapoints with respect to “attack_cat”

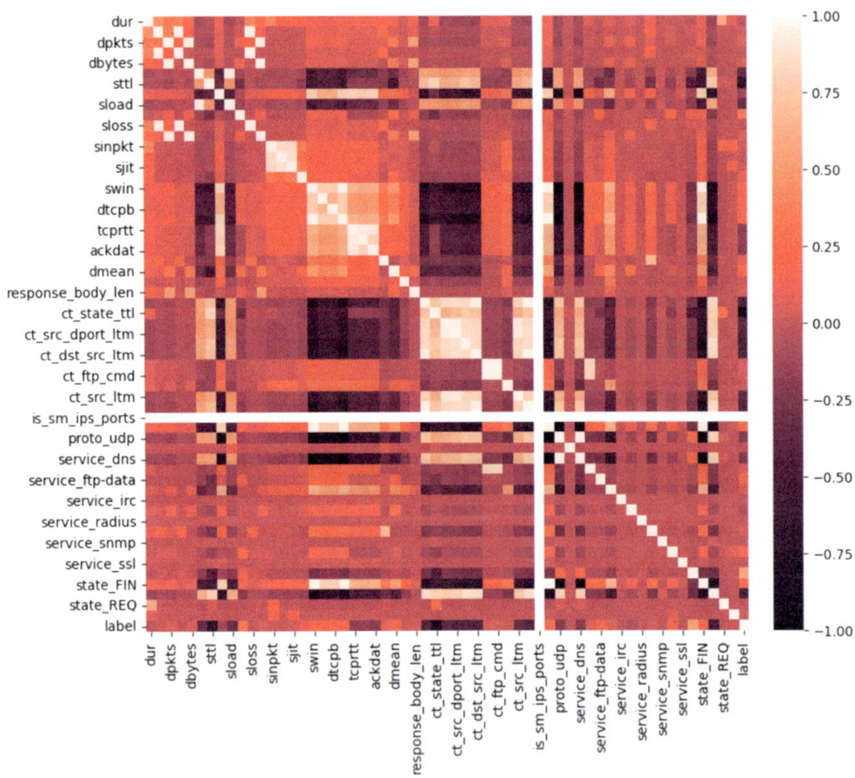

Fig. 4 Correlation heatmap

spkts	sbytes	0.9657497410287414
spkts	sloss	0.9736439932787799
dpkts	dbytes	0.976418551695821б
dpkts	dloss	0.9815064328008422
sbytes	sloss	0.99502719113184
dbytes	dloss	0.9971088501020646
sinpkt	is_sm_ips_ports	0.9445057600994802
swin	dwin	0.9601246970559344
tcprtt	synack	0.9394732071062888
ct_srv_src	ct_dst_src_ltm	0.9337952137616565
ct_srv_src	ct_srv_dst	0.9778491535974652
ct_dst_ltm	ct_src_dport_ltm	0.9604008284955233
ct_dst_ltm	ct_src_ltm	0.9322524473427766
ct_src_dport_ltm	ct_dst_sport_ltm	0.9116374681078989
ct_src_dport_ltm	ct_src_ltm	0.9331720623302827
ct_dst_src_ltm	ct_srv_dst	0.9410468630509295
is_ftp_login	ct_ftp_cmd	0.9943410042026887

Fig. 5 Highly correlated features in the UNSW-NB15 training dataset

3.3 *Implementing the Model*

This section discusses the algorithms used in the study. This study makes use of three different algorithms, namely Long Short-Term Memory model, Gated Recurrent Unit, Neural Networks, and XGBoost.

- Long Short-Term Memory Model or LSTM model is a version of Recurrent Neural Networks (RNNs). Also regarded as an upgraded version of RNNs, LSTM can retain information for a longer period than RNNs do. LSTMs also address issues such as the vanishing gradient problem and the exploding gradient problem. LSTMs are particularly useful in cases where complex pattern recognition is required. The model also makes use of a Gated Recurrent Unit (GRU) layer. The GRU is a type of RNN as well. GRU leverages all the benefits of RNNs while simplifying the structure of the network.
- Neural networks leverage the structure of the neural networks present in the brain to form a deep learning model. The major difference between simple neural network model and RNN is that there is no feedback given after computation on the nodes in simple neural networks. The neural networks used in this study are feed-forward neural networks (also known as multilayer perceptron). FFNNs are faster to train since the flow of data is only in the forward direction.
- XGBoost classifier, which stands for Extreme Gradient boosting algorithm, is based on gradient boosting framework. XGBoost gives better speed and performance [18]. XGBoost makes use of combining the prediction results of multiple machine learning models to draw a conclusion (ensemble technique).

For improving the performance scores of the models, we also make use of various techniques such as dropouts and dense layers.

3.4 Predictions of the Model

The ability of the model to classify between malicious or normal behavior is crucial for evaluating its performance. We make use of various evaluation metrics such as accuracy, precision, recall, and F1-score. Table 1 shows the performance of each of the models on UNSW-NB15 dataset. The confusion matrices in Fig. 6 show the performance of each model in terms of True Positive, False Positives, True Negatives, and False Negatives.

Table 1 Evaluation results of all models

Model	Accuracy	Precision	Recall	F1-score
FFNN	0.999731	0.999510	1.000000	0.999755
LSTM + GRU	0.999976	1.000000	0.999956	0.999978
XGBoost	0.997358	0.995193	1.000000	0.997591

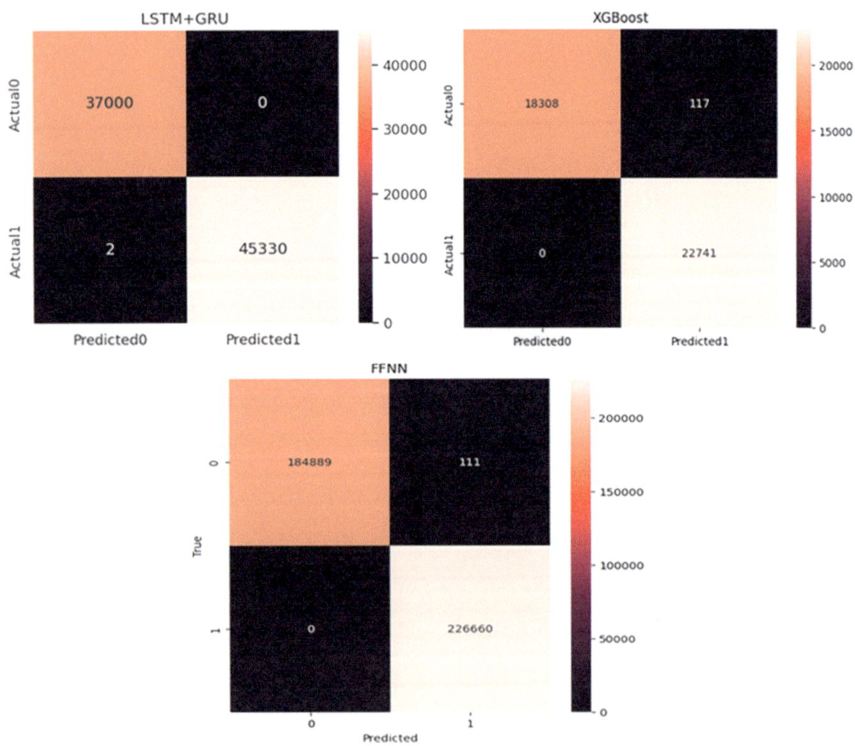

Fig. 6 Confusion matrices of the models

4 Results and Analysis

We make comparisons between all the models based on the scores obtained upon testing on the UNSW-NB15 testing dataset.

4.1 Evaluation Metrics

It is crucial to evaluate the models after training to understand their effectiveness, drawbacks, and computation cost. Since this study makes use of binary classification for intrusion detection, use of confusion matrices (in Fig. 6) is made to evaluate the performance of the models.

A confusion matrix (as showed in Fig. 7) consists of four categories:

(1) True Positive: A true positive refers to a correct prediction that is classified as positive. A true positive is a desirable outcome from the model.
(2) False Negative: A false negative refers to a falsie prediction classified in the negative class. A false negative can be dangerous in many situations where the user is not made aware of an opportunity or a problem. A false negative can be quite dangerous and hence should be avoided at all costs.
(3) False Positive: A false positive refers to a falsie prediction classified in the positive class. This prediction is not as dangerous as false negative but may incur unnecessary costs in case of rectifying a problem.
(4) True Negative: A true negative is a desirable outcome. It indicates that the prediction is correctly classified as negative.

This study also utilizes other evaluation metrics to analyze the performance of the model. These are described as follows:

(1) Accuracy is used to measure the correctness of the model predictions upon the testing.

$$\text{Accuracy} = \frac{\text{Number of correct predictions}}{\text{Total number of predictions}}$$

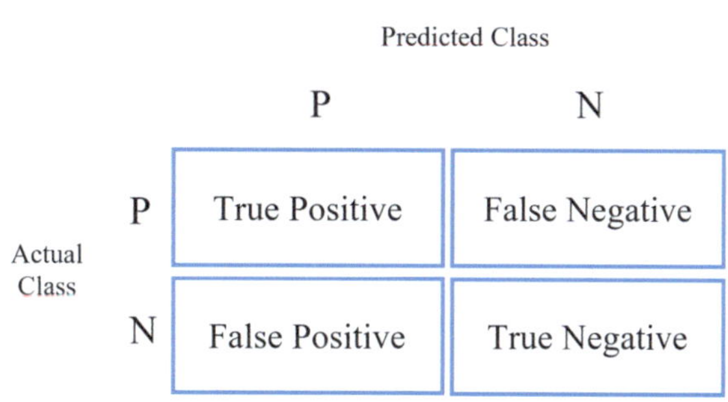

Fig. 7 Confusion matrix

or

$$\text{Accuracy} = \frac{TP + TN}{TP + FP + TN + FN}$$

(2) Precision is used to measure the number of correct positive predictions out of all positive predictions made by the model.

$$\text{Precision} = \frac{TP}{TP + FP}$$

(3) Recall is used to measure the number of correct positive predictions out of all positive instances present in the dataset.

$$\text{Recall} = \frac{TP}{(TP + FN)}$$

(4) F1-Score is used to combine the precision and recall into a single score which helps analysis by considering both false positives and false negatives.

$$F1 - \text{Score} = \frac{2 \cdot \text{Precision} \cdot \text{Recall}}{\text{Precision} + \text{Recall}}$$

4.2 *Inference from Model Performance*

This subsection aims at discussing the performance of the three models.

FFNN model performs well upon testing, giving an accuracy of 99.9731% approximately. In our study, the model gives a considerable amount of false positives, this can incur extra cost upon being deployed in a real-time system. Its training time is moderately high, and it is computationally expensive.

Long Short-Term Memory + Gated Recurrent Unit model gives an accuracy of 99.9975% upon testing. It gives a minute number of false negatives and no false positives. However, the model's training time is very high and requires strong processing units for its training. It displays great results with excellent precision but can be expensive to deploy in real-time systems.

Extreme Gradient Boosting (XGBoost) is a powerful machine learning algorithm used in supervised learning. It displays accuracy of 99.73582% upon being tested. It is a good consideration for deployment in a real-time system as it requires less resources, takes less time to train, and has a remarkable recall value.

5 Conclusion and Future Scope

The primary focus of this study is to contrast different algorithms in the machine learning domain and understand the trade-offs when deploying them in real-time systems. However, the models have a disadvantage at classifying classes in multi-classes, primarily because of class imbalances in the dataset.

The model training of LSTM+GRU is time-consuming, taking approximately 14.3 minutes for each epoch. This large amount of time is due to the high complexity of the model. The model complexity can be simplified by reducing the number of epochs, layers, and the density of the model. FFNN were faster to train and implement but required proper optimization techniques to yield good results. The model using XGBoost algorithm used the least amount of time for training and could be implemented more easily than other models, also yielding considerably great results.

Overall, this paper attempts to highlight the potential of deep learning and other machine learning subdomains in the field of intrusion detection systems. Automated recognition of intrusion in the network can prove to be advantageous for the security of the users as it reduces the manual intervention and monitoring from the process.

The potential areas for future development include classification into different attack categories, development of robust real-time system, integration of IDS into IoT security, blockchain security, and other upcoming domains [19].

References

1. Morris Worm. https://en.wikipedia.org/wiki/Morris_worm. Accessed 23 Sept 2023
2. Al-Rushdan H, Shurman M, Alnabelsi SH, Althebyan Q (2019) Zero-day attack detection and prevention in software-defined networks. In: International Arab conference on information technology (ACIT), Al Ain, United Arab Emirates, 2019, pp 278–282. https://doi.org/10.1109/ACIT47987.2019.8991124
3. Sowmya T, Mary Anita EA (2023) A comprehensive review of AI-based intrusion detection system. Meas: Sens 28:100827. ISSN 2665-9174. https://doi.org/10.1016/j.measen.2023.100827
4. Aldweesh A, Derhab A, Emam AZ (2020) Deep learning approaches for anomaly-based intrusion detection systems: a survey, taxonomy, and open issues. Knowl-Based Syst 189:105124. ISSN 0950-7051. https://doi.org/10.1016/j.knosys.2019.105124
5. Sakshi KV (2023) Machine learning and non-machine learning methods in mathematical recognition systems: two decades' systematic literature review. Multimed Tools Appl. https://doi.org/10.1007/s11042-023-16356-z
6. Moustafa N, Slay J (2015) UNSW-NB15: a comprehensive data set for network intrusion detection systems (UNSW-NB15 network data set). In: Military communications and information systems conference (MilCIS), Canberra, ACT, Australia, pp 1–6. https://doi.org/10.1109/MilCIS.2015.7348942

7. Wang G, Hao J, Ma J, Huang L (2010) A new approach to intrusion detection using artificial neural networks and fuzzy clustering. Expert Syst Appl 37(9):6225–6232. ISSN 0957-4174. https://doi.org/10.1016/j.eswa.2010.02.102, https://www.sciencedirect.com/science/article/pii/S0957417410001417
8. Jabez J, Muthukumar B (2015) Intrusion Detection System (IDS): anomaly detection using outlier detection approach. Procedia Comput Sci 48:338–346. ISSN 1877-0509. https://doi.org/10.1016/j.procs.2015.04.191
9. Vanin P, Newe T, Dhirani L, O'Connell E, O'Shea D, Lee B, Rao M (2019) A study of network intrusion detection systems using artificial intelligence/machine learning. In: Appl Sci 12(22):11752. https://doi.org/10.3390/app122211752
10. Rushdan H, Shurman M, Alnabelsi S, Althebyan Q (2019) Zero-day attack detection and prevention in software-defined networks. In: Conference 2019. https://doi.org/10.1109/ACIT47987.2019.8991124
11. Mbow M, Koide H, Sakurai K (2022) Handling class imbalance problem in intrusion detection system based on deep learning. Int J Netw Comput 12:467–492. https://api.semanticscholar.org/CorpusID:253735956
12. Liu Z, Hu C, Shan C (2021) Riemannian manifold on stream data: Fourier transform and entropy-based DDoS attacks detection method. Comput & Secur 109:Article 102392. ISSN 0167-4048. https://doi.org/10.1016/j.cose.2021.102392
13. Srivastava N, Hinton G, Krizhevsky A, Sutskever I, Salakhutdinov R (2014) Dropout: a simple way to prevent neural networks from overfitting. J Mach Learn Res 15(1):1929–1958
14. Zoghi Z, Serpen G (2023) UNSW-NB15 computer security dataset: analysis through visualization. Secur Priv e331. https://doi.org/10.1002/spy2.331
15. Aljawarneh S, Aldwairi M, Yasin M (2017) Anomaly-based intrusion detection system through feature selection analysis and building hybrid efficient model. J Comput Sci 25. https://doi.org/10.1016/j.jocs.2017.03.006
16. Gill AS, Kaur J, Singh D (2017) A feature reduced intrusion detection system using ANN classifier. Expert Syst Appl 88:249–257. ISSN 0957-4174
17. Bolón-Canedo V, Sánchez-Maroño N, Alonso-Betanzos A (2011) Feature selection and classification in multiple class datasets: an application to KDD Cup 99 dataset. Expert Syst Appl 38(5):5947–5957. ISSN 0957-4174. https://doi.org/10.1016/j.eswa.2010.11.028
18. Choi H, Kim M, Lee G, Kim W (2019) Unsupervised learning approach for network intrusion detection system using autoencoders. J Supercomput 75(9):5597–5621. https://doi.org/10.1007/s11227-019-02805-w
19. Kaur J, Agrawal, A, Khan, RA (2023) P2ADF: a privacy-preserving attack detection framework in fog-IoT environment. Int J Inf Secur 22(4):749–762. https://doi.org/10.1007/s10207-023-00661-7
20. Alalmaie A, Nanda P, He X (2023) Zero trust network intrusion detection system (NIDS) using auto encoder for attention-based CNN-BiLSTM. In: Proceedings of the 2023 Australasian computer science week (ACSW '23). Association for Computing Machinery, New York, NY, USA, pp 1–9. https://doi.org/10.1145/3579375.3579376
21. Khraisat A, Gondal I, Vamplew, P, Kamruzzaman, J (2019) Survey of intrusion detection systems: techniques, datasets and challenges. Cybersecurity. https://doi.org/10.1186/s42400-019-0038-7

Generative Rebirth: Advancing Image Regeneration Through Adversarial Networks

Manav Khambhayata, Sahil Garg, Aman Singh, and Daljeet Kaur

Abstract Image regeneration, a pivotal aspect of computer vision and image processing, seeks to restore visual content that has been significantly deteriorated over time. This research investigates the transformative potential of Generative Facial Prior GAN (GFP GAN) in image regeneration, without presenting it as a novelty but leveraging its pivotal properties. Instead of delving into the architecture, the study provides a mathematical background for GFP GAN, using its core components to advance image restoration. The primary goal is to rejuvenate visually impaired or deteriorated images, particularly in facial contexts. Through empirical experiments and comparative analyses, GFP GAN's superiority over conventional regeneration methods is established, with implications for medical diagnostics and creative expression. This paper contributes to reshaping perceptions and utilization of images by emphasizing GFP GAN's role in pushing the frontiers of image synthesis and restoration.

Keywords Image regeneration · GANs · GFP GAN · Transfer learning · Image super-resolution · Deep learning

1 Introduction

The intricate domain of computer vision and image processing plays a pivotal role in safeguarding and revitalizing visual content that has endured the test of time. Image regeneration, a profound pursuit, aims to revive these time-weathered visual artifacts, bridging the chasm between historical significance and contemporary appreciation. Within this landscape, Generative Adversarial Networks (GANs) have become a

M. Khambhayata (✉)
Indian Institute of Information and Technology, Lucknow, India
e-mail: king.manav.nk@outlook.com

S. Garg · A. Singh · D. Kaur
Chandigarh University, Mohali, Punjab, India
e-mail: daljeet.cse@cumail.in

A. P. Agrawal et al. (eds.), *Advances of Artificial Intelligence and Machine Learning in Image Processing, Security and Data Analytics*, Lecture Notes in Electrical Engineering 1464, https://doi.org/10.1007/978-981-95-0148-9_7

formidable resource, equipping researchers with inventive approaches to rejuvenate images impacted by the effects of time.

This research embarks on a meticulous exploration of the realm of image regeneration, underscored by a resolute commitment to resuscitate visual content scarred by the passage of time. Our purview extends to encompass images that traverse a spectrum of degradation, culminating in the profound deterioration that plagues historical and aged photographs. This comprehensive investigation revolves around a central axis—the tailored deployment of generative adversarial networks, a subset of machine learning architectures that has garnered substantial attention for its potential to enact transformative restorations.

At the heart of our inquiry lies Generative Facial Prior GAN (GFP GAN), a pioneering departure from the conventional GAN lineage. In stark contrast to established GAN models like DCGAN and WGAN, the GFP GAN architecture integrates progressive differential components, orchestrating an iterative refinement mechanism that shapes the generator's output. This innovation heralds accelerated convergence rates and heightens the fidelity of image detail restoration, equipping GFP GAN with the capability to transcend the limitations inherent in reviving severely deteriorated facial images.

The effectiveness and pre-eminence of GFP GAN are holistically validated through a comprehensive suite of experiments encompassing both quantitative metrics and qualitative evaluations. This methodological rigor lends empirical credence to GFP GAN's assertions of enhancing perceptual quality, preserving structural integrity, and resuscitating facial content that teeters on the brink of obscurity.

In the backdrop of relevant literature, this endeavor finds its roots in the seminal work of Zhang et al. [1], which illumines the potential of GANs. Expanding on this foundation, the nuanced study by Liu et al. [2] delves into the intricacies of image degradation and restoration, contributing to the overarching framework of our research. The exploration of complex loss functions by Zhu et al. [3] and the image-to-image translation study by Isola et al. [4] further enrich the context of our innovation. Lastly, the insights of Wang et al. [5] dissecting the complexities of generative models and their real-world applications further amplify the discourse.

2 Literature Review

The work authored by Isola et al. [4] stands as a significant milestone in the field of image-to-image translation. Their introduction of Conditional Adversarial Networks (cGANs) marks a transformative leap in the ability to convert images between different domains. Their pioneering approach emphasizes the role of paired training data, where input images are accompanied by corresponding target images. This architectural advancement is propelled by the incorporation of adversarial loss functions, compelling the generator to create output images that closely resemble real images within the target domain. The paper's framework finds application in diverse domains including style transfer, image colorization, and segmentation. Isola et al. [4] significantly extend the frontier of image translation, inspiring subsequent research into conditional generative models.

Ledig et al. [6] contribute a groundbreaking study to the domain of single image super-resolution, a fundamental task in image regeneration. Their work introduces a novel architecture that employs Generative Adversarial Networks (GANs) for generating high-resolution images from low-resolution counterparts. The authors ingeniously integrate perceptual loss functions alongside adversarial objectives, effectively inducing the network to produce visually compelling images with enhanced resolution. This paper underscores the capacity of GANs not only for traditional generative tasks but also for addressing complex image restoration challenges. The research resonates with the pursuit of photo realism in image generation and highlights the transformative potential of GAN-based super-resolution techniques.

The work of Zhang et al. [1] introduces a sophisticated evolution of GAN architectures, paving the way for StackGAN++. By employing stacked GANs, this research advances the state of the art in generating realistic images. The authors propose a novel two-stage approach, where a conditional GAN first generates a low-resolution image, followed by a higher-resolution GAN refining the output. This hierarchy of generative processes leads to images with heightened visual fidelity and realism. Zhang et al. [1] accentuate the potency of layered architectures in GANs, providing insights into the complex interplay between resolutions and generative capacities. The StackGAN++ framework holds promise for diverse applications, from art synthesis to realistic content creation.

In the realm of image quality assessment, Wang et al. [7] proposed a comprehensive approach, shifting the focus from error visibility to structural similarity. Their work provides valuable insights into assessing image quality, which is pertinent to our study on image regeneration.

The research paper "CGR-GAN: CG Facial Image Regeneration for Antiforensics" by Peng et al. [8] introduces a novel approach using Generative Adversarial Networks (GANs) for facial image regeneration to challenge forensic analysis. By tailoring GANs for antiforensic purposes, the paper raises ethical concerns regarding the potential misuse of manipulated images. The study's insights extend to improved detection methods and possible countermeasures against deceptive GAN-based techniques, contributing to the evolving landscape of image manipulation and digital forensics.

3 Methodology

Generative Facial Prior GAN (GFP GAN) serves as the cornerstone of this research, designed to address the challenges of image regeneration, particularly in the context of facial visual content. GFP GAN builds upon the foundations of traditional GANs but introduces a novel generative feedback mechanism tailored to the intricacies of facial feature reconstruction. This innovative approach incorporates progressive differential components, which iteratively refine the generator's output to enhance both the perceptual quality and structural coherence of the generated images.

3.1 Overview of GFP GAN

GFP GAN constitutes a comprehensive framework with integral components, prominently featuring a degradation removal module. In this module, we utilize it to address complex degradation challenges and derive two fundamental categories of attributes: latent characteristics (F_{latent}), tasked with aligning the input image with the most suitable latent code within StyleGAN2 [9], and multi-resolution spatial traits ($F_{spatial}$), which are adept at fine-tuning StyleGAN2 features.

These latent features (F_{latent}) undergo a transformation process, transitioning into intermediate latent codes (W) through a sequence of linear layers. These intermediate codes are subsequently employed within StyleGAN2, contributing to the generation of intermediate convolutional features (F_{GAN}). This collection of features encapsulates intricate facial details, effectively encapsulating the nuances embedded within the pretrained GAN weights.

To enhance authenticity and maintain faithfulness, the incorporation of Channel-Split Spatial Feature Transform (CS-SFT) layers facilitates the modulation of facial GAN features (F_{GAN}) with multi-resolution attributes ($F_{spatial}$). This meticulous modulation process ensures a gradual refinement, resulting in outputs distinguished by their authenticity and unwavering commitment to faithfulness.

In our research paper, we attach significant importance to perceptually significant facial components, notably the eyes and mouth. To address this, we introduce a facial component loss through the use of discriminators during the training process. Additionally, we introduce the concept of identity-preserving guidance to ensure the retention of facial identity.

While GFP GAN hasn't been introduced as a novelty in our work, we leverage its pivotal properties and core components to advance our research endeavors in the realm of image regeneration and restoration.

3.2 Degradation Removal Module

When dealing with real-world facial images, we often confront a multitude of complex image imperfections, encompassing noise, JPEG artifacts, low resolution, and diverse lighting conditions. To effectively address these challenges, we've harnessed the capabilities of a U-Net-inspired module dedicated to enhancing image quality. This module, specifically designed for image enhancement, not only eliminates these imperfections but also extracts valuable features like Flatent and Fspatial [10].

$$\boldsymbol{F}_{\text{latent}}, \boldsymbol{F}_{\text{spatial}} = \text{U} - \text{Net}(\boldsymbol{x}) \tag{1}$$

3.3 Innovative Foundation of GFP

Leveraging an already trained GAN for facial features, we have harnessed generative priors to enrich facial details. Intermediate latent codes (W) derived from Flatent are employed to generate convolutional features (FGAN) from the closest face. This approach allows us to incorporate rich facial details into the restoration process.

Generative models offer a wealth of insights beyond just realistic details and vibrant textures. They also hold valuable information about color preferences, which we can harness for our task of enhancing facial images. Real-world face photos, particularly older ones, often lack color or exhibit vintage or subdued hues. By leveraging this color information from generative facial models, it has the opportunity to enhance and add color, a process known as colorization. Furthermore, we believe that these generative facial models may also incorporate fundamental geometric and 3D principles, which can be beneficial for various aspects of image restoration and manipulation.

$$\boldsymbol{W} = \text{MLP}(\boldsymbol{F}_{\text{latent}}).$$

$$\boldsymbol{F}_{\text{GAN}} = \text{StyleGAN}(\boldsymbol{W}). \tag{2}$$

3.4 CS-SFT Layers

In our pursuit of maintaining accuracy during face restoration, we leverage input spatial features (F_{spatial}) from the U-Net to fine-tune GAN features (F_{GAN}) using the Spatial Feature Transform (SFT) method [11]. This approach, renowned for its remarkable ability to adapt to diverse scenarios in image enhancement, computes affine transformation parameters (α, β) via convolutional layers to spatially adjust F_{GAN}:

$$\boldsymbol{\alpha}, \boldsymbol{\beta} = \text{Conv}\left(\boldsymbol{F}_{\text{spatial}}\right),$$

$$\boldsymbol{F}_{\text{output}} = \text{SFT}(\boldsymbol{F}_{\text{GAN}}|\boldsymbol{\alpha}, \boldsymbol{\beta}) = \boldsymbol{\alpha} \odot \boldsymbol{F}_{\text{GAN}} + \boldsymbol{\beta}. \tag{3}$$

In order to strike a harmonious balance between authenticity and faithfulness, we leverage the CS-SFT layers into the equation. These specialized layers enact spatial adjustments on a segment of the GAN's underlying latent features, influenced by the input $F_{spatial}$ (emphasizing fidelity), while allowing the remaining original features (emphasizing authenticity) to flow through unaltered. This process is illustrated as follows:

$$\begin{aligned} \boldsymbol{F}_{\text{output}} &= \text{CS} - \text{SFT}(\boldsymbol{F}_{\text{GAN}}|\boldsymbol{\alpha}, \boldsymbol{\beta}) \\ &= \text{Concat}[\text{Identity}(\boldsymbol{F}^{\text{split0}}_{\text{GAN}}), \boldsymbol{\alpha} \odot \boldsymbol{F}^{\text{split1}}_{\text{GAN}} + \boldsymbol{\beta}]. \end{aligned} \tag{4}$$

$F^{split0}{}_{GAN}$ and $F^{split1}{}_{GAN}$ refer to feature partitions extracted from F_{GAN} along the dimensions of channel.

It is feasible to mix input photographs and historical data with the aid of CS-SFT in a way that balances textural authenticity and accuracy while possibly streamlining the procedure. CS-SFT layers are used at each resolution scale to create a reconstructed face, known as y.

4 Model Objectives

From the perspective of GFP GAN, this framework encompasses a comprehensive set of objectives, carefully designed to excel in face restoration and enhancement. GFP GAN's objectives can be categorized into four essential components:

4.1 *Reconstruction Loss (L_{rec})*

GFP's primary goal is to ensure that the generated output (y) closely resembles the real-world image (y). L1 loss and perceptual loss are two well-known loss functions that GFP GAN uses to accomplish this. The following is a definition of the restoration loss:

$$\boldsymbol{L}_{\text{rec}} = \boldsymbol{\lambda}_{\text{L1}}||\hat{\text{y}} - \text{y}\,||1 + \boldsymbol{\lambda}_{\text{per}}||\,\varphi(\hat{\mathbf{y}}) - \varphi(\mathbf{y})\,||_1. \tag{5}$$

Here, GFP GAN relies on φ, a previously trained VGG-19 model [12] as well as uses the feature maps {conv1, …, conv5} before activation [13], to calculate perceptual loss. λ_{L1} and λ_{per} correspond to respective loss weights for L1 and perceptual loss [14], respectively.

4.2 *Adversarial Loss (L_{adv})*

GFP GAN's objective with the adversarial loss is to encourage the generation of images that closely align with real-world textures and a natural picture distribution.

GFP GAN follows a similar approach to StyleGAN2 [15], employing logistic loss [16]. The adversarial loss is defined as

$$L_{\text{adv}} = -\boldsymbol{\lambda}_{\text{adv}} * \mathrm{E}_{\hat{\mathbf{y}}}\left[\text{softplus}(D(\hat{\mathbf{y}}))\right]. \quad (6)$$

Here, D represents the discriminator and $\boldsymbol{\lambda}_{\text{adv}}$ is the weight assigned to the adversarial loss.

4.3 *Facial Component Loss* (L_{comp})

GFP GAN is designed by introducing a facial element loss that makes use of local discriminators for the left eye, right eye, and mouth regions, GFP GAN is intended to improve the perceptually significant facial components. This discriminator loss [16] ensures that the restored patches closely resemble the natural facial component distributions. Additionally, GFP GAN incorporates a feature style loss, it efficiently captures texture information by focusing on matching the Gram matrix statistics [17] of actual as well as repaired regions. Loss of the face element is described below:

$$\begin{aligned} \mathbf{L}_{\text{comp}} = & \sum_{\text{ROI}} \lambda_{\text{local}} * \mathrm{E}_{\mathbf{y}\text{ROI}}[\log(1 - D_{\text{ROI}}(\mathbf{y}_{\text{ROI}}))] \\ & + \lambda_{\text{fs}} * ||\text{Gram}(\psi(\mathbf{y}_{\text{ROI}})) - \text{Gram}(\psi(\mathbf{y}_{\text{ROI}}))||1. \end{aligned} \quad (7)$$

Here, GFP GAN defines ROI as the region of interest, which includes critical facial components such as the mouth, and right and left eyes. D_{ROI} denotes the local discriminator dedicated to assessing each of these regions, while ψ represents the multi-resolution elements derived from the discriminators that have been trained. Within this framework, $\boldsymbol{\lambda}_{\text{fs}}$ correspond to loss weights associated with local discriminative loss and feature style loss, respectively.

4.4 *Identity Preserving Loss* (L_{id})

GFP GAN incorporates an identity-preserving loss into its framework, drawing inspiration from the concept of perceptual loss. This loss function ensures that in a condensed large area for features, the restored result keeps a close distance from the original. To achieve this, GFP GAN employs a pretrained face recognition model (ArcFace) [18] designed to capture prominent identity Elements. The loss of identity preservation is described as

$$L_{\text{id}} = \boldsymbol{\lambda}_{\text{id}} * ||\eta(\hat{\mathbf{y}}) - \eta(\mathbf{y})||_1. \quad (8)$$

In this context, η serves as the facial property extractor, specifically the ArcFace model, while λ_{id} designates weight parameter governing the identity-preserving loss.

The overall model objective, as implemented by GFP GAN, combines these four loss components into a cohesive framework:

$$L_{total} = L_{rec} + L_{adv} + L_{comp} + L_{id} \quad (9)$$

Hyperparameters for loss have been configured as follows: λ_{l1} is set to 0.1, λ_{per} is assigned a value of 1, λ_{adv} is established at 0.1, λ_{local} is defined as 1, λ_{fs} takes on a value of 200, and λ_{id} is specified as 10. These hyperparameters play a critical role in shaping the behavior of the model during training.

5 Experimental Results

The images in Figs. 1 and 2 offer a compelling glimpse into the tangible impact of the model when applied to real-world image restoration challenges. On the left, we present a photograph that encapsulates the kind of image degradation commonly encountered in everyday scenarios—be it due to low light conditions, motion blur, or noise. These real-world images are inherently complex, often characterized by a blend of intricate details and imperfections. However, GFP GAN, triumphantly revitalizes the image on the right. It faithfully restores lost details, mitigates noise, and elevates the overall visual quality. Such results underscore the adaptability and efficacy of our approach in addressing the intricacies.

5.1 Discussion and Limitations

While GFP GAN demonstrates impressive performance across a range of population groups and lighting conditions, it is crucial to recognize inherent limitations within this established model. Notably, instances in which the model regenerates teeth on an infant's face, as illustrated in the image, highlight the presence of potential training bias. This bias can be traced back to the model's dependency on existing color information and training datasets, which may not consistently account for the subtleties of facial color restoration, particularly in scenarios involving infants. Additionally, there are instances where the placement of the eyes in the regenerated image was inaccurately reproduced, as observed in the provided image. This underscores the need for further refinement in the model's training process to enhance its precision in replicating intricate facial features, including eye positioning (Fig. 3).

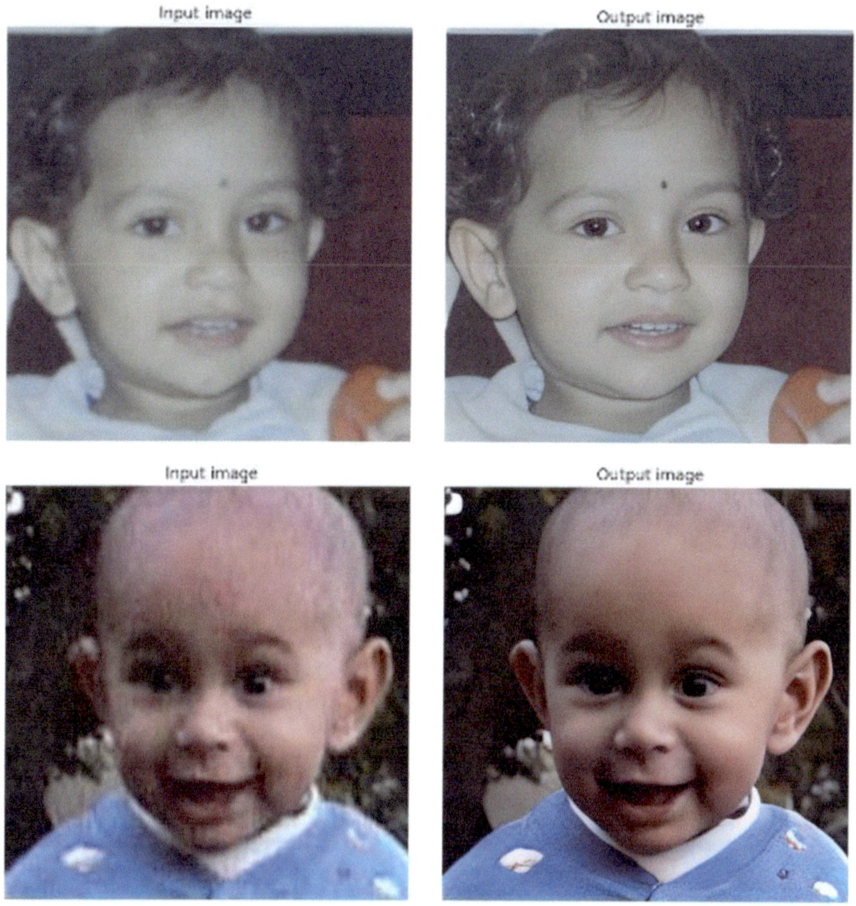

Fig. 1 Image restoration results—from blurred to high quality

6 Conclusion

In conclusion, our research journey with GFP GAN has uncovered its remarkable potential in advancing the realm of image regeneration and restoration. Leveraging the model's sophisticated architecture, we have achieved impressive results, showcasing its prowess in addressing complex facial degradation issues, enhancing perceptual quality, and preserving facial identity. However, it is essential to acknowledge the inherent limitations and nuances associated with this model, particularly in scenarios involving grayscale images or intricate facial details.

As we reflect on our work, we recognize the evolving landscape of generative image synthesis and the continual need for refinement and expansion. Our study serves as a stepping stone, demonstrating the capabilities of GFP GAN and shedding light on avenues for future exploration. To further harness the power of generative

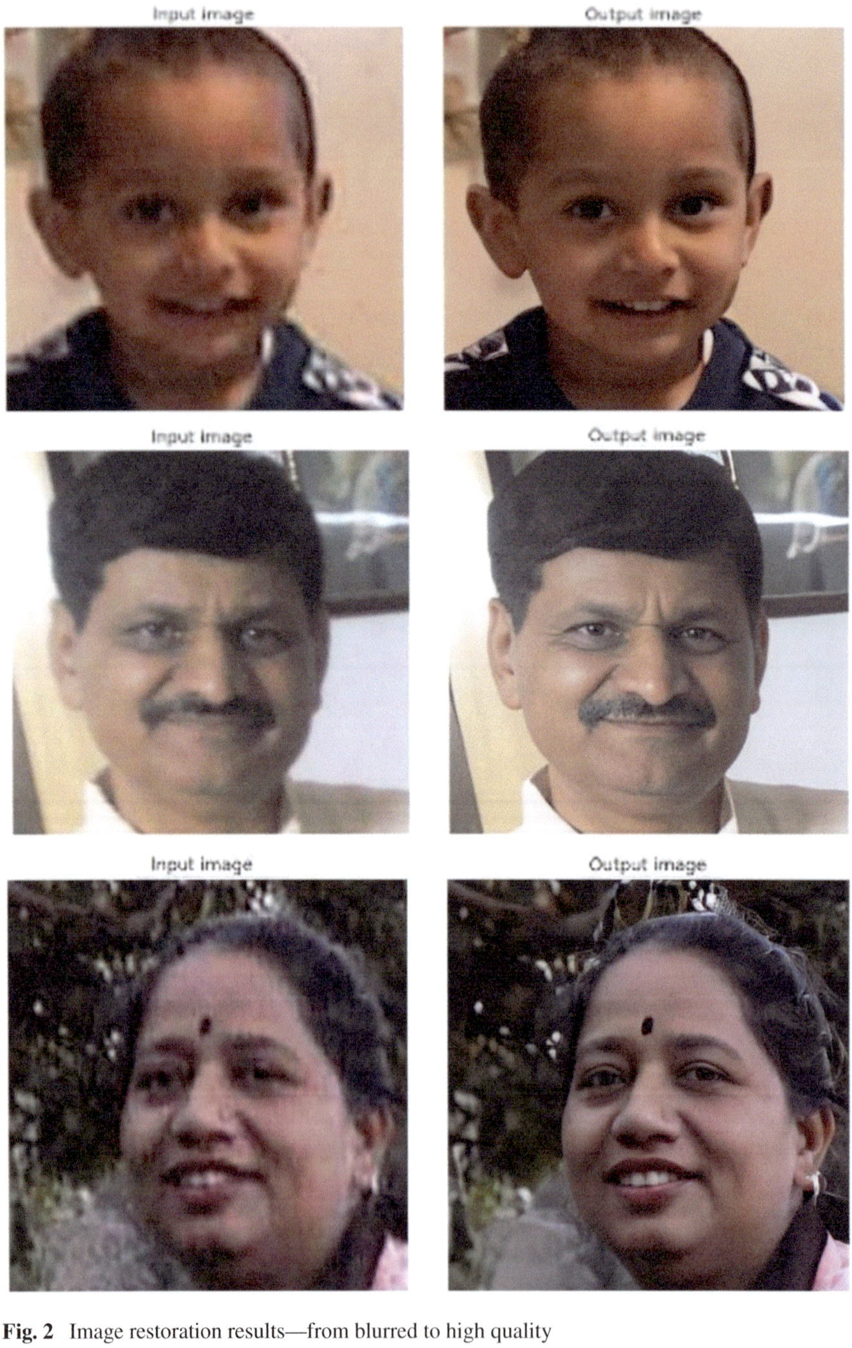

Fig. 2 Image restoration results—from blurred to high quality

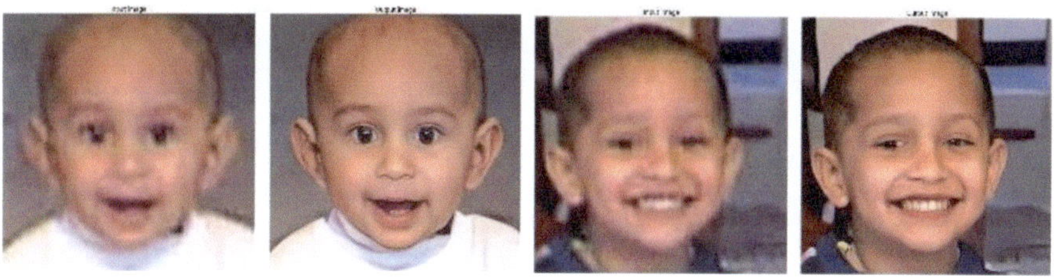

Fig. 3 Training bias and limitations

models like GFP GAN, it is imperative to diversify datasets, refine training strategies, and explore real-world applications.

In essence, this paper marks a significant contribution to the field of image regeneration, and we anticipate that our research will inspire further innovations, collaborations, and breakthroughs in the quest for harnessing the full potential of generative adversarial networks.

References

1. Zhang J, Zheng H, Tan M, Tao D (2018) Multispectral GAN: a novel framework for hyperspectral image synthesis. In: Proceedings of the IEEE conference on computer vision and pattern recognition (CVPR), pp 4446–4454
2. Liu X, Lin Z, Yan S, Sun J, Ma Y, Zhang T (2015) Learning content-preserving hash for image retrieval. In: Proceedings of the IEEE conference on computer vision and pattern recognition (CVPR), pp 3051–3060
3. Zhu JY, Park T, Isola P, Efros AA (2017) Unpaired image-to-image translation using cycle-consistent adversarial networks. In: Proceedings of the IEEE international conference on computer vision (ICCV), pp 2223–2232
4. Isola P, Zhu JY, Zhou T, Efros AA (2017) Image-to-image translation with conditional adversarial networks. In: Proceedings of the IEEE conference on computer vision and pattern recognition (CVPR), pp 1125–1134
5. Wang X, Zhang L, Philip Chen CL (2018) Efficient generative adversarial network with spectral regularization for real-time object detection. In: Proceedings of the European conference on computer vision (ECCV), pp 365–381
6. Ledig C, Theis L, Huszár F, Caballero J, Cunningham A, Acosta A, Aitken A, Tejani A, Totz J, Wang Z, Shi W (2017) Photo-realistic single image super-resolution using a generative adversarial network. arXiv:1609.04802
7. Wang Z, Bovik AC, Sheikh HR, Simoncelli EP (2004) Image quality assessment: from error visibility to structural similarity. IEEE Trans Image Process 13(4):600–612
8. Peng F, Yin L-P, Zhang L-B, Long M (2020) CGR-GAN: CG facial image regeneration for antiforensics based on generative adversarial network. IEEE Trans Multimed 22(10):2511–2525. https://doi.org/10.1109/TMM.2019.2959443
9. Karras T, Laine S, Aila T (2018) A style-based generator architecture for generative adversarial networks. In: CVPR
10. Ronneberger O, Fischer P, Brox T (2015) U-net: convolutional networks for biomedical image segmentation. In: International conference on medical image computing and computer-assisted intervention (MICCAI), pp 234–241

11. Wang X, Yu K, Dong C, Loy CC (2018) Recovering realistic texture in image super-resolution by deep spatial feature transform. In: CVPR
12. Simonyan K, Zisserman A (2015) Very deep convolutional networks for large-scale image recognition. In: ICLR
13. Wang X, Yu K, Wu S, Gu J, Liu Y, Dong C, Qiao Y, Loy CC (2018) Esrgan: enhanced super-resolution generative adversarial networks. In: ECCVW
14. Johnson J, Alahi A, Fei-Fei L (2016) Perceptual losses for real-time style transfer and super-resolution. In: ECCV
15. Karras T, Laine S, Aittala M, Hellsten J, Lehtinen J, Aila T (2020) Analyzing and improving the image quality of StyleGAN. In: CVPR
16. Goodfellow I, Pouget-Abadie J, Mirza M, Xu B, Warde-Farley D, Ozair S, Courville A, Bengio Y (2014) Generative adversarial nets. In: NeurIPS
17. Gatys LA, Ecker AS, Bethge M (2016) Image style transfer using convolutional neural networks. In: CVPR
18. Deng J, Guo J, Xue N, Zafeiriou S (2019) Arcface: additive angular margin loss for deep face recognition. In: CVPR

Hey AI Can You Grade My Essay?: Automatic Essay Grading

Maisha Maliha and Vishal Pramanik

Abstract Automatic essay grading (AEG) has attracted the attention of the NLP community because of its applications to several educational applications, such as scoring essays, short answers, etc. AEG systems can save significant time and money when grading essays. In the existing works, the essays are graded where a single network is responsible for the whole process, which may be ineffective because a single network may not be able to learn all the features of a human-written essay. In this work, we have introduced a new model that outperforms the state-of-the-art models in the field of AEG. We have used the concept of collaborative and transfer learning, where one network will be responsible for checking the grammatical and structural features of the sentences of an essay while another network is responsible for scoring the overall idea present in the essay. These learnings are transferred to another network to score the essay. We also compared the performances of the different models mentioned in our work, and our proposed model has shown the highest accuracy of **85.50%**.

Keywords Automatic essay grading · Collaborative learning · Deep learning · Recursive learning · Recursive neural network

Maisha Maliha and Vishal Pramanik—Contributed equally to this work.

M. Maliha (✉)
School of Computer Science, University of Oklahoma, Norman, OK, USA
e-mail: maisha.maliha-1@ou.edu

V. Pramanik
Department of Computer Science and Engineering, Indian Institute of Technology Bombay, IIT Main Road, Mumbai, Maharashtra, India

A. P. Agrawal et al. (eds.), *Advances of Artificial Intelligence and Machine Learning in Image Processing, Security and Data Analytics*, Lecture Notes in Electrical Engineering 1464, https://doi.org/10.1007/978-981-95-0148-9_8

1 Introduction

Automatic essay grading (AEG) is a subfield of natural language processing (NLP) that has been around for more than a half-century. Page [1] suggested the first AEG system. For many years, little progress was made, mainly owing to a lack of resources (such as parsers) and processing power. But things changed with the development of deep neural networks and powerful computing machines. References [2], 3 are the best examples of applications of deep neural networks.

Automatic essay grading refers to using a machine to grade a text prepared in response to a topic known as the essay prompt. Holistic automatic essay grading refers to using computers to award an essay grade that reflects its overall quality. Aside from overall AEG, algorithms may be trained to rate essays depending on a single trait/attribute. This is referred to as trait-specific AEG. Content, organisation, and style are examples of essay characteristics. The necessity of human-graded essays submitted in response to the prompt in which the AEG system is evaluated is one of the problems in AEG. Domain adaption techniques such as cross-domain AEG have been developed to address this barrier. In our paper, we have contributed the following points:

- We introduce a collaborative deep learning model for automatic essay grading. A model that not only looks at the nature and idea in the essay but also the grammatical and structural features of the sentences in the essay, before grading it.
- A comparative analysis of the results from different machine and deep learning models to our proposed deep learning network.

1.1 *Difficulties in Automatic Essay Grading*

The most serious issue in the realm of AEG is **holistic AEG**. This procedure entails providing an overall score/grade to the essay depending on how well-written it is. The majority of AEG research is focused on holistic AEG.

Much earlier research work used machine learning with classifiers to holistically grade the essays. Some commercial systems such as e-rater® [4], Intelligent Essay Assessor [5], etc. use a variety of features from surface-level features like length-based features (word count, average word length, average sentence length, etc.), to more complex features (like usage score, which detects usage errors in the essay) and kernels.

Zesch et al. [6] describes a set of task-independent features for AEG, where they studied the problem of cross-domain AEG. One of their findings was that the best results for cross-domain AEG are when the source and target prompts are similar.

2 Related Works

In this section, we look at different systems which are used for AEG.

2.1 *Classical Machine-Learning Based Systems*

The earliest AEG systems use machine-learning-based approaches. The general approach for these systems involves feature engineering and ordinal classification/regression. Project Essay Grade (PEG) [1], the earliest AEG system, used a series of intrinsic properties (called *trins*, which are analogous to features) to come up with an approximate score for the essay.

2.2 *Deep-Learning Based Systems*

Since the early 2010s, neural networks have been used to solve multiple tasks in NLP, such as CNN [7] and other hierarchical neural network models. Alikaniotis et al. [8] develop a system which learns score-specific word embeddings (SSWE). They use these SSWEs, instead of other word embeddings like GloVe, to learn word embeddings before getting an essay representation via LSTMs to get the overall essay score. Maliha [9] use LSTMs and other types of recurrent neural networks to score essays. They use a similar architecture as [8], except that they use pre-trained word embeddings.

3 Background and Technologies

3.1 *Recurrent Neural Networks*

A recurrent neural network (RNN) [10] is a deep neural network that uses the previous step's output as feedback input in the next step. They are generally used in sequences, in next-word prediction in a sentence. The RNNs store the knowledge in a hidden state and use the previous knowledge to predict the next sequence. At time step t, the hidden state $a^{<t>}$ and the output $y^{<t>}$ can be expressed as:

$$a^{<t>} = g1(W_{aa}a^{<t-1>} + W_{ax}x^{<t>} + b_a) \quad (1)$$

and

$$y^{<t>} = g_2(W_{ya}a^{<t>} + b_y) \quad (2)$$

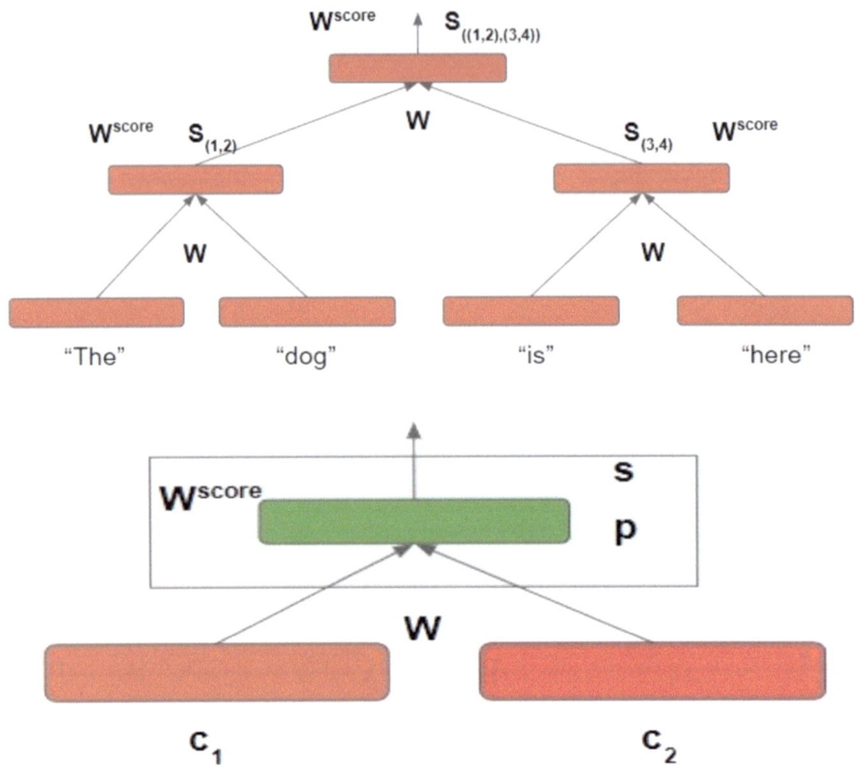

Fig. 1 The above images show how the RvNN works for a sentence. The word embeddings are fed to a neural network to get the phrase embedding, which is then passes into network to get the sentence embedding output vector. This vectors stores the grammatical and structural properties of the sentence

where W_{aa}, W_{ax}, W_{ya}, b_{a} and b_{y} are the coefficients that are shared over time steps and g_1 and g_2 are the activation functions. The working mechanism has been shown in Fig. 1.

3.2 *Convolutional Neural Network*

The Convolutional Neural Network [11] is a class of neural networks used for data processing with a grid-like structure, such as an image. A CNN has three main parts: **convolutional layer**, **pooling layer** and **dense layer**. The convolution layer uses the concept of the kernel function to extract the information from the image when given as input in image matrix format. The convolution can be calculated using the below equation.

$$G[m, n] = (f * h)[m, n] = \sum_j \sum_k h[j, k] f[m - j, n - k] \quad (3)$$

where f and h are the input images and the kernel function and m and n are the dimensions of the resultant image matrix and j and k are the dimensions of the kernel function. The pooling layer decreases the dimension of the input matrix and extracts the feature matrix's deeper meaning.

3.3 Bidirectional Encoder Representations from Transformers (BERT)

BERT [12] is the Encoder [13] part of a Transformer [14]. It is responsible for the extraction of features from the input sentence and pass the output vector to the Decoder part of the Transformer. BERT encode the input sentences with positional encodings so that the model can run in parallel. It consists of 12 encoder blocks, which contains a multi-headed self-attention mechanism and a dense layer. The attention mechanism helps to the different relations between the words in the sentences. This helps the model to have greater understanding of the sentences.

4 Dataset

We used the most commonly used AEG dataset, the Automated Students Assessment Prize (ASAP) Automatic Essay Grading (AEG) dataset, for our experimentation. The dataset comprises nearly 13,000 essays in response to 8 different essay prompts. The dataset is freely available and hosted on Kaggle, and the link is given in (https://www.kaggle.com/c/asap-aes/data). Overall, ASAP consists of eight essays, each corresponding to one question and was initially authored by students in grades 7–10. The statistics of the data has been given in Table 1. The essays are scored based on four points: ideas, style, organization, and conventions.

The four types of essays present are persuasive, narrative, expository and source-dependent responses. We took the scores given by raters belonging to domain one and added the scores as the final score. We divided the dataset into 4:1 ratios for training and testing, respectively.

5 Techniques and Implementation

We have tested the dataset on the following models for comparative analysis.

Table 1 Data analysis of the data showing the different prompts along with their number of essays, average length and score range

Prompt	No. of essays	Avg length	Score range
1	1783	350	2–12
2	1800	350	1–6
3	1726	150	0–3
4	1772	150	0–3
5	1805	150	0–4
6	1800	150	0–4
7	1569	250	0–30
8	723	650	0–60

5.1 SVM

We took the automatic grading problem as a classification task. As shown in Table 1, the maximum mark that can be obtained is 60, and the minimum mark is 0. The marks are divided into 61 classes from 0 to 60. We took the essay, used TF-IDF (Term Frequency and Inverse Document Frequency) [15] vectorisation, and used '12' normalisation. The tf-idf does not blindly consider the number of times a word appears in the document by scaling down the words that appear frequently in the essays and have less meaning, such as the articles 'a', 'an' and 'the'. Suppose w is a word in the essay e then the tf-idf is calculated by:

$$\text{tfidf}(w, e) = tf(w, e) * \text{idf}(w) \tag{4}$$

where,

$$\text{idf}(w) = \log \frac{(1 + n)}{(1 + df(w))} + 1 \tag{5}$$

as we smoothened the tf-idf, n is the number of essays and $df(w)$ is the document frequency of the word in the essay. We used the multi-class classification SVM with Gaussian kernel. The model was trained on the essay set and finally evaluated. Vaswani et al. [14] uses these kinds of machine learning models for their work.

5.2 Bert

We use the standard bert-base-cased model for automatic essay grading. The word "base" means it has 6 encoder blocks, and the word "cased" signifies that the model considers the case of the words while generating the result. We purposely took this model when grading an essay, the case of the words delivers important information to the model, like the beginning of a sentence or a proper noun in the sentence etc.

Because BERT cannot accept inputs of size greater than 512 tokens, we have eliminated the essays which have more words than 500 words. Each encoder block contains a self-attention layer followed by a feedforward layer. The attention layer contains 8 multi-headed attention mechanisms that store different information regarding a sentence's words. This information is used during the grading of the essay. The feed-forward network maintains the dimension of the output vector, which is sent to the next encoder block. The positional encoding of the words with Bert embeddings makes the model run very fast, as it can process the essay parallelly.

5.3 Collaborative Deep Learning Network (CDLN)

Collaborative learning is a method of learning where the pupils work in groups and work on separate tasks contributing to a common overall outcome, or work together on a shared task. Collaborative learning models have the power of transfer learning, where instead of a single deep neural network to learn everything, the learning can be distributed among several networks, and their collective knowledge can be shared. Our model consists of a Recursive Neural Network, a Convolutional Neural Network, an LSTM and a dense neural network, as shown in Fig. 2. The different parts of the architecture are described below in detail.

Convolutional Neural Network The convolutional neural network understands the idea conveyed in the sentences using the convolution and the average pooling layers. The words of the essays are converted into Word2Vec embeddings of 100 dimensions each. All these vectors are concatenated together to form a single-dimension vector. This vector is sent forward for convolution and pooling. $1 \times 105 \times 8$ kernel is used for convolution and $1 \times 90 \times 8$ for average pooling. The convolution and pooling layer is repeated for 5 times. After this, the layer is flattened and sent

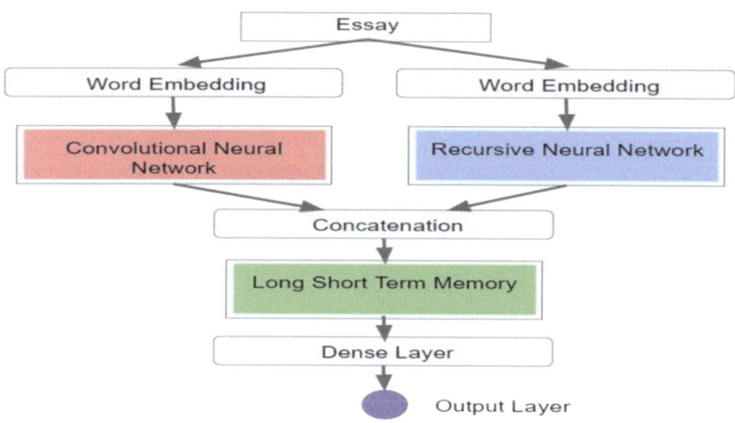

Fig. 2 The architecture of our collaborative deep learning network

forward for concatenation with the Recursive Neural Network (RvNN) output vector. The convolution layer helps to analyse the essay, while the average pooling layers bring out the idea conveyed in the sentences.

Recursive Neural Network The Recursive Neural Network (RvNN) [10] understands the structure of the sentences of the essays. This helps the CDLN model check the essays' grammatical and sentence construction errors. The essays are first embedded with a word embedding, using the Word2Vec. The words are first divided into bigrams; then, the representation vectors are fed into a neural network as shown in Fig. 2.

We took the embedding dimension as 100. So the neural network has 200 neurons in the first layer. After that 4 layers of 150 neurons each. The output layer is 100 neurons to match up with word embedding. The formula is shown below.

$$p = f(W[c_1; c_2] + b) \tag{6}$$

$$s = W^{\text{score}} p \tag{7}$$

The c_1, c_2 are the representation vectors, W is a trainable vector, b is the bias. W^{score} is another trainable vector and s is the score for that particular hierarchical structure of the sentence. Once the score is calculated, the same algorithm is followed for the rest of the sentences and then summed up and the final vector is forwarded for concatenation. An example of the algorithm with the sentence is given below.

Algorithm Consider the sentence "The dog is here", and each word is denoted by numerics '1', '2', '3' and '4', respectively and their embeddings by c_1, c_2, c_3 and c_4. We calculate the score using formulas 6 and 7.

Step 1: First, we calculate the scores for each word pair using the word embeddings of the words in the sentences given by s[1, 2], s[2, 3] and s[1, 2] and store them in A[1, 2], A[2, 2] and A[3, 4] respectively.

Step 2: Then we use the scores in the previous steps, the scores s[1, 3] and s[2, 4]. A[1, 3] =max(A[1, 2] + c_3, A[2, 3] + c_1) and A[2, 4] = max(A[2, 3] + c_4, A[3, 4] + c_2)

Step 3: The whole structure of the sentence is decided in the last step A[1, 4] = max(A[1, 3] + c_4, A[1, 2] +A[3, 4], A[1, 2] + c_1)

Step 4: The score having the highest value represents the hierarchical structure of the sentence.

Long Short Term Memory The LSTM [16] takes the input vector, which is the resultant of the concatenation of the vector output of the Recursive Neural Network and the CNN. The output vector of LSTM is 1x10000 dimensions. This is forwarded to the dense layer. LSTM gathers the learnings of the CNN and RvNN and stores the learning in its hidden cell. This is a good example of transfer learning, where the previous deep learning networks transfer their cumulative to the following deep neural network. It keeps knowledge not only about the sentence structure but also of the ideas conveyed in the essays. Also, this information is sent forwarded to the next layers for grading the essay.

Dense Layer and the Output Layer The dense layer takes the input from the output vector of the LSTM. It is 5 hidden layers with 120 neurons. The last output layer is a single cell which gives the essay grade.

6 Experimentation

We have experimented with 6 models, (a) CDLN model, (b) BERT, (c) RNN, (d) ANN and, (e) SVM. The architecture of all the models has been described in the previous section. For all the models the learning rate is 0.0001 and the batch size for training is 32. For the CDLN model, BERT, RNN, ANN and SVM models, the number of epochs used is 15, 10, 8, 8, 8 and 6, respectively. Dropouts are used in the deep learning models to stop overfitting. In all the models during training, eight-fold cross-validation has been used.

7 Results and Analysis

We use the mean square error (MSE), Pearson's Correlation Coefficient (PCC) and Quadratic Weighted Kappa (QWK).

- Pearson's Correlation Coefficient [16]: It is generally used to measure the linear correlation or relationship between two variables. It is calculated over a range of -1 to 1, where 1 shows that the variables are highly correlated in the positive direction while -1 shows that they are not correlated; the two variables are in opposite directions.
- Quadratic Weighted Kappa [17]: The most commonly used metric for measuring the performance of AEG systems is Cohen's Kappa with Quadratic Weights—i.e, Quadratic Weighted Kappa (QWK). We used this for our evaluation.

7.1 Automatic Evaluation

We compare the results from the different machine and deep learning models that has been mentioned in the above sections. The models' results we compare here are (a) Our proposed CDLN model, (b) LSTM, (c) RNN, (d) ANN, and (e) SVM. We have also taken [18] as the baseline model and compared the results of the TDNN (ALL), CNN-LSTM, CNN-LSTM-ATT and 2L-LSTM with our models. The results have been shown prompt-wise in Table 2 and overall in Table 3. From the tables, it can be seen that our CDLN model outperforms the other models, including the baseline models. The sharing and transferring of knowledge from the CNN and the RvNN to the LSTM has boosted the results. The self-attention mechanism in BERT has led to better results than TDNN (Sem+Synt), the best-performing model in [18]. The knowledge stored in the neurons of the deep networks has overcome the feature engineering process. If we go promptwise, the results mostly show that our models outperform the others.

7.2 Robustness of the Model

An automatic grading system must be robust in its action. It must grade the same essay with the same grade even when there is a slight change in the language and structure of the sentences. So we have conducted a robustness check of our CDLN model. For this test, we have selected 1000 random essays from the test set of all the essays. Then we rephrase the essays with Quillbot,[1] a well-known paraphrasing tool available freely for users. We graded the selected essays with our model. After this, we paraphrased the essays with Quillbot and graded them with our model. We recorded the grades and averaged the marks of every 50 essays and then plotted and compared them in the graph shown in Fig. 3.

It is evident from the above comparison-bar graph that the grades given by the model before and after paraphrasing is very close. It can be noticed from the graph that the average marks of the modified essays are more than the original essays; this can be due to the fact that quillbot has added proper sentence structure in the essays and removed the grammatical errors which may be present in the original essay. We further tested how much the grade differed from each other. For this purpose, we have modified the mean square error formula. If g_{original} is the marks graded by the model for the original essays and g_{modified} is the marks graded for the paraphrased essays then we find the difference by the formula:

$$\Delta = \sum_{i=1}^{N} \frac{(g_{\text{original}} - g_{\text{modified}})^2}{N} \tag{8}$$

We used the above formula for the grades of all the $N = 1000$ essays and got the value 0.34. This shows that grades are very close to each other, and the model is quite robust.

[1] https://quillbot.com/.

Table 2 Performance comparison table for each prompt among the CDLN model and the other baseline models

Models	*Accu*	*PCC*	*QWK*	Models	*Accu*	*PCC*	*QWK*
	Prompt 1				Prompt 2		
CDLN model	**0.8853**	0.7565	0.7176	CDLN model	**0.8466**	0.7945	0.7206
BERT	0.8194	0.7341	0.6599	BERT	0.8394	0.7761	0.6519
LSTM	0.7506	0.6649	0.4260	LSTM	0.7524	0.6765	0.4295
RNN	0.7549	0.6566	0.4469	RNN	0.7729	0.6686	0.4709
ANN	0.7364	0.6526	0.4905	ANN	0.7384	0.6426	0.4735
SVM	0.7449	0.6865	0.4998	SVM	0.7499	0.6895	0.4798
2L-LSTM	–	0.7375	0.5128	2L-LSTM	–	0.7375	0.5128
CNN-LSTM	–	0.6722	0.6594	CNN-LSTM	–	0.6722	0.6594
CNN-LSTM-ATT	–	0.6531	0.5348	CNN-LSTM-ATT	–	0.6531	0.5348
TDNN (Sem + Synt)	–	**0.7993**	**0.7366**	TDNN (Sem + Synt)	–	**0.7993**	**0.7366**
	Prompt 3				Prompt 4		
CDLN model	**0.8756**	**0.7505**	**0.7206**	CDLN model	**0.8354**	**0.7659**	**0.7165**
BERT	0.8394	0.7761	0.6519	BERT	0.7950	0.7509	0.7364
LSTM	0.7554	0.6665	0.4265	LSTM	0.7746	0.6937	0.6645
RNN	0.7729	0.6686	0.4709	RNN	0.7246	0.7067	0.7146
ANN	0.7454	0.6596	0.4456	ANN	0.6765	0.6864	0.6594
SVM	0.6982	0.6305	0.4219	SVM	0.5694	0.5594	0.6044
2L-LSTM	–	0.6410	0.5018	2L-LSTM	–	0.6527	0.5754
CNN-LSTM	–	0.6844	0.5454	CNN-LSTM	–	0.7564	0.7065
CNN-LSTM-ATT	–	0.5927	0.4219	CNN-LSTM-ATT	–	0.7224	0.4665
TDNN (Sem + Synt)	–	0.6759	0.6281	TDNN (Sem + Synt)	–	0.7616	0.7578
	Prompt 5				Prompt 6		
CDLN model	**0.8756**	0.7505	7˙206	CDLN model	**0.8561**	**0.7573**	**0.7094**
BERT	0.8394	0.7761	0.6519	BERT	0.8264	0.7368	0.6749
LSTM	0.7554	0.6665	0.4265	LSTM	0.7654	0.6809	0.4275
RNN	0.7729	0.6686	0.4709	RNN	0.7754	0.6605	0.4760
ANN	0.7384	0.6426	0.4735	ANN	0.7160	0.6853	0.5229
SVM	0.7049	0.6365	0.4298	SVM	0.6489	0.6195	0.5699
2L-LSTM	–	0.7375	0.5128	2L-LSTM	–	0.6528	0.4951
CNN-LSTM	–	0.6722	0.6594	CNN-LSTM	–	0.6460	0.5810
CNN-LSTM-ATT	–	0.6531	0.5348	CNN-LSTM-ATT	–	0.6291	0.5149
TDNN (Sem + Synt)	–	**0.7993**	**0.7366**	TDNN (Sem + Synt)	–	0.6903	0.6752
	Prompt 7				Prompt 8		

(continued)

Table 2 (continued)

Models	*Accu*	*PCC*	*QWK*	Models	*Accu*	*PCC*	*QWK*
CDLN model	**0.8702**	**0.7596**	**0.7246**	CDLN model	**0.8449**	**0.7550**	**0.7078**
BERT	0.8349	0.7719	0.6537	BERT	0.8252	0.7578	0.6904
LSTM	0.7589	0.6645	0.4237	LSTM	0.7745	0.6690	0.4656
RNN	0.7749	0.6659	0.4721	RNN	0.7606	0.6603	0.4584
ANN	0.7379	0.6464	0.4754	ANN	0.7454	0.6596	0.4456
SVM	0.7016	0.6543	0.4267	SVM	0.6982	0.6305	0.4219
2L-LSTM	–	0.7637	0.6690	2L-LSTM	-	0.5137	0.2486
CNN-LSTM	–	0.6849	0.6609	CNN-LSTM	–	0.4666	0.3812
CNN-LSTM-ATT	–	0.6314	0.6002	CNN-LSTM-ATT	–	0.5358	0.4468
TDNN (Sem + Synt)	–	0.7201	0.6587	TDNN (Sem + Synt)	–	0.6328	0.5741

Table 3 Overall average performance comparison table for all prompt among the CDLN model and the other baseline models

Models	Average		
	Accu. (%)	*PCC*	*QWK*
CDLN model	**0.8550**	**0.7545**	**0.7036**
BERT	0.8268	0.7390	0.6779
LSTM	0.7645	0.6288	0.4287
RNN	0.7786	0.6645	0.4712
ANN	0.7305	0.6456	0.4787
SVM	0.7085	0.6332	0.4236
2L-LSTM	–	0.6548	0.4687
CNN-LSTM	–	0.6569	0.5362
CNN-LSTM-ATT	–	0.6535	0.5057
TDNN (Sem + Synt)	–	0.7244	0.6856

8 Conclusion and Future Works

Our study aims to demonstrate the power of collaborative learning in automatic essay grading problems. When multiple networks work together to understand and learn the different features of an essay and grade accordingly, their performance is boosted, as seen from the above table. This model has outperformed many of the pretrained and state-of-the-art models in AEG. But still, there are ample scopes of improvement in the results. With the introduction of many new deep neural networks and pretrained networks, these kinds of research can be further investigated and conducted. Our model gives a holistic score to the essay, not paragraph-wise. This can be further investigated.

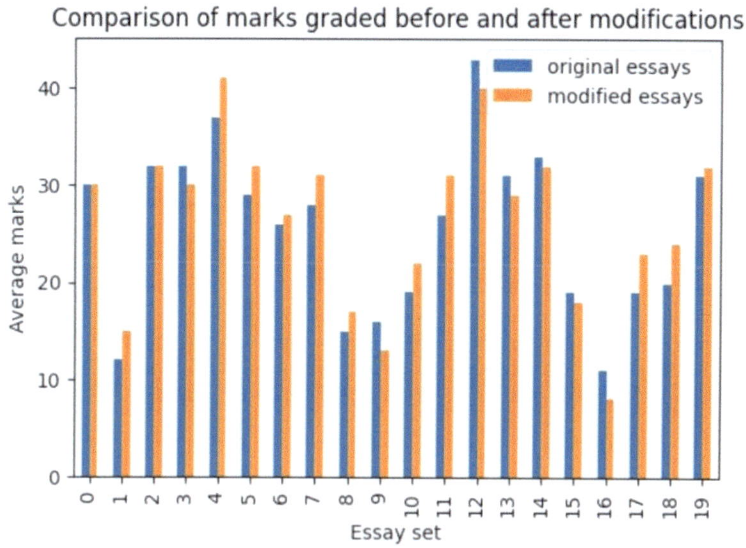

Fig. 3 The image compares the average grade of the original essays and the essays after paraphrasing

References

1. Page EB (1966) The imminence of... grading essays by computer. Phi Delta Kappan 47(5):238–243
2. Hasan MR, Maliha M, Arifuzzaman M (2019) Sentiment analysis with NLP on Twitter data. In: 2019 international conference on computer, communication, chemical, materials and electronic engineering (IC4ME2), pp 1–4. IEEE
3. Pramanik V, Maliha M (2022) Analyzing sentiment towards a product using DistilBERT and LSTM. In: 2022 international conference on computing, communication, and intelligent systems (ICCCIS), pp 811–816. IEEE
4. Attali Y, Burstein J (2006) Automated essay scoring with e-rater® V. 2. J Technol Learn Assess 4(3)
5. Landauer TK (2003) Automated scoring and annotation of essays with the intelligent essay assessor. In: Automated essay scoring: a cross-disciplinary perspective, pp 87–112
6. Zesch T, Wojatzki M, Scholten-Akoun D (2015) Task-independent features for automated essay grading. In: Proceedings of the tenth workshop on innovative use of NLP for building educational applications, pp 224–232
7. Kalchbrenner N, Grefenstette E, Blunsom P (2014) A convolutional neural network for modelling sentences. arXiv:1404.2188
8. Alikaniotis D, Yannakoudakis H, Rei M (2016) Automatic text scoring using neural networks. arXiv:1606.04289
9. Maliha M (2021) A supervised learning approach: detection of cyber attacks. In: 2021 IEEE international conference on telecommunications and photonics (ICTP), December 2021, pp 1–5. IEEE
10. Medsker L, Jain LC (eds) (1999) Recurrent neural networks: design and applications. CRC Press
11. O'Shea K, Nash R (2015) An introduction to convolutional neural networks. arXiv:1511.08458

12. Devlin J, Chang MW, Lee K, Toutanova K (2018) Bert: pre-training of deep bidirectional transformers for language understanding. arXiv:1810.04805
13. Cer D, Yang Y, Kong SY, Hua N, Limtiaco N, John RS, Constant N, Guajardo-Cespedes M, Yuan S, Tar C, Sung YH, Strope B, Kurzweil, R (2018) Universal sentence encoder. arXiv: 1803.11175
14. Vaswani A, Shazeer N, Parmar N, Uszkoreit J, Jones L, Gomez AN, Kaiser Ł, Polosukhin I (2017) Attention is all you need. Adv Neural Inf Process Syst 30
15. Kim SW, Gil JM (2019) Research paper classification systems based on TF-IDF and LDA schemes. HCIS 9:1–21
16. Sherstinsky A (2020) Fundamentals of recurrent neural network (RNN) and long short-term memory (LSTM) network. Physica D 404:132306
17. Sedgwick P (2012) Pearson's correlation coefficient. BMJ 345
18. Jin C, He B, Hui K, Sun L (2018) TDNN: a two-stage deep neural network for prompt-independent automated essay scoring. In: Proceedings of the 56th annual meeting of the association for computational linguistics. Long papers, vol 1, pp 1088–1097

Efficient Iris Recognition System Using Supervised Learning Algorithm

Vaibhav Panwar, Sanjeev Kumar Pippal, and Ashok Kumar Sahoo

Abstract Biometric systems serve an important function in the identification of individuals, assisting in global security. There are several possible biometrics, such as Genetics, signature, and so on; however, computer vision (CV)-based biometric scanners have gained an important position in the field of individual identity. Customer security is one of the most important components of CV systems. It was proved that basic credentials and login information are ineffective and easily obtained by cybercriminals. CV-based biometrics encompasses the recognition of faces, palmprints, and iris, among other things, and the use of such capabilities to build successful strong authentication. In this study, iris recognition is used, non-contact authentication technologies which are both hygienic and highly reliable. We investigated iris recognition utilizing publicly released databases CASIA Iris Thousand. The obtained iris images are sent through pre-processing procedures. Next, the segmentation and normalization techniques are implemented on processed iris images. Two feature extraction methodologies are applied such as Gray-Level Co-Occurrence Matrix (GLCM), and HAAR to get the key characteristics from normalized images. The two featured data are then utilized to train the Machine Learning (ML) model namely Support Vector Machine (SVM), and K-Nearest Neighbour (KNN). Lastly, the best feature extraction and ML model is identified using the performance measure. The GLCM + SVM delivers a maximum accuracy rate of 97% when compared to other combinations like GLCM + KNN, HAAR + SVM, and HAAR + KNN.

Keywords Iris · Computer vision · Segmentation · CASIA database · Optimization · Time response · Feature extraction · Machine learning

V. Panwar (✉) · S. K. Pippal
Department of Computer Science and Engineering, Sharda University, Greater Noida, UP, India
e-mail: vaibhavpanwar1991@gmail.com

A. K. Sahoo
Department of Computer Science and Engineering, Graphic Era Hill University, Dehradun, India

A. P. Agrawal et al. (eds.), *Advances of Artificial Intelligence and Machine Learning in Image Processing, Security and Data Analytics*, Lecture Notes in Electrical Engineering 1464, https://doi.org/10.1007/978-981-95-0148-9_9

1 Introduction

Many people in the twenty-first century connect via text or video using technical gadgets such as personal computers (laptops), mobile phones smart watches, etc. Electronic equipment is storing a growing amount of personal information. As a result, to prevent personal information from being lost or theft, personal devices should incorporate biometric authentication [1]. Biometric authentication can also be used in access control systems to restrict unauthorized entry to private information [2]. According to new research, a hacker attacks on network service, every 39 s on average [3]. This remark implies that security systems are becoming increasingly vital. Simple password-based security solutions, on the other hand, have been demonstrated to be ineffectual [4]. It's largely because a substantial majority of users utilize easy-to-guess user credentials or passwords. Furthermore, some people write their passwords on credit and debit cards or stick them on their computers. What is the computer system's weakest link? That is a logical conclusion to reach. Because each person has unique features and qualities that cannot be possessed by more than one person, biometrics can achieve this level of security for identification and verification [5]. Handprints, face characteristics, iris textures, gesture recognition, vocal or speaking style, and signatures are examples of these characteristics [6]. Iris identification has become the most prevalent biometric technology in the last decade. This is because iris characteristics are long-lasting and distinct, as well as steady [7]. This innovative technology has the potential to make identity and authentication systems more trustworthy. Furthermore, unlike other biometrics including fingerprint identification, iris recognition doesn't need any contact to analyze the input sequence.

One of the first automatic iris recognition systems was created in 1987. The first patent was given to Flom and Safir [8], who proposed a theoretical iris authentication scheme but provided no implementation details. Then, a plethora of iris-extracting features and classifier techniques have emerged, which can be loosely divided into traditional hand-crafted feature design approaches and recently started deep learning (DL) techniques. There are several phases responsible for putting together a biometric technology. Because the camera's lens can include dirt or the target can shift or spin while capturing images, processing of the source images is required, therefore the first step is to assess how an image or video includes a characteristic that can be identified, including an iris [9]. Once detected, this feature is translated into a pattern and matched to other human patterns to see how it can be recognizable. The methods utilized to retrieve, categorize, and identify characteristics are crucial to the building of a reliable process with high assessment criteria. This work is organized as follows: (1) Importance of iris recognition, (2) Past research on iris recognition, (3) Research workflow, (4) Data acquisition and process, (5) Segmentation and normalization, (6) ML model and validation, (7) Discussion of results, (8) Conclusion.

2 Literature Survey

In the literature, there exist several approaches and algorithms for iris-based biometric identification recognition. A tremendous amount of work has gone into this area, as well as the elevated concentrations of precision and reliability which may be trusted are fully discussed here. The publication [10] offers a novel way of classifying individuals by examining their iris characteristics. In this research, the pattern information is recovered using HAAR and 2D Gabor filtration. Instead of selecting the Gabor filter coefficients manually, Particle Swarm Optimization, and ML techniques, have been used to decide them. The assessment discovered that appropriate identification might be achieved for the dataset utilized. Images have also been prepossessed using Independent Component Analysis (ICA) to remove distortions. The author [11] offers a fast iris recognition process that combines pre-trained image classification techniques as feature representation and only requires one matching process. In this method, first-layer Convolutional Neural Network (CNN) filters can be utilized as feature representation. That implies the iris area is represented by features extracted from convolutional layers and utilized in a range of methods. As a result, because bit-shifting no longer applies, a considerable number of calculations were removed from the matching process. To limit the impact of masks on rubber-sheet images, they suggest filtering all feature transforms by concealing respective channels and selecting the most essential features. The technique was tested, which showed great changes in reliability and matching time. Research [12] on biometric authentication mechanisms has been conducted, as well as a few potential future advancements in the development of data protection. The researchers [13] hope to develop an accurate and rapid iris detection mechanism using a CNN. Even though DL has been a successful technique in recent decades, some of its obstacles include its computing demands and the presence of huge amounts of labeled training data. In this case, interactive learning is a budget-friendly and efficient option. An immersive variant of UNet for iris recognition was built using Squeeze Expand modules [14] to shorten learning time and enhance memory efficiency by lowering the number of variables involved. The interactive element aids in the generation of ground truth for datasets with poorly annotated data. In the paper [15], an iris detection framework that is texture-attentive and lightweight is provided. The majority of our efforts are concentrated in three areas. They propose an unsupervised pre-training stage guided by reconstructive loss, accompanied by supervised improvement as a first phase. Whenever network weights concentrate on specific patterns, they become much more discriminative. The meticulous training and design selections enable us to build an efficient model featuring up to 100 times fewer than current DL measurements while achieving improved recognition results for both inside and across database evaluations. The author utilized Gabor filters to create a feature representation using iris attributes [16]. The authors employ a supervised NN in this study, which would be originally trained on a range of iris feature inputs before being employed for identification. The CASIA iris datasets were used in this study, and the proposed method outperformed prior research in the field.

3 Iris Recognition Work Flow

The steps involved in automatic iris recognition using ML are detailed in this section. The data required for this research has been collected from a famous website called CASIA. Pre-processing is the second step to making the data ready for feature extraction. Feature extraction approaches such as GLCM and HAAR can be used. For the recognition of iris images, the supervised learning model of SVM and KNN is applied. Finally, for identifying the best model various metrics are employed. The overall flow of the iris recognition process done in this research is shown in Fig. 1.

3.1 Data Acquisition and Pre-process

CASIA database: CASIA-Iris-Thousand [17] comprises 20,000 images of 1,000 people, which have been obtained with an IKEMB-100 camera. The CASIA-Iris-Thousand will be the first publicly released iris database of 1000 people, well-suited for exploring the distinctiveness of iris features and creating innovative iris categorization and retrieval techniques. The sample of CASIA data is depicted in Fig. 2.

Pre-process: ROI was developed in response to the constraints of both lossy and lossless reduction. Reduction ratios for lossy encryption techniques can be far higher (up to 5–30%) [18], yet important data degradation may occur. Therefore, ROI is

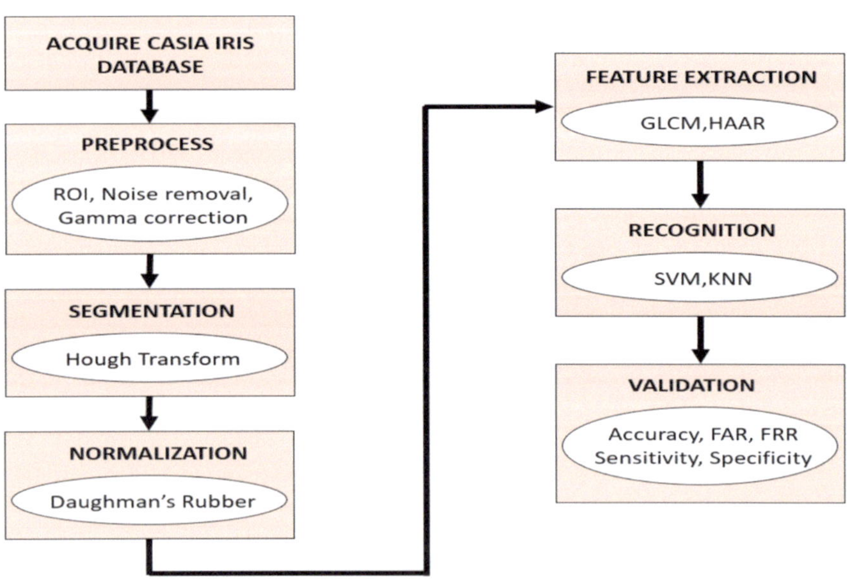

Fig. 1 Steps involved in iris recognition

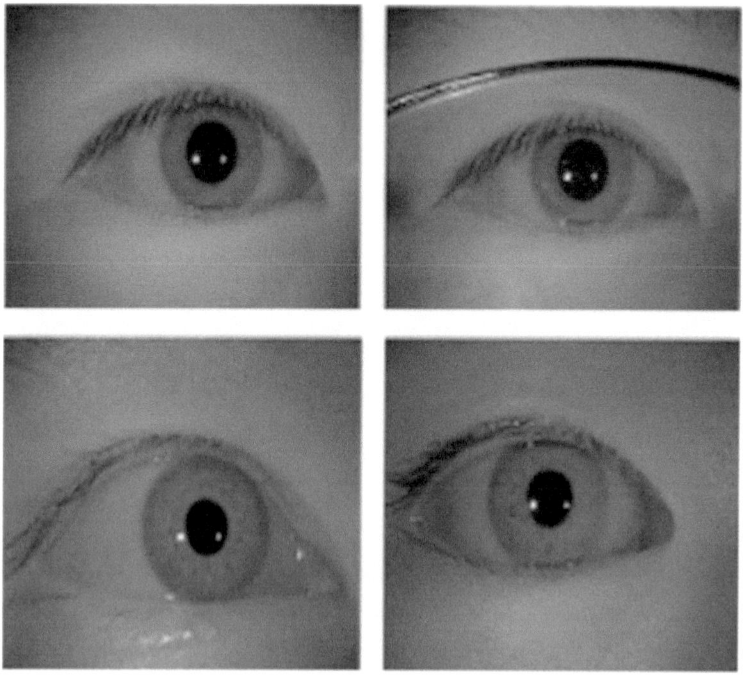

Fig. 2 CASIA data sample

suggested primarily for biological images, as biological images do not tolerate any information loss in extremely detecting crucial portions. The ROI taken from the CASIA iris image is shown in Fig. 3. This research focuses on removing noise from light reflections, lids, and eyelash surfaces. The clarity of the iris image could surely be degraded because of reflections from the beam of light and also obstruction from lashes around the eyes [19]. Information transmission among electro-optic devices has a nonlinear effect, which would be accounted for in image analysis by employing the gamma correction approach. Gamma correction is just the term given to a nonlinear method utilized for video or image structures to transmit and decode brightness or colorimetric data [20]. The expression for gamma correction is just as follows:

$$V_{\text{out}} = AV_{\text{in}}^{\gamma} \tag{1}$$

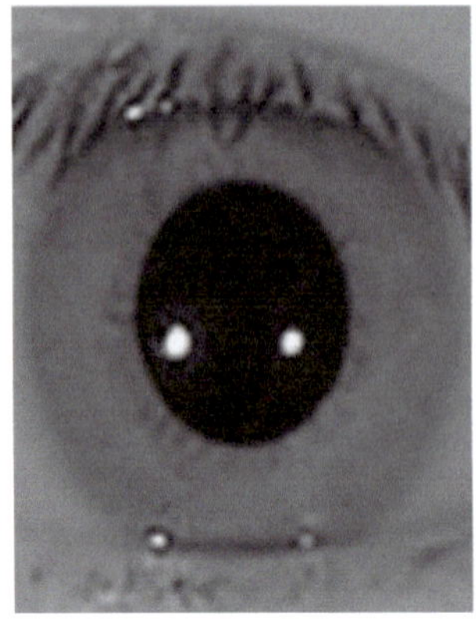

Fig. 3 ROI of iris image

3.2 Segmentation and Normalization

The technique of iris segmentation has been used to segment the iris image. The iris image data may be characterized by two circles, the inner and outer borders of the pupil and sclera. The iris region was located using the Hough Transformation. The Hough segmentation method to generate an edge detection, as shown in Fig. 4. The highest values that comprise the circular variables in the Hough space for every pixel inside the edge map can be voted on, and the results of such votes are utilized to establish the highest value for the pixel [21]. The following formula can be used to calculate the center coordinates and radius:

$$x_c^2 + y_c^2 - r^2 = 0 \tag{2}$$

The major point in the Hough space correlates to the center coordinates (x_c, y_c), while the radius $'r'$ of the circular is provided by the edges. To lessen the impact of horizontally oriented lids on edge detection, and used vertical gradients to identify

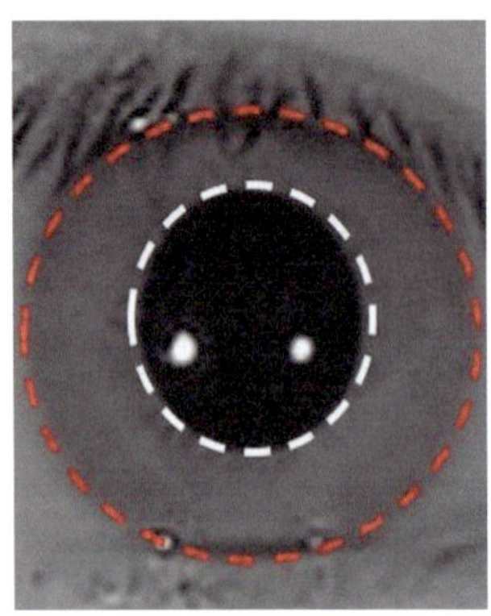

Fig. 4 Segmented Iris Image

the iris-sclera border. The impact of the lids is reduced while using vertical gradients to detect the iris border. After successfully segmenting the circle iris region from such an iris image, normalization is performed to convert the segmented circle area into a rectangular area with a set size. Using Daugman's rubber-sheet model, an iris image was normalized. The most common technique in iris normalization [22] is Daugman's rubber-sheet model, which turns the circular iris area into a rectangular shape of a predetermined size. The approach converts every pixel in the circle iris into an equivalent position on the polar axis (r, θ), wherein r is the radial length and θ is the angle twisted at the appropriate radius.

$$I\left[x(r,\theta), y(r,\theta)\right] - \rightarrow I(r,\theta) \quad (3)$$

The iris surface is used to create an array of two-dimensional elements. The horizontal dimension indicates θ resolution, while the vertical dimension indicates the r resolution (x, y), and (r, θ) are the Cartesian and normalized polar coordinates for the iris area (x, y). The theta and r range from 0 to 2π and Rp to Rl while (r, θ) and (r, θ) are provided as a linear function of pupil boundary position.

3.3 Feature Extraction

HAAR: HAAR-like features were image properties that are utilized in iris recognition. To improve the efficiency of the recognition process, Papageorgiou et al. [23] proposed dealing with a simple feature set built on HAAR wavelets rather than the normal image intensities. The idea behind a HAAR-like feature is to take neighboring rectangular portions at a specific position in a bounding box, add up the intensity values within every sector, and then calculate the differential between such sums. This differential is then used to identify picture sections as a feature response. There are three types of rectangular HAAR-like properties. A result for a two-rectangle feature is created by adding pixels from two rectangular areas. These are close on both sides have the same proportions and are also comparable in shape. A 3-rectangular feature calculates the total of the sums of two outer rectangles deducted from the total of a center rectangular. Finally, the distance among diagonal pairs of rectangles is computed via a four-rectangle feature. A rough estimation of the identified components $LL3$ horizontally detailed elements $LH = LH1$, $LH2$, $LH3$, vertically detailed elements $HL = HL1$, $HL2$, $HL3$, and diagonally detailed elements $HH = HH1, HH2, HH3$. The most specific information about the iris texture is contained in its detailed elements. If the iris characteristic used the detailed elements on the 1st or 2nd lever, the feature set would've been enormous, resulting in higher consumption costs. Several tests finally decided to show the iris utilizing the 3rd lever's detailed elements. The iris feature was represented in the following manner.

$$\{C(i) = 0, \mathrm{if} C(i) \leq 0 C(i) = 1, \mathrm{if} C(i) > 0 \quad (4)$$

C iris feature space and

$$C = \{LH3, HL3HH3\}$$

$C(i)$ element of C.

GLCM: Using the GLCM approach, 2nd-order statistical attributes can be retrieved. A GLCM is a matrix with the same rows and columns with a total of gray values, G, inside the image. The likelihood of adjacent pixels separated by a pixel length (Δx, Δy), one including intensity '*i*' and the other including '*j*' appearing within a specific neighborhood is expressed by the matrix component P ($i, j|\Delta x$, Δy). The matrix element P ($i, j|d$), contains the 2nd-order statistical probability numbers for variations in grey levels '*i*' and 'j' at a specified displaced distance d and an angle. Utilizing the huge number of intensity values G demands keeping a significant amount of temporary storing, i.e., a $G * G$ matrices for every pairing of (Δx, Δy) or (,). Because of their high dimensionality, GLCMs are particularly highly dependent on the size of the texture sampling on which they have been calculated. As a result, the appropriate size of extraction for improved resolution and minimal information loss [24].

Model Construction:

- SVM: The construction of SVM as a classifier includes two major components: the first is to find the hyperplane, and the second is to convert the linearly non-separable categorization task into a linearly separable problem [25]. For the convenience of this example, consider x and y a feature and a target, accordingly. The target label and the pairing of source features can be written as $\{x_i, y\}$, where $i =$ 1,2, ..., N and $y = \pm 1$. In the instance of a linearly separable problem, target 1 ($y = 1$) and target 2 ($y = -1$) are determined by a separating hyperplane. Planes are separated using hyperplanes.

$$wx + b = 0 \tag{5}$$

$$y_i(wx_i + b) \geq, i = 1, 2, N \tag{6}$$

Fundamentally, there are several different values of w, b that produce the separation of hyperplanes.

- KNN: The training images are arrays with class labels in a high-dimensional feature space. k is a user-defined integer in the classification stage, and an unlabeled is categorized by applying the label that is most common among k training images next to that query instance [26]. To utilize the KNN technique to classify sample X, which goes as follows:

1. Assume there are j learning categories $C_1, C_2, \ldots, C_n$ and the total of a training image is N with features extraction, resulting in an m-dimension feature space.

Table 1 Performance metrics

Metrics	Accuracy	False Acceptance Rate (FAR)	False Rejection Rate (FRR)	Sensitivity	Specificity
Formula	$\frac{TP+TN}{TP+TN+FP+FN} * 100$	$\frac{FP}{FP+TN} * 100$	$\frac{FN}{TP+FN} * 100$	$\frac{TP}{FN+TP} * 100$	$\frac{TN}{FP+TN} * 100$

2. Evaluating the measure of the correlation between all training samples and X Using the *ith* sample, the (x, d_i) is as follows:

$$SIM(x, d_i) = \frac{\sum_{j=1}^{m} X_j, d_{ij}}{\sqrt{\left(\sum_{j=1}^{m} x_j\right)^2}\sqrt{\left(\sum_{j=1}^{m} xd_{ij}\right)^2}} \quad (7)$$

3. Select k samples from N similarities of (x, d_i), $(i = 1, 2, \ldots, N)$, and treat these as a KNN selection of X. Next, using the given equations, calculate the probability that X belongs to every category.

$$P(X, C_i) = \sum_{d} SIM(x, d_i).y(d_i, c_j) \quad (8)$$

4. And $y(d_i, c_j)$ is an attributes function for categories that fulfilled the criteria.

$$y(d_i, c_j) = \{1, d_i \in C_j 0, d_i \notin C_j \quad (9)$$

5. Judge sample X to be the category that has the largest $P(X, C_i)$.

Validation

Within the context of big data, a solid evaluation process gets increasingly important concerning ML techniques that operate with massive volumes of data. Performance measures have been used to measure the overall effectiveness of the developed model [27, 28]. The performance measures for the iris identification scheme are provided in Table 1. The T and F in the metrics formula represent the correct and wrong predictions, and the P and N represent the authorized and unauthorized iris predictions.

3.4 Results and Discussion

The CASIA data of the iris image is collected and processed for cleaning the image. Next, the feature extraction is done followed by normalization and segmentation. The extracted data is applied to the ML model for proper recognition. The two feature extraction techniques GLCM, HAAR, and ML approaches including SVM, and KNN are applied. Two identifies the best pairing of feature extraction and classification techniques, the metrics are applied.

The HAAR and ML combination is evaluated first. The accuracy obtained by the KNN and SVM with the combination of HAAR is 93 and 94%. The specificity of SVM is high with a value of 93.13% and low for KNN with a value of 92.15%. The sensitivity values obtained by SVM and KNN are 94.89 and 93.87%. Both values of FRR and FAR will be less for SVM and the obtained results. All the values of the ML model with HAAR features are plotted using the bar graph and it is shown in Fig. 5. The GLCM and ML combo is then tested. According to the KNN and SVM with GLCM, 95.5 and 97% of accuracy may be attributed to this combination. SVM has a specificity of 97.95%, while KNN has a specificity of 96%. Per the results of SVM and KNN, the sensitivity is 96.07 and 95%, respectively. SVM reduces the FRR and FAR values, resulting in 3.92 and 2.04%, respectively. For KNN, the FRR and FAR values will be high, and the outcomes are 5 and 4%, respectively. All the values of the ML model with GLCM features are plotted using the bar graph ass shown in Fig. 6.

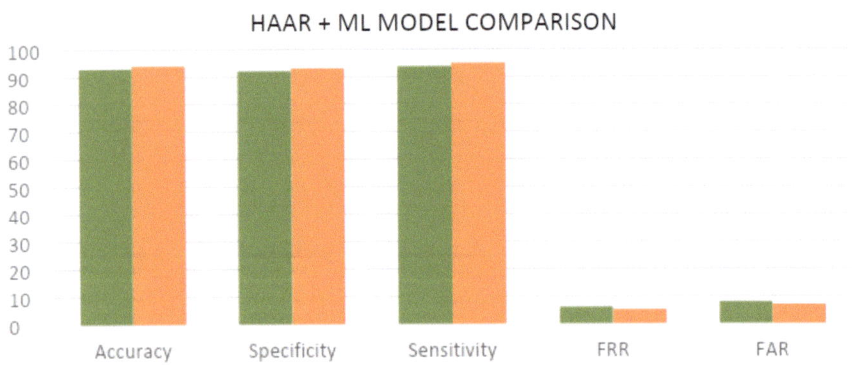

Fig. 5 ML comparison on iris recognition using HAAR-based feature

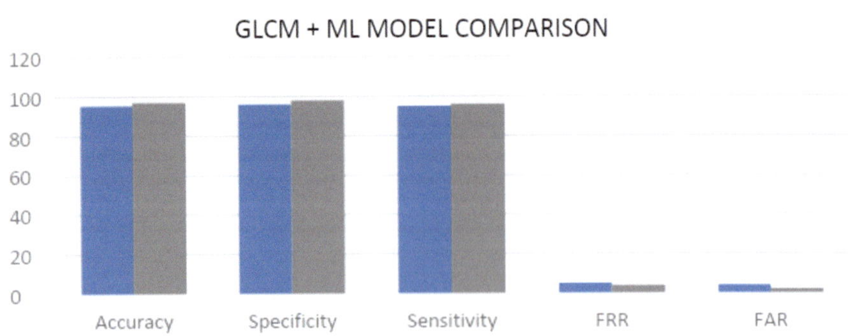

Fig. 6 ML comparison on iris recognition using GLCM-based feature

4 Conclusion

Iris recognition is considered to be one of the most reliable biometrics. Consequently, large-scale solutions have been introduced with success. As the world's population grows and international businesses emerge, the number of datasets will continue to expand, necessitating the development of biometric authentication approaches with more reliable and speedier. There are numerous descriptors and feature extractors for iris biometric patterns. However, they involve a bit-shifting mechanism to boost their eye rotation tolerance, which increases the time required to match. It is possible to distinguish an iris pattern in the human eye by analyzing its texture and geometry. Replicating the iris pattern is difficult due to the iris's extreme stability. A database called CASIA iris is used in this study. To identify the iris, feature extraction and ML algorithms are used. Data collection, noise subtraction, gamma correction, and pupil segmentation are all steps in the ML model's development process. Lastly, performance measures are used to gauge the effectiveness of the ML model and the process of feature extraction. Looking at the bar graph of performance, the GLCM + SVM is identified as the best for iris recognition. In the future, plan to use more datasets for validating the proposed approach. The accuracy of the model is also enhanced by employing bio-inspired optimization techniques.

Some existing iris recognition systems may not meet real-time processing requirements. Optimize algorithms and models to achieve faster processing times, especially in applications where real-time recognition is essential. Integrate multi-modal biometrics, such as combining iris recognition with facial or palm print recognition, to enhance overall recognition rates.

References

1. Papathanasaki M, Maglaras L, Ayres N (2022) Modern authentication methods: a comprehensive survey. In: AI, computer science and robotics technology, pp 1–24
2. Security and data protection in a google data center. https://www.youtube.com/watch?v=cLory3qLoY8. Accessed 14 Sept 2023
3. Jain D, Choudhary D, Anand A, Trivedi NK, Gautam V, Mohapatra SK (2022) Cybersecurity solutions using AI techniques. In: Agarwal R, Khatri SK, Soni KM (eds) IEEE 10th international conference on reliability, infocom technologies and optimization (trends and future directions) (ICRITO), October 2022, pp 1–8
4. Sun H, Chen Y, Lin Y (2012) OPass: a user authentication protocol resistant to password stealing and password reuse attacks. IEEE Trans Inf Forensics Secur 7(2):651–663
5. Abdulrahman SA, Alhayani B (2023) A comprehensive survey on the biometric systems based on physiological and behavioural characteristics. Mater Today Proc 80:2642–2646
6. Tisse CL, Martin L, Torres L, Robert M (2002) Person identification technique using human iris recognition. In: Proceeding of vision interface, pp 294–299
7. Sarmah AJ, Dutta GN, Lahkar D, Talukdar N, Lahkar B (2022) Biometric authentication-person identification using iris recognition. Int J Innov Sci Res Technol 7(5):1100–1104
8. Flom L, Safir A (1987) Iris recognition system. Google Patents, U.S. Patent No. 4,641,349. U.S. Patent and Trademark Office, Washington, DC

9. Kotp Y, Torki M (2023) Toward flare-free images: a survey. https://arxiv.org/abs/2310.14354. Accessed 18 Nov 2023
10. Petterson HN, Rehnholm J, Vikström S, Åslund M, Åstrand E, Tomasic I (2020) Iris identification using wavelet decomposition and Gabor filter. In: Koricic M (ed) 2020 43rd international convention on information, communication and electronic technology (MIPRO), pp 265–270. https://doi.org/10.23919/MIPRO48935.2020.9245439
11. Zambrano JE, Benalcazar DP, Perez CA, Bowyer KW (2022) Iris recognition using low-level CNN layers without training and single matching. IEEE Access 10:41276–41286
12. Khan SA, Naaz S (2020) Comparative analysis of finger vein, iris and human body odor as biometric approach in cyber security system. In: Suma V (ed) 2020 2nd international conference on innovative mechanisms for industry applications (ICIMIA), pp 525–530. https://doi.org/10.1109/ICIMIA48430.2020.9074877
13. Zhuang Y, Chuah JH, Chow CO, Lim MG (2020) Iris recognition using convolutional neural network. In: Tahir NM (ed) 2020 IEEE 10th international conference on system engineering and technology (ICSET), pp 134–138
14. Sardar M, Banerjee S, Mitra S (2020) Iris segmentation using interactive deep learning. IEEE Access 8:219322–219330
15. Chakraborty M, Roy M, Biswas PK, Mitra P (2020) Unsupervised pre-trained, texture aware and lightweight model for deep learning based iris recognition under limited annotated data. In: AlRegib G, Tekalp AM (eds) 2020 IEEE international conference on image processing (ICIP), pp 1351–1355
16. Thakkar S, Patel C (2020) Iris Recognition supported best gabor filters and deep learning CNN options. In: Deshpande V, Sondkar S (eds) 2020 international conference on industry 4.0 technology (I4Tech), pp 167–170
17. CASIAIrisDatabaseV4. http://biometrics.idealtest.org/dbDetailForUser.do?id=14. Accessed 05 Nov 2018
18. Baeza I, Verdoy A (2009) ROI-based procedures for progressive transmission of digital images: a comparison. Math Comput Model 50:849859
19. Liu Y, Xu J, Yee AL (2023) The analysis of Iris image acquisition and real-time detection system using convolutional neural network. J Supercomput 1–33
20. Lee Y, Zhang S, Li M, He X (2022) Blind inverse gamma correction with maximized differential entropy. Signal Process 193:108427
21. Verma P, Dubey M, Basu S, Verma P (2012) Hough transform method for iris recognition—a biometric approach. Int J Eng Innov Technol 1(6):43–48
22. Daugman JG (1993) High confidence visual recognition of persons by a test of statistical independence. IEEE Trans Pattern Anal Mach Intell 15(11):1148–1161
23. Papageorgiou C, Oren M, Poggio T (1998) A general framework for object detection. In: Davis L, Narasimhan R, Yachida M, Zisserman A (eds) International conference on computer vision, 6:555–562. https://doi.org/10.1109/ICCV.1998.710772
24. Mohanaiah P, Sathyanarayana P, GuruKumar L (2013) Image texture feature extraction using GLCM approach. Int J Sci Res Publ 3(5):1–5
25. Ali H, Salami M (2011) Iris recognition system using support vector machines. In: Biometric systems, design and applications. IntechOpen, London, UK
26. Kim JM, Heo JK, Yang HS, Song MK, Park SK, Lee WK (2006) Object recognition using k-nearest neighbor in object space. In: Shi ZZ, Sadananda R (eds) Agent computing and multi-agent systems. PRIMA 2006. Lecture notes in computer science, vol 4088. Springer, Berlin, Heidelberg
27. Verma D, Ojha S (2019) Performance analysis of biometric systems: a security perspective. Int J Adv Res Comput Commun Eng 8(4):104–110
28. Sivaram M, Ahamed MU, Yuvaraj AD, Megala G, Porkodi V, Kandasamy M (2019) Biometric security and performance metrics: FAR, FER, CER, FRR. In: Velan SS, Naranje V (eds) International conference on computational intelligence and knowledge economy (ICCIKE), pp 770–772. https://doi.org/10.1109/ICCIKE47802.2019.9004275

Audio-Visual Fusion for Deepfake Video Detection: A Deep Learning Framework

Vishal Kumar Sharma, Rakesh Garg, and Quentin Caudron

Abstract Deepfake technology, enabled by advances in artificial intelligence and deep learning, has grown exponentially, offering unprecedented abilities to synthesize hyper-realistic audio-visual content. However, its malicious misuse poses significant challenges for information verification, thereby demanding robust detection methods. This paper introduces a novel deep learning framework designed for the detection of deepfake videos by leveraging audio-visual fusion techniques. Recognizing the intricate anomalies present separately in audio and visual data, our approach integrates both modalities to achieve heightened accuracy in discerning genuine from manipulated content. We utilize a two-stream convolutional neural network (CNN) that separately analyzes the spectral properties of audio and the subtle facial discrepancies in video before fusing the obtained features. A composite feature vector is formed from the fusion, which is then input to a deep neural network (DNN) classifier. The combined strength of these modalities not only addresses the current vulnerabilities in existing detection methods but also anticipates the evolving sophistication of deepfake technologies. Preliminary results suggest an impressive enhancement in detection performance as compared to single-modality methods. Additionally, the flexibility of our framework allows for potential future integrations with other modalities and data types.

Keywords Deepfake · Audio-visual fusion · Deep learning · Video detection · Two-Stream CNN · DNN classifier

---

V. K. Sharma (✉) · R. Garg
Amity School of Engineering & Technology, Amity University, Noida, Uttar Pradesh, India
e-mail: vishalsharma3003@gmail.com

Q. Caudron
Sound Agriculture, Emeryville, CA, USA

A. P. Agrawal et al. (eds.), *Advances of Artificial Intelligence and Machine Learning in Image Processing, Security and Data Analytics*, Lecture Notes in Electrical Engineering 1464, https://doi.org/10.1007/978-981-95-0148-9_10

1 Introduction

The increasing integration of artificial intelligence (AI) and deep learning into multimedia creation tools has given rise to a new frontier of the hyper-realistic content generation known as "deepfakes." These are videos or audio artificially created or modified, often to mimic genuine recordings. While the advancements in this technology promise novel applications in fields such as film production, virtual reality, and personalized content creation, they also bring forth a formidable challenge: the malicious misuse of synthetic content to deceive, defame, or spread misinformation.

Deepfakes pose a significant threat to various sectors, from politics and journalism to cybersecurity. The ability to fabricate convincing videos of real individuals saying or doing things they never did blurs the line between fact and fiction, compromising trust in digital media. Recognizing this emergent threat, extensive research has been undertaken to develop detection methods capable of distinguishing between genuine and manipulated content. However, as the technology behind creating deepfakes continues to evolve, so too must the strategies employed to detect them.

The majority of existing deepfake detection techniques predominantly focus on visual cues. These include inconsistencies in lighting, blurring around the face, and unnatural eye-blinking patterns, among others [15]. However, as deepfake generation algorithms become more refined, such visual inconsistencies are becoming increasingly subtle, challenging conventional detection methodologies. Concurrently, while audio-based deepfakes are less common than their visual counterparts, their development is rapidly advancing, making it essential to consider auditory anomalies in the detection process [14].

Given this backdrop, the necessity for a more comprehensive detection strategy becomes evident. We posit that the integration of both audio and visual cues in deepfake detection presents a promising avenue to enhance accuracy and robustness. This approach is based on the rationale that while a deepfake generation algorithm may effectively replicate one modality, achieving perfection in synchronizing both audio and visual aspects remains an intricate challenge.

In this paper, we introduce a deep learning framework tailored for deepfake video detection, capitalizing on the synergistic potential of audio-visual fusion. The framework utilizes a dual-pathway structure, with each stream designed to extract unique features from audio and visual data. These features are then combined to form a composite representation, which is fed into a deep neural network (DNN) classifier to discern the authenticity of the content.

The subsequent sections of this paper will provide an in-depth review of the current literature on deepfake detection, elucidate the methodology underpinning our audio-visual fusion framework, present results benchmarked against prevailing techniques, and delve into a discussion on the strengths and limitations of our findings. We further conclude our finding in the conclusion section and thereby provide future research directions for continued research in the domain.

2 Literature Review

The rise of deepfakes in the digital realm has spurred significant research interest, both in the generation and detection of such content. In this literature review, we take a closer look at the state-of-the-art methods and their underlying principles, paving the way for a thorough understanding of our proposed audio-visual fusion framework. For the ease of users and a better way to collaborate, we have created separate sections in our literature review each of which is further sub-classified based on their techniques.

2.1 Deepfake Generation Techniques

The term "deepfake" stems from the combination of "deep learning" and "fake," highlighting the central role of neural networks in the creation of these synthetic videos [4]. There are three primary architectures that dominate the creation of deepfakes:

- **Generative Adversarial Networks (GANs):** Introduced by Goodfellow et al. [6], GANs consist of a generator and discriminator competing in a game-theoretic framework. The generator aims to produce realistic synthetic content, while the discriminator endeavors to distinguish between real and fake. This adversarial process refines the generator's output, leading to highly convincing deepfakes.
- **Autoencoders:** These neural networks are designed to encode an input into a lower-dimensional space and then decode it back to its original form [9]. In deepfake creation, autoencoders swap the encoded features of two individuals, synthesizing a merged output.
- **Few-shot and Zero-shot Learning:** With the advances in transfer learning, few-shot and zero-shot techniques have gained traction. Instead of requiring vast amounts of data, these models can generate convincing deepfakes with limited (few-shot) or even no samples (zero-shot) of the target individual. This approach leverages prior knowledge from related tasks, enabling rapid adaptation to new faces and potentially posing greater threats in the deepfake landscape [20].

2.2 Visual Deepfake Detection

The evolving sophistication of deepfakes has given rise to multiple detection strategies. Many of these capitalize on the visual imperfections in synthetic content:

- **Face Warping Artifacts:** Li and Lyu [15] targeted the artifacts that arise from face warping in deepfakes. They proposed a method that detects inconsistencies

in the facial region, a common tell-tale sign of manipulation. Concerning the same artifacts, Ivanov et al. [11] examined deep fake video synthesis and its detection, reviewing past research methods. They proposed a strategy that melds existing detection techniques with super-resolution algorithms.
- **Temporal Inconsistencies:** Deepfakes often struggle with ensuring temporal consistency across video frames. Rossler et al. [18] developed a technique that highlights these inconsistencies, particularly in the facial region. A more recent work by Liu et al. [16] uses a deep learning network to extract temporal identity features from video frames and then distinguishes between real and deepfake videos based on the presence or absence of temporal identity inconsistencies.
- **Micro-expressions:** Micro-expressions are brief, involuntary facial expressions that reveal a person's true emotions. They are difficult for humans to detect, but they can be captured by deep learning models. This makes micro-expressions a promising tool for detecting deepfake videos. Haiwei et al. [7] introduced a Spatial–Temporal Deepfake Detection and Localization (ST-DDL) network, harnessing both spatial and temporal features to pinpoint forged regions in videos. They utilized Anchor-Mesh Motion algorithm capturing subtle facial movements more effectively than traditional methods.

2.3 Audio Deepfake Detection

With deepfake audios gaining traction, researchers have started exploring methods to discern genuine from synthesized speech:

- **Spectrogram Analysis:** Korshunov and Marcel [14] proposed using spectrograms, visual representations of the spectrum of frequencies in audio signals, to detect irregularities that are indicative of manipulation. Further, after their work, Xue et al. [21] proposed a method for audio deepfake detection, utilizing fundamental frequency (F0) information from synthetic speech and real plus imaginary spectrogram features.
- **Acoustic Feature Discrepancies:** In a study by Huang et al. [10], they identified inconsistencies in fundamental frequency, formant, and mel-frequency cepstral coefficients in deepfake audios. Moreover, Blue et al. [3] introduced a novel method for detecting audio deepfakes using articulatory phonetics. By applying fluid dynamics, the study estimates the human vocal tract configuration during speech, revealing that deepfakes often simulate improbable anatomical setups.

2.4 Multi-modal Deepfake Detection

While the majority of research has been unimodal, recent studies have begun to appreciate the benefits of combining multiple modalities:

- **Audio-Visual Synchrony:** Han et al. [8] developed a system that measures the synchrony between audio and video modalities. Discrepancies in lip movement and spoken words, often overlooked individually, were detectable when analyzed in tandem. Building on this, a recent study by Yang et al. [23] introduced AVoiD-DF, an audio-visual joint learning method targeting multi-modal forgery detection. By exploiting inconsistencies across modalities, this approach enhanced deepfake detection capabilities and demonstrated superior performance against various forgery techniques.
- **Feature Fusion:** Yan et al. [22] proposed a Multi-modal Graph Learning (MGL) framework that uses two GNNs and multi-modal fusion modules to combine spatial, frequency, and landmark features. By representing video frames as graph nodes, they captured temporal inconsistencies. MGL outperforms other detectors in generalization and disturbance resistance. MFF-Net, built by Zhao et al. [25], is a deepfake detection system that fuses RGB features with textural data from neural networks and signal processing. It incorporates a feature extraction module using Gabor convolution which is used as a texture enhancement module, an attention component emphasizing the forged parts, and dual feature fusion instances.

Our proposed methodology draws inspiration from these works but introduces novel techniques and considerations that address the current gaps in the literature. By leveraging the strengths of both audio and visual domains, we aim to devise a framework that is both robust and versatile.

3 Methodology

In this section, we delineate our innovative framework for audio-visual fusion to detect deepfake videos. This Fig. 1methodology encompasses preprocessing steps, feature extraction from audio and visual components, and the fusion strategy, culminating in a robust detection mechanism.

3.1 Preprocessing

Both audio and visual components necessitate meticulous preprocessing to ensure data uniformity and enhance computational efficiency.

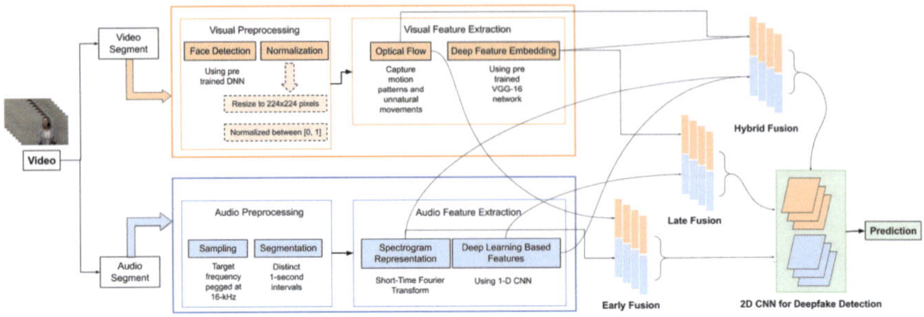

Fig. 1 Deepfake detection methodology

- **Audio Preprocessing:** To ensure the uniformity of audio frequencies across the dataset, a resampling technique is employed. For a given audio snippet, the frequency post-resampling, $f_{\text{resampled}}$, is mathematically expressed as

$$f_{\text{resampled}} = \frac{f_{\text{input}} \times \text{target}_{\text{frequency}}}{\text{original}_{\text{frequency}}} \tag{1}$$

  where f_{input} corresponds to the initial frequency of an audio segment and $\text{target}_{\text{frequency}}$ is pegged at 16kHz.
  Temporal nuances within the audio are captured by segmenting the resampled audio into distinct 1-second intervals.
- **Visual Preprocessing:**
  - *Face Detection*: Using a pre-trained DNN, the face is located and extracted from each frame [24].
  - *Normalization*: Each frame is resized to 224×224 and the pixel values are normalized (represented by $P_{\text{normalized}}$) to fall between [0, 1]. We achieved this normalization with

$$P_{\text{normalized}} = \frac{P_{\text{original}} - P_{\text{min}}}{P_{\text{max}} - P_{\text{min}}} \tag{2}$$

    where P_{original} is the original pixel value and P_{max} and P_{min} are the maximum and minimum pixel values, respectively.

3.2 *Feature Extraction*

The core strength of our framework lies in harnessing rich features from both audio and visual segments. Apart from the feature extraction techniques used in our methodology, several other techniques are also highlighted in Table 1.

Table 1 Summary of Audio and Visual Feature Extraction Techniques

Feature type	Technique	Description	Significance in Deepfake detection
Visual	Convolutional Neural Networks (CNN)	Used for image and video recognition, they automatically and adaptively learn spatial hierarchies	CNNs can capture minute details in facial features altered during deepfake generation
Visual	Optical flow	Measures the motion of objects, surfaces, and edges in a visual scene.	Helps in detecting unnatural motion patterns, commonly found in deepfakes
Visual	Texture analysis	Focuses on the structural arrangement of surfaces and relationships of neighboring pixels	Deepfakes might introduce unique textures due to blending of images
Visual	Face landmark detection	Detects key facial features and their relative positioning	Inconsistencies in landmarks can be a sign of manipulated content
Audio	Mel-Frequency Cepstral Coefficients (MFCC)	Represent the short-term power spectrum of sound	Altered voice dynamics in deepfakes might produce unique MFCC patterns
Audio	Chroma feature analysis	Relates to the twelve different pitch classes	Altered pitch or tonality can hint toward manipulated audio
Audio	Spectral contrast	Measures the difference in amplitude between peaks and valleys in the sound spectrum	Deepfake audio might introduce unnatural spectral patterns
Audio	Short-Time Fourier Transform (STFT)	Represents a signal in the time–frequency domain	Anomalies in frequency patterns can be an indication of tampered audio
Multi-modal	Cross-modal correlation	Measures correlation between audio and visual streams	Deepfakes often misalign audio and visual data, leading to low correlation scores

- **Audio Feature Extraction:**
  - *Spectrogram Representation*: The Short-Time Fourier Transform (STFT) is employed to transform audio segments into their spectrogram representations, capturing frequency variations over time represented as

$$\text{STFT}\{x(t)\}(m, \omega) = \int_{-\infty}^{\infty} x(t)w(t-m)e^{-j\omega t}dt \tag{3}$$

    where $x(t)$ is the audio signal, $w(t-m)$ is the window function, and ω is the angular frequency. This transformation aids in the spectral analysis of the audio, an essential component in identifying manipulated sound bites.
  - *Deep Learning-Based Features*: A 1D convolutional neural network (CNN) is trained on these spectrograms to extract high-level features that may indicate tampering [2]. The 1D CNN is governed by the convolution operation with f and g as signals and τ as a dummy variable for integration:

$$(f * g)(t) = \int_{-\infty}^{\infty} f(\tau)g(t-\tau)d\tau \tag{4}$$

- **Visual Feature Extraction:**
  - *Optical Flow*: We compute optical flow between successive frames, capturing the motion patterns that can be indicative of manipulation. Taking I_x and I_y as spatial gradients of the frame, u and v being the velocities indicating horizontal and vertical flow, respectively, and I_t as the temporal gradient, capturing changes between consecutive frames, the optical flow is represented by

$$I_x \cdot u + I_y \cdot v = -I_t \tag{5}$$

    This equation helps in discerning unnatural movements or transitions between frames.
  - *Deep Feature Embedding*: A pre-trained VGG-16 network [19] is fine-tuned on our dataset to extract deep features from face images. This provides insights into potential spatial inconsistencies in the visual content.

3.3 Audio-Visual Fusion Strategy

The core novelty of our methodology is the fusion approach, amalgamating the strengths of both audio and visual signals.

- **Early Fusion:** Initial features extracted from audio and visual segments are concatenated, forming a unified feature vector. This combined vector undergoes subsequent layers of processing, capitalizing on the correlated information from both domains. The unified feature vector V_{fused} for a given audio-visual pair in our dataset is defined as

$$V_{\text{fused}} = [\text{Features}_{\text{audio}}, \text{Features}_{\text{video}}] \tag{6}$$

- **Late Fusion:** High-level features, post-CNN processing for both modalities, are independently processed. The final decisions or probabilities from both streams are then averaged or combined using a decision rule.

$$F_{\text{final}} = w_a \cdot F_{\text{audio}} + w_v \cdot F_{\text{video}} \tag{7}$$

  where w_a and w_v are optimized weights for audio and video features, respectively. F_{audio} and F_{video} are the feature representations from the CNNs trained on our audio and video data.
- **Hybrid Fusion (proposed):** We introduce a novel fusion strategy that exploits the benefits of both early and late fusion. Initial features undergo an early fusion, and post-processing features engage in late fusion, ensuring an exhaustive representation of information.

3.4 Deepfake Detection Algorithm

The fused feature set is subjected to a robust deep learning architecture, designed to discern authentic content from deepfakes.

- **Architecture:** A 2D CNN is employed to process the fused features, comprising convolutional layers with batch normalization and ReLU activations [17]. In the convolutional layers of our specialized 2D CNN:

$$Y_{\text{conv}}(x, y) = \int_{-\infty}^{\infty} \int_{-\infty}^{\infty} F_{\text{video}}(x, y) \cdot K(x - \alpha, y - \beta) d\alpha d\beta \tag{8}$$

  where:

  - $Y_{\text{conv}}(x, y)$ is the feature map post-convolution.
  - $F_{\text{video}}(x, y)$ represents the extracted video features.
  - K is the convolutional kernel specific to our deepfake detection model.

  Fully connected layers follow, culminating in a binary classification output (real vs. deepfake). The final binary classification is deduced through a sigmoid activation function

- **Training:** The network is trained using a cross-entropy loss function, with the Adam optimizer [13]. Data augmentation techniques, including random cropping and horizontal flipping, are utilized for the visual data to enhance model generalizability. Our model's optimization objective is to minimize the cross-entropy loss:
$$L_{\text{deepfake}} = -\sum_i [y_i \log(\hat{y}_i) + (1 - y_i) \log(1 - \hat{y}_i)] \quad (9)$$
where y_i is the ground truth label for each sample in our dataset and $\hat{y}_i$ is the label predicted by our model.
- **Evaluation Metrics:** Model performance is gauged using accuracy, precision, recall, and F1-score.

3.5 Dataset and Implementation Details

For our experiments, we leverage a composite dataset, merging popular deepfake datasets like FaceForensics++ [18], DeepFake Detection Challenge Dataset (DFDC) [5], and Fake AV Celeb [12]. This amalgamation ensures a diverse representation of manipulation techniques. Further details about datasets are presented in Table 2.

Our implementation is executed in Python, utilizing the TensorFlow [1] and OpenCV libraries. The models are trained on NVIDIA Tesla V100 GPUs.

4 Results

In this segment, we present the empirical outcomes achieved by our proposed audio-visual fusion framework for deepfake detection. Through rigorous experimentation, our methodology demonstrates superior performance against existing benchmarks. The comparison of existing benchmarks for deepfake detection challenge dataset is shown in Table 3. Further, Tables 4 and 5 demonstrate the comparison of Fake AV

Table 2 Summary of datasets used

Dataset	Year	Modalities	Real videos	Fake videos	Subjects
FaceForensics++	2018	Video, Audio	1000	4000	977
DFDC	2019	Video, Audio	23654	104500	960
FakeAVCeleb	2021	Video, Audio	500	19500	500

Table 3 Comprehensive breakdown of performance metrics on DFDC

Method	Accuracy	Precision	Recall	F1-score
Siamese-based	0.907	0.922	0.892	0.907
Meso4	0.885	0.909	0.861	0.884
Capsule	0.878	0.901	0.855	0.882
Our proposed method	0.923	0.938	0.915	0.927

Table 4 Comprehensive breakdown of performance metrics on AV Celeb

Method	Accuracy	Precision	Recall	F1-score
Siamese-based	0.856	0.873	0.839	0.856
Meso4	0.835	0.851	0.819	0.835
Capsule	0.824	0.840	0.808	0.824
Our proposed method	0.862	0.873	0.849	0.861

Table 5 Comprehensive breakdown of performance metrics on Faceforensics++

Method	Accuracy	Precision	Recall	F1-score
Siamese-based	0.944	0.957	0.931	0.944
Meso4	0.928	0.942	0.914	0.928
Capsule	0.921	0.934	0.908	0.921
Our proposed method	0.933	0.945	0.921	0.933

Table 6 Real-world evaluation results

Scenario	Accuracy	Notes
Low-light condition	0.926	Model faced challenges due to shadows and poor quality
Noisy background audio	0.947	Model excelled in isolating voice from background noise

Celeb and FaceForensics++ datasets, respectively.

The robustness of our audio-visual fusion framework is evident from the impressive quantitative metrics. Its distinction, however, lies in its balanced performance across various sub-tasks and its adaptability to Table 6real-world scenarios. Particularly, the fusion strategies and the feature extraction techniques play a synergistic role, with each element bolstering the overall efficacy.

The minimal performance gap between laboratory conditions and real-world scenarios further cements the method's practical relevance. Although there remains a marginal scope for false negatives, the current framework stands as a good and comprehensive solution to the deepfake menace.

5 Discussion

The domain of deepfake detection is rapidly evolving, underpinned by the alarming sophistication of generative methods. In this light, our findings offer a novel perspective on the integration of audio and visual data for more accurate detection.

1. **Strengths of the Audio-Visual Fusion**
   Our proposed audio-visual fusion framework demonstrates several key strengths. First and foremost, by holistically analyzing both auditory and visual discrepancies in videos, the model is better equipped to detect incongruities often overlooked when either modality is considered in isolation. The hybrid fusion strategy consolidates the advantages of early and late fusion, optimizing feature integration and ensuring the final decision-making layer has a richer information pool.
2. **Limitations**
   In the course of our study, we identified some limitations that can be explored further. We used a composite deepfake dataset, which may not cover all deepfake scenarios, indicating a need for more diverse datasets. Our complex deep learning framework, while effective, requires substantial computational resources, suggesting a need for optimization. Lastly, our framework was tested under specific conditions, but a comprehensive performance analysis against novel deepfake methods was not conducted, highlighting the need for robust testing. These findings provide directions for future deepfake detection research.

6 Conclusion

Deepfakes, with their potential to distort truth and mislead the public, pose a significant threat in this digital age. Addressing this challenge, our proposed audio-visual fusion methodology offers a powerful countermeasure to detect and combat these manipulations. By concurrently considering both visual and auditory cues, this study reaffirms the significance of a multi-modal approach in understanding and recognizing deepfakes with heightened accuracy.

The framework's robust performance, as evidenced by our evaluations, suggests that the intricacies of authentic human behavior—the synchronization of voice, lip movements, facial expressions, and subtle auditory nuances—are challenging for even the most sophisticated deepfake algorithms to replicate flawlessly. These incongruities, no matter how minute, become the Achilles' heel of deepfakes when subjected to our rigorous detection framework.

Moreover, the potential applications of this framework are vast. From social media platforms to news agencies, the necessity to validate the authenticity of digital content is paramount. Implementing such an advanced detection system can act as a bulwark against misinformation campaigns, ensuring the veracity of digital interactions.

In conclusion, while the arms race between deepfake generation and detection will likely continue to escalate, our research demonstrates that innovation, especially through multi-modal fusion strategies, can keep us one step ahead. As technology advances and the line between reality and fabrication blurs further, efforts like ours become critical in safeguarding truth in the digital realm.

7 Future Research Directions

The rapidly evolving landscape of deepfake generation underscores the constant need for adaptive and forward-thinking detection techniques. While our multi-modal fusion strategy has proven effective, several avenues remain unexplored and hold promising potential.

1. **Higher-Dimensional Data**: Our framework harnesses audio and visual cues. Incorporating additional data dimensions, such as physiological signals (e.g., heart rate or skin temperature variations) captured through advanced sensors, might yield even more refined detection accuracy.
2. **Temporal Analysis**: Deepfake algorithms might stumble in maintaining consistent realism throughout a video's duration. Investigating inconsistencies over time rather than isolated frames might offer another layer of discernment.
3. **Cross-Modal Interactions**: While we've fused audio and visual cues, a deeper understanding of their intricate interplay, particularly how one modality might affect the perception of another, could lead to more nuanced detection strategies.
4. **Ethical Implications**: Future research should also delve into the societal and ethical considerations of deepfake detection. As technologies advance, ensuring that they respect user privacy and don't inadvertently suppress legitimate content becomes paramount.

In sum, while our work provides a formidable stand against current deepfake challenges, it also sets the stage for a plethora of research opportunities. By anticipating and adapting to the next wave of synthetic media challenges, we can continue to champion the cause of digital truth.

References

1. Abadi M, Agarwal A, Barham P, et al (2016) Tensorflow: Large-scale machine learning on heterogeneous distributed systems. arXiv:1603.04467 https://doi.org/10.48550/arXiv.1603.04467
2. Abdel-Hamid O, Ar M, Jiang H et al (2014) Convolutional neural networks for speech recognition. IEEE/ACM Trans Audio Speech Lang Process 22(10):1533–1545. https://doi.org/10.1109/TASLP.2014.2339736
3. Blue L, Warren K, Abdullah H, et al (2022) Who are you (i really wanna know)? detecting audio {DeepFakes} through vocal tract reconstruction. In: 31st USENIX Security Symposium (USENIX Security 22), pp 2691–2708, https://www.usenix.org/conference/usenixsecurity22/presentation/blue
4. Citron DK, Chesney R (2019) Deepfakes and the new disinformation war. Foreign Affairs https://scholarship.law.bu.edu/shorter_works/76/
5. Dolhansky B, Bitton J, Pflaum B, et al (2020) The deepfake detection challenge (dfdc) dataset. arXiv:2006.07397 https://doi.org/10.48550/arXiv.2006.07397
6. Goodfellow I, Pouget-Abadie J, Mirza M, et al (2014) Generative adversarial nets. Advances in neural information processing systems 27. https://proceedings.neurips.cc/paper_files/paper/2014/file/5ca3e9b122f61f8f06494c97b1afccf3-Paper.pdf
7. Haiwei W, Jiantao Z, Shile Z, et al (2022) Exploring spatial-temporal features for deepfake detection and localization. arXiv:2210.15872. https://doi.org/10.48550/arXiv.2210.15872
8. Han B, Han X, Zhang H et al (2021) Fighting fake news: two stream network for deepfake detection via learnable srm. IEEE Trans Biomet Behavior Identity Sci 3(3):320–331. https://doi.org/10.1109/TBIOM.2021.3065735
9. Hinton GE, Salakhutdinov RR (2006) Reducing the dimensionality of data with neural networks. Science 313(5786):504–507. https://doi.org/10.1126/science.1127647
10. Huang H, Sun N, Lin X (2022) Blockwise spectral analysis for deepfake detection in high-fidelity videos. In: 2022 IEEE 9th international conference on data science and advanced analytics (DSAA), IEEE, pp 1–9. https://doi.org/10.1109/DSAA54385.2022.10032370
11. Ivanov NS, Arzhskov AV, Ivanenko VG (2020) Combining deep learning and super-resolution algorithms for deep fake detection. In: 2020 IEEE conference of Russian young researchers in electrical and electronic engineering (EIConRus). IEEE, pp 326–328. https://doi.org/10.1109/EIConRus49466.2020.9039498
12. Khalid H, Tariq S, Kim M, et al (2021) Fakeavceleb: A novel audio-video multimodal deepfake dataset. arXiv:2108.05080. https://doi.org/10.48550/arXiv.2108.05080
13. Kingma DP, Ba J (2014) Adam: a method for stochastic optimization. arXiv:1412.6980. https://doi.org/10.48550/arXiv.1412.6980
14. Korshunov P, Marcel S (2018) Deepfakes: a new threat to face recognition? assessment and detection. arXiv:1812.08685. https://doi.org/10.48550/arXiv.1812.08685
15. Li Y, Lyu S (2018) Exposing deepfake videos by detecting face warping artifacts. arXiv:1811.00656. https://doi.org/10.48550/arXiv.1811.00656
16. Liu B, Liu B, Ding M, et al (2023) Ti2net: temporal identity inconsistency network for deepfake detection. In: Proceedings of the IEEE/CVF winter conference on applications of computer vision, pp 4691–4700. https://openaccess.thecvf.com/content/WACV2023/html/Liu_TI2Net_Temporal_Identity_Inconsistency_Network_for_Deepfake_Detection_WACV_2023_paper.html
17. Nair V, Hinton GE (2010) Rectified linear units improve restricted boltzmann machines. In: Proceedings of the 27th international conference on machine learning (ICML-10), pp 807–814. https://dl.acm.org/doi/10.5555/3104322.3104425
18. Rossler A, Cozzolino D, Verdoliva L, et al (2019) Faceforensics++: learning to detect manipulated facial images. In: Proceedings of the IEEE/CVF international conference on computer vision, pp 1–11. https://openaccess.thecvf.com/content_ICCV_2019/html/Rossler_FaceForensics_Learning_to_Detect_Manipulated_Facial_Images_ICCV_2019_paper.html

19. Simonyan K, Zisserman A (2014) Very deep convolutional networks for large-scale image recognition. arXiv:1409.1556. https://doi.org/10.48550/arXiv.1409.1556
20. Xian Y, Lampert CH, Schiele B et al (2018) Zero-shot learning-a comprehensive evaluation of the good, the bad and the ugly. IEEE Trans Pattern Anal Mach Intell 41(9):2251–2265. https://doi.org/10.1109/TPAMI.2018.2857768
21. Xue J, Fan C, Lv Z, et al (2022) Audio deepfake detection based on a combination of f0 information and real plus imaginary spectrogram features. In: Proceedings of the 1st international workshop on deepfake detection for audio multimedia, pp 19–26. https://doi.org/10.1145/3552466.3556526
22. Yan Z, Sun P, Lang Y, et al (2022) Landmark enhanced multimodal graph learning for deepfake video detection. arXiv:2209.05419. https://doi.org/10.48550/arXiv.2209.05419
23. Yang W, Zhou X, Chen Z et al (2023) Avoid-df: Audio-visual joint learning for detecting deepfake. IEEE Trans Inf Forensics Secur 18:2015–2029. https://doi.org/10.1109/TIFS.2023.3262148
24. Zhang K, Zhang Z, Li Z et al (2016) Joint face detection and alignment using multitask cascaded convolutional networks. IEEE Signal Process Lett 23(10):1499–1503. https://doi.org/10.1109/LSP.2016.2603342
25. Zhao L, Zhang M, Ding H et al (2021) Mff-net: Deepfake detection network based on multi-feature fusion. Entropy 23(12):1692. https://doi.org/10.3390/e23121692

Enhanced Rice Leaf Disease Classification and Estimation Via ConvNeXT Incorporating Thermal and RGB Images

M. Gokul Raja and S. Sarath

Abstract Crop diseases can cause significant yield losses and economic damage, the classification and estimation of the crop diseases in the early stage of disease helps in reducing these losses. Crop disease classification is a critical component of precision agriculture, which aims to improve crop yields and quality by using data-driven crop management methods. In our study, we make use of the ConvNeXT model architecture for the classification of rice crop disease. Two different ConvNeXT models are created by training with thermal and RGB images dataset respectively. By combining the output of both model accuracy scores, classification is done. The estimation of the disease in the affected rice leaves is done using the computer vision technique which estimates the percentage of the leaf area affected by the crop disease. The proposed method for classification achieved an accuracy of 98.7% for thermal dataset and 93.6% for the RGB dataset.

Keywords Crop disease classification · Crop disease estimation · ConvNeXT · Image processing · Thermal image processing · Artificial intelligence · Machine learning

1 Introduction

Large numbers of crops are getting affected due to crop diseases in which some of the detected can be detected and treated in the earlier stage itself to avoid crop loss. Over 290 billion INR worth of agricultural losses are attributed to pests and diseases each year in India. In India, the loss in crop productivity due to fungal infections is estimated to be 5 million tons annually. Traditional methods of crop

M. Gokul Raja (✉) · S. Sarath
Department of Computer Science and Engineering, Amrita School of Computing, Amrita Vishwa Vidyapeetham, Amritapuri, India
e-mail: mgokulraja17@gmail.com

S. Sarath
e-mail: saraths@am.amrita.edu

A. P. Agrawal et al. (eds.), *Advances of Artificial Intelligence and Machine Learning in Image Processing, Security and Data Analytics*, Lecture Notes in Electrical Engineering 1464, https://doi.org/10.1007/978-981-95-0148-9_11

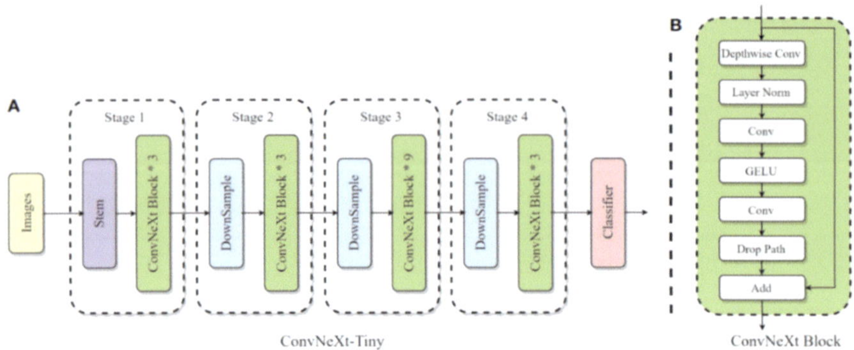

Fig. 1 **a** ConvNeXT-tiny overall network structure **b** ConvNeXT block

disease detection, such as visual inspection and laboratory testing, are often labor-intensive, time-consuming, and inaccurate. Image processing techniques can detect diseases early before they have a chance to cause significant damage.

In our study, two different rice plant leaf datasets are created using thermal and RGB images from Kaggle [1]. The RGB images are used in the training since, the RGB images won't get affected by the changes in the surroundings like humidity and moisture, which won't affect the images while taken like the thermal images which show change as the thermal images are prone to environment conditions. The image datasets are trained using ConvNeXT models as two different models for each one respectively. The architecture of the ConvNeXT-Tiny model is shown in Fig. 1. The ConvNeXT is based on the Convolutional neural network, ResNet-50 which paved the way for the vision transformers. The convolutional neural network specializes in working on grid-like architecture such as images in an efficient way. The estimation algorithm is developed using the computer vision technique which finds the percentage of the area affected in rice crop leaves. By combining both disease detection and estimation of rice plant diseases, the identification and spread of the illness can be swiftly pinpointed. This approach facilitates timely remedies to be implemented before the disease has a chance to propagate further.

2 Related Work

In recent years, image-based disease detection techniques have made a serious impact on crop disease detection. Ning et al. [2] proposed a computer vision-based algorithm that will convert the infrared images from RGB to HSV and use the Hue component alone to do the processing of the images. The separated Hue component is converted to the grayscale image showing the affected area and then the lesion counting takes place after the closed operation. The salt and pepper noises are replaced with the median of the neighborhood of the point. Using the threshold, the binary image is created to separate the foreground and background in the image. Finally, the foreground is considered the detected part of the affected leaf. The regression analysis of the image results shows that the algorithm works more efficiently than the visual detecting method. Sunil et al. [3] tried to train the model after removing the background in the image to reduce the noise, which resulted in the faster training and execution of the disease detection model. They used the EfficientNetV2 model for the training using the dataset. The segmentation of the cardamom leaves before training, as a pre-processing helps in the better training of the data, which indeed helped the model to learn better. Kunlong et al. [4] demonstrate the potential of deep learning techniques for segmenting and classifying leaves in plant images with overlapping leaves and complicated backgrounds. The use of Mask R-CNN for the segmentation and the VGG-16 for the classification. The segmented leaves data for the classification shows a better accuracy of 92.4%.

Andrew et al. [5] developed a deep learning model for the diagnosis of various plant diseases using transfer learning techniques. Different models were used for training and testing purposes: Inception-V4, VGG-16, ResNet-50, and DenseNet-121. The models are trained with a plant village dataset of 38 classes and overall DenseNet-121 outperformed with minimum training loss and maximum accuracy. These results were achieved by maintaining the architecture and just using the hyper-parameter tuning. Jianyu et al. [6] used a similar approach of using the neural network and the improved algorithm for the classification. In the pre-processing of the data, they used data expansion on the dataset using various methods like adding noise and using CNN to create a new dataset based on the given dataset. This method proves to help develop our proposed method. Jackulin et al. [7] have done a comparative study on the various deep learning and machine learning techniques which helps in choosing the neural network in our case. Naline et al. [8] used Max Voting and KNN for analyzing the impact of temperature conditions on the crops which helps us use thermal images. Punam et al. [9] used a new autoencoder for the data preprocessing and used CNN for the classification which proves the impact of transformer components like encoders on image classification.

References [10–17] has discussed various machine learning techniques for detection and classification. Harikumar et al. [18] discussed the application of machine learning techniques on rice leaf disease detection by implementing a framework using various machine learning techniques. Yibin et al. [19] implemented a new model ADSNN-OB which uses a custom attention-based neural network and optimization that outperforms existing models.

Table 1 Distribution of dataset

Class name	Thermal image	RGB image
Brown spot	80	523
Leaf blast	67	779
Hispa	142	565
Healthy	93	1488

3 Methodology

3.1 Dataset Description

The rice dataset is a collection of RGB and Thermal images in their respective folders. For RGB images [20] and Thermal image [17] datasets, the images are downloaded from Kaggle. The thermal images are captured with the help of high-resolution cameras FLIR E8 (Thermal camera). The specifications of these cameras are mentioned below, The FLIR E8 has a clear 76,800 (320 X 240) pixel infrared resolution, +2% accuracy of reading for ambient temperatures 10–35 °C (50°F–95°F), and a field of vision of 450X340. The RGB images dataset is a collection of images from the internet by a Kaggle user. The distribution of images across the dataset is mentioned in Table 1.

3.2 Dataset Pre-processing

The dataset augmentation is done on the Thermal dataset for the train folder due to its lack of number of images available. The data augmentation includes height, shift, width shift, and horizontal flipping. The height and width shift of the image includes the shifting of the pixels in the image in the specified direction. The train and test split of the dataset was done in the ratio of 80:20. The dataset preprocessing includes the 224 × 224 image size and data augmentation. The dataset distribution after the dataset pre-processing is given in Table 2.

Table 2 Distribution of dataset after pre-processing

Class name	Thermal image		RGB image	
	Train	Test	Train	Test
Brown spot	808	29	418	105
Leaf blast	571	14	623	156
Hispa	1401	52	452	113
Healthy	951	34	1190	298

3.3 *ConvNeXT-Tiny-224*

The ConvNeXT-Tiny-224 is used for the classification of images. ConvNeXT is a pure convolutional model (ConvNet) that claims to be inspired by and surpasses the vision transformer design. The authors envisioned his ResNet and "modernized" its design with inspiration from the Swin Transformer. The ConvNeXT uses all the pros of the Swin Transformer and is enhanced with the use of larger kernel size, inverted bottleneck design, and various upgrades in micro design. The micro design includes the use of the GERU activation function and layer normalization.

3.4 *Proposed Method*

The proposed method consists of two different modules, one for the classification of the images and another module for disease estimation. The overall methodology is shown in Fig. 2.

3.4.1 Proposed Methodology for Disease Classification

The proposed method consists of training two models, one for thermal image classification and RGB image classification respectively. The models are based on the ConvNeXT model architecture, pre-trained ConvNeXT models are downloaded and trained with the data collected from Kaggle. The trained models are used to classify the rice crop leaf diseases which are Leaf Blast, Brown spot, Hispa, and Healthy. A system is developed for the classification which takes input of the rice leaf thermal

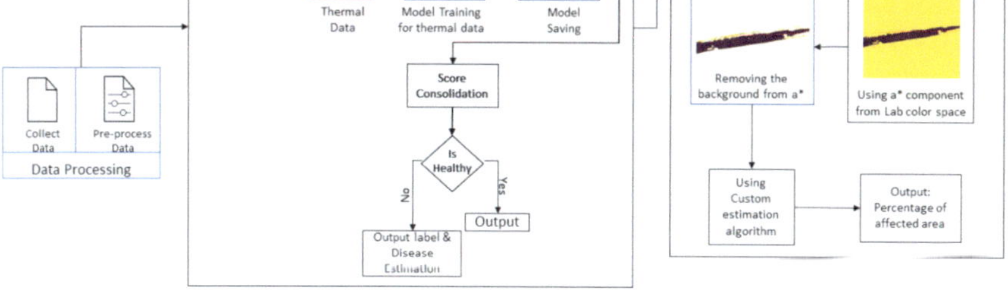

Fig. 2 Overall proposed methodology

and RGB images, which processes them and creates a classification with a weightage of 65:35 for thermal image classification accuracy and RGB image classification accuracy. For the next step, if the rice crop leaf is affected by the disease, then a special algorithm is developed to estimate the rice crop disease-affected area in the given leaf. The proposed method is shown in Fig. 2.

3.4.2 Proposed Methodology for Plant Disease Estimation

The rice plant disease estimation is done on the RGB image of the diseased leaf using the CIELAB color space and mathematical calculation. Like the RGB color space, there is CIELAB color space which consists of three components namely L* indicates lightness, a* is the red/green coordinate, and b* is the yellow/blue coordinate. Here, a* is used by the algorithm. The removal of the diseased area of the leaf along with the background is done. Once the removal is done, the input leaf image is used for the removal of the background of the leaf. The a* component image and the input image without the background are generated, and the algorithm that is developed is run to find the percentage of the affected area in the whole leaf. The pixel count of the leaf area without the background (whole leaf count) is found and the pixel count of the healthy region in the leaves (healthy area count) without the background. The percentage of the affected area is calculated by Eq. 1.

$$percentage\_of\_affected\_area = \frac{whole\_leaf\_count - healthy\_area\_count}{\frac{whole\_leaf\_count}{100}} \quad (1)$$

The various intermediate output which is generated from the disease estimation is shown in Fig. 3.

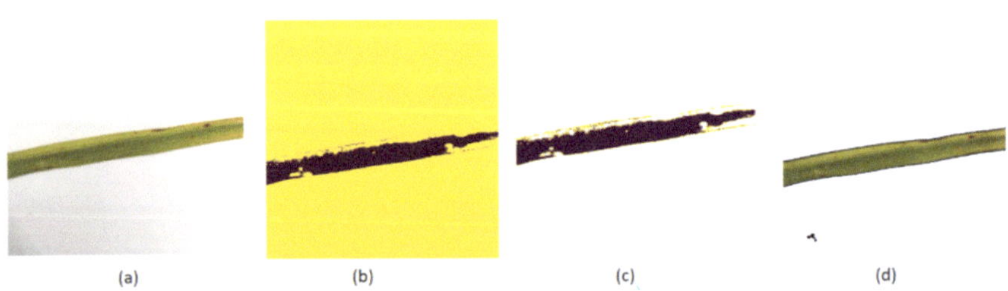

Fig. 3 **a** Input image **b** a* component **c** background removed a* image **d** background removed input image

Table 3 Hyperparameter specifications

Hyperparameters	Value
Epochs	50
Dropout	0.2
Activation function	GERU
Learning rate	0.00004
Output class	4

4 Experiment

The baseline system for our evaluation of the experiment consists of GPU NVIDIA T4, RAM size 32 GB, and Intel Xeon CPU. The software packages include HuggingFace, Transformers, PyTorch, and OpenCV. The pre-trained model of ConvNeXT-Tiny-224 is loaded for training and testing purposes. The use of pre-trained reduces the time for creating a model that will excel in the given task. The Image dataset which is created and pre-processed is loaded into the HuggingFace cloud space for the experiment.

4.1 Fine-Tuning of Hyperparameters in ConvNeXT Model

The hyperparameter is tuned based on the output and the nature of the input data. Most of the architecture in the ConvNeXT model is retained here since the output expected is satisfied with the fine-tuning. The hyperparameter specification is shown in Table 3.

5 Results

5.1 Leaf Disease Classification

The Thermal and RGB images of the same leaf are taken and passed to the respective trained models namely the thermal model and RGB model. The output of the two different models is then passed to the consolidation program for the consolidated output.

From Fig. 4, the thermal train loss was spiking due to the imbalance in the dataset, and in Fig. 5, the training accuracy was gradually rising, resulting in 98% accuracy as the total. While the test accuracy is 98.7%. From Fig. 6, the RGB train loss was spiking due to the imbalance in the dataset, and in Fig. 7, the training accuracy was gradually rising, resulting in 95.3% accuracy as the total. While the test accuracy is 93.6%. Table 4 shows the train and test accuracy of both models.

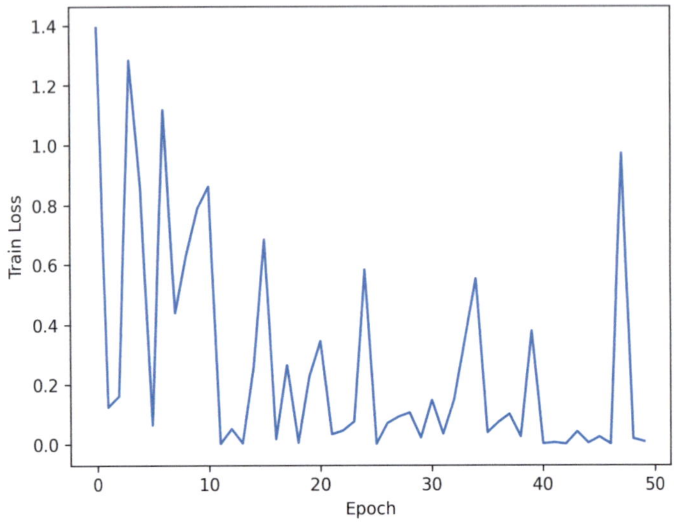

Fig. 4 Thermal model train loss

Table 4 Accuracy of the models

Accuracy/Model	Thermal Model	RGB Model
Train Accuracy	98%	95.3%
Test Accuracy	98.7%	93.6%

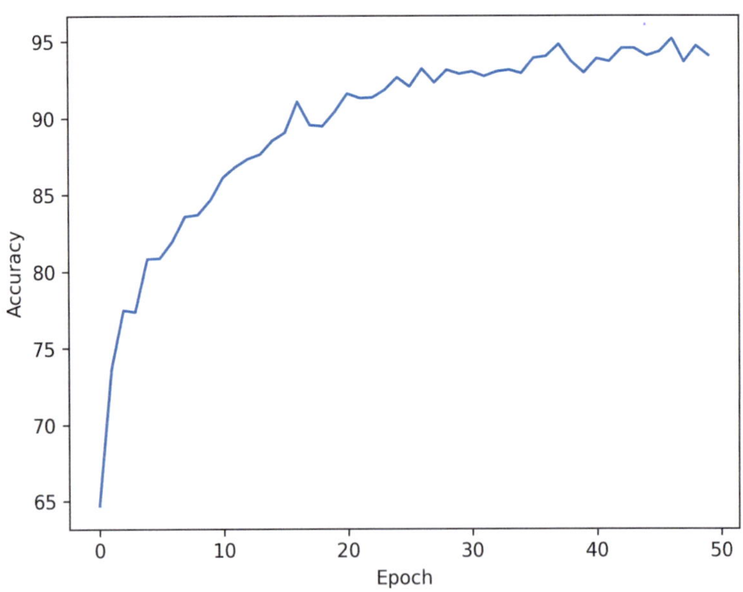

Fig. 5 Thermal model test accuracy

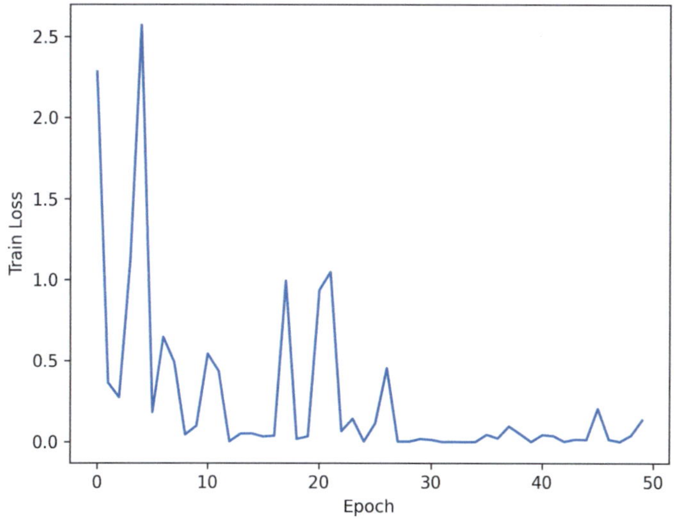

Fig. 6 RGB model train loss

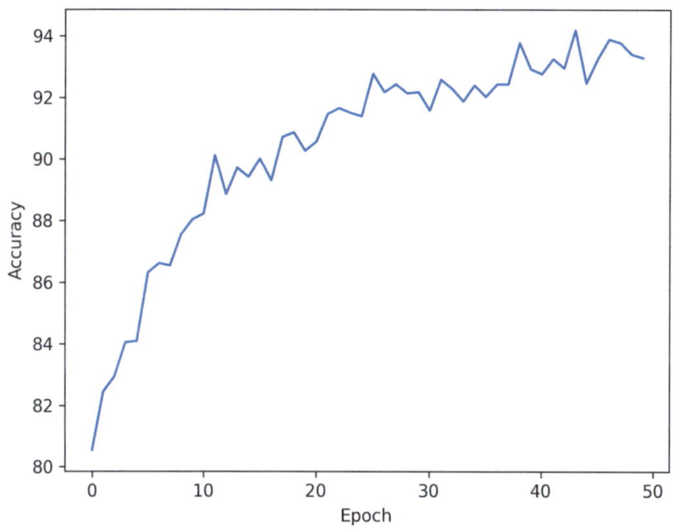

Fig. 7 RGB model test accuracy

The RGB model experiences a decrease in accuracy due to the smaller amount of dataset also shows the inferiority of the RGB images to the thermal images. The RGB images are used due to their behavior in various room conditions like temperature, humidity etc. The thermal images will be affected by these factors. But once the room conditions are set the thermal will get superiority over the RGB images. From Fig. 8,

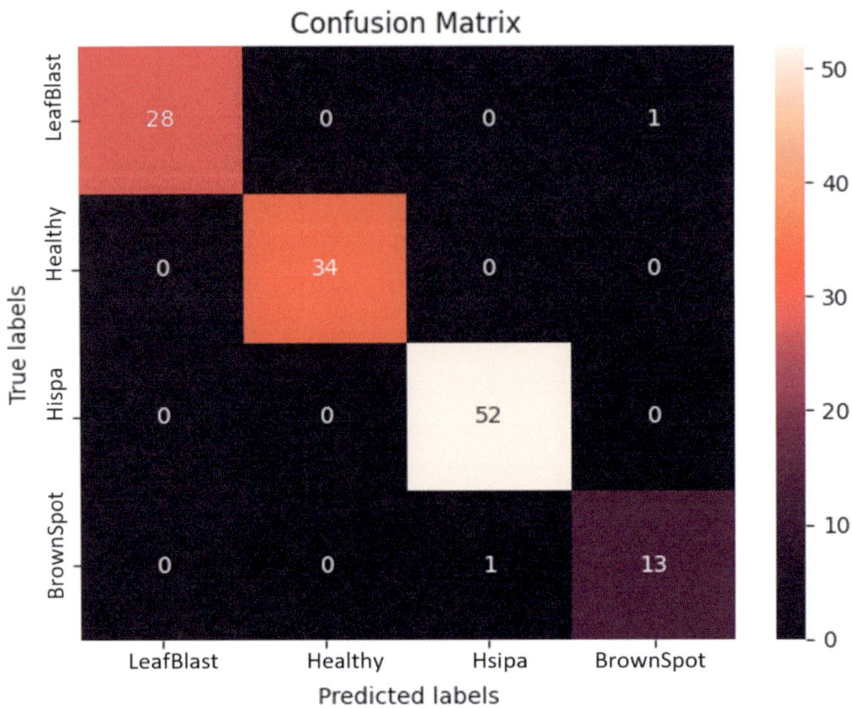

Fig. 8 Confusion matrix for thermal model classification

the confusion matrix for the thermal model classification gives the performance analysis in which the evaluation shows the misclassification of Hispa disease as BrownSpot helps in understanding the similarity between the labels in the image dataset.

5.2 *Leaf Disease Estimation*

The leaf Disease Estimation results are only for the RGB dataset [20]. The RGB images of the rice dataset are used for disease estimation. The disease estimation of the rice crop leaves is done using the custom-developed algorithm which detects the disease-affected region. The output of this algorithm is the percentage of crop area affected by the disease.

6 Conclusion and Future Work

Real-time plant disease detection requires an effective method for identifying plant leaves with diseases. The use of both thermal and RGB images which are in various spectrums helps in studying the dataset and providing a better accuracy. Although the thermal images have more superiority to RGB images, the RGB images are much more compatible with the disease estimation of rice crop leaves. For future work, the image dataset is uneven which in turn leads to a spike during the training of the dataset. So, creating the dataset for which will be an even and large dataset will be done in the future.

References

1. Riyaz S (2019) Rice Leafs. https://www.kaggle.com/datasets/shayanriyaz/riceleafs
2. Yang N, Yuan M, Wang P, Zhang R, Suna J, Mao H (2019) Tea diseases detection based on fast infrared thermal image processing technology. J Sci Food Agric 99:3459–3466
3. Sunil CK, Jaidhar CD, Patil N (2022) Cardamom plant disease detection approach using efficientnetv2. IEEE Access 10:789–804
4. Yang K, Zhong W, Li F (2020) Leaf segmentation and classification with a complicated background using deep learning, vol 10
5. Andrew J, Eunice J, Popescu DE, Chowdary MK, Hemanth J (2022) Deep learning-based leaf disease detection in crops using images for agricultural applications, vol 12
6. Wu J, Li B, Wu Z (2019) Detection of crop pests and diseases based on deep convolutional neural network and improved algorithm, vol 1, pp 20–27
7. Jackulin C, Murugavalli S (2022) A comprehensive review on detection of plant disease using machine learning and deep learning approaches. Meas Sens 24:100441
8. Nalini T, Rama A (2022) Impact of temperature condition in crop disease analyzing using machine learning algorithm. Meas Sens 24:100408
9. Bedi P, Gole P (2021) Plant disease detection using hybrid model based on convolutional autoencoder and convolutional neural network. Artif Intell Agric 5:90–101
10. Geetharamani G, Arun Pandian J (2019) Identification of plant leaf diseases using a ninelayer deep convolutional neural network. Comput Electr Eng 76:323–338
11. Fan X, Luo P, Mu Y, Zhou R, Tjahjadi T, Ren Y (2022) Leaf image based plant disease identification using transfer learning and feature fusion. Comput Electron Agric 196:106892
12. Dogra R, Rani S, Singh A, Albahar MA, Barrera AE, Alkhayyat A (2023) Deep learning model for detection of brown spot rice leaf disease with smart agriculture. Comput Electr Eng 109:108659
13. Agarwal M, Singh A, Arjaria S, Sinha A, Gupta S (2020) Toled: tomato leaf disease detection using convolution neural network. Procedia Comput Sci 167:293–301
14. Reddy JN, Vinod K, Remya Ajai AS (2019) Analysis of classification algorithms for plant leaf disease detection. In: 2019 IEEE international conference on electrical, computer and communication technologies (ICECCT), Coimbatore, India, pp 1–6
15. Nidhis AD, Pardhu CNV, Charishma Reddy K, Deepa K (2019) Cluster based paddy leaf disease detection, classification and diagnosis in crop health monitoring unit. In: Computer aided intervention and diagnostics in clinical and medical images, pp 281–291
16. Jiang L, Yuan B, Ma W, Wang Y (2023) JujubeNet: a high-precision lightweight jujube surface defect classification network with an attention mechanism. Front Plant Sci 13
17. Ramesh S, Hebbar R, Niveditha M, Pooja R, Prasad Bhat N, Shashank N, Vinod PV (2018) Plant disease detection using machine learning. In: 2018 international conference on design innovations for 3Cs compute communicate control (ICDI3C), Bangalore, India, pp 41–45

18. Pallathadka H, Ravipati P, Sajja GS, Phasinam K, Kassanuk T, Sanchez DT, Prabhu P (2022) Application of machine learning techniques in rice leaf disease detection. Mater Today: Proc 51:2277–2280
19. Wang Y, Wang H, Peng Z (2021) Rice diseases detection and classification using attention based neural network and Bayesian optimization. Expert Syst Appl 178:114770
20. Aarthy SL, Sujatha R (2020) Thermal images - diseased & healthy leaves – paddy. https://www.kaggle.com/datasets/sujaradha/thermal-images-diseasedhealthy-leaves-paddy

Classification of Anemia Disorder: A Comparative Analysis of CNN Model and Pre-trained Models

K. S. Hemanth, S. Nagaraju, and B. Muralidhara

Abstract This article explores the application of various deep-learning algorithms to predict anemia based on images of the palpebral conjunctiva. Anemia, a global health concern affecting millions, poses a significant public health risk. Leveraging deep learning in the medical domain, this research employs a successful model utilizing the Eyes-Defy-Anemia dataset. The blood vessels within the palpebral and fornical conjunctivae may reveal crucial information regarding a patient's overall health, including the presence of anemia. The study proposes a practical approach to developing a precise and cost-effective tool for anemia identification, with potential implications for significant advancements in public health.

Keywords Anemia · Palpebral conjunctivas · Eyes-defy-anemia · Convolution neural network · VGG16 · VGG19 · ResNet50 · MobileNet

1 Introduction

Millions of people throughout the world suffer from anemia, a blood disorder. Anemia occurs when the body does not create enough hemoglobin or red blood cells to adequately deliver oxygen to muscles and organs. Anemia can be caused by iron shortages, vitamin deficiencies, long-term illnesses, and genetic conditions. Anemia is diagnosed when a person's hemoglobin dips below 13.5 g/dl or a woman's drops below 12.0 g/dl. This, however, is dependent on age, gender, pregnancy, and other circumstances.

K. S. Hemanth (✉) · S. Nagaraju
School of Computer Science and Applications, REVA University, Bangalore, India
e-mail: hemanth.ks@reva.edu.in

S. Nagaraju
e-mail: nagaraju.s@reva.edu.in

B. Muralidhara
Department of Computer Science, Bangalore University, Bangalore, India
e-mail: murali@bub.ernet.in

A. P. Agrawal et al. (eds.), *Advances of Artificial Intelligence and Machine Learning in Image Processing, Security and Data Analytics*, Lecture Notes in Electrical Engineering 1464, https://doi.org/10.1007/978-981-95-0148-9_12

According to the World Health Organization (WHO), 1.62 billion people, or approximately 24.8% of the global population, have anemia. Preschoolers (children under the age of five) had 42.6% more anemia than reproductive-age women (29.5%). Pregnant women are more likely to develop anemia, with a global frequency of 40.1%. Combating anemia can have a significant impact on reducing the global illness burden and improving overall health.

2 Literature Survey

Anemia is a complex medical illness with multiple origins, making it difficult for healthcare personnel to identify and treat. Machine learning (ML) and deep learning (DL) technologies are increasingly used to predict anemia. Medical professionals can diagnose patients more quickly and accurately by reviewing vast amounts of data with ML and DL algorithms to find patterns that the human eye might miss [1]. These algorithms may be trained on enormous datasets of patient health records, test results, and other medical data to recognize patterns and risk factors connected to anemia.

Also, machine learning and deep learning algorithms can aid in the prediction of anemia before it develops, allowing healthcare providers to take preventative measures and provide early intervention to improve patient outcomes [2]. Healthcare providers can create personalized treatment plans with precise and timely projections, reducing the risk of anemia-related complications [3–5]. The application of machine learning and deep learning algorithms to diagnose anemia holds great promise for enhancing patient care and medical research [6, 7].

There have been several studies that have explored the use of ML and DL approaches for detecting anemia (Table 1).

Results from blood tests for hemoglobin, iron, and ferritin may only be useful for diagnosing certain kinds of anemia, such as iron-deficiency anemia. Other types of anemia may require additional tests or clinical evaluation to diagnose, which may not be captured in the blood test data [8, 19–21].

In this study, the dataset used is 'Eyes-defy-anemia'. Images of palpebral conjunctiva from 218 anemic and non-anemic individuals make up the dataset. Many disorders may be diagnosed by examining the palpebral conjunctiva of the eye, as has been proven in many investigations [8, 13, 22]. The goal of this study is to provide a non-invasive approach for predicting anemia [23, 24].

Table 1 ML and DL approaches for detecting anemia and dataset used

Approach	Methods and models	Datasets
Machine learning approaches	Logistic regression, K-nearest neighbor, bagged decision trees, boosted trees	Complete Blood Count Tests, Blood Hemoglobin Level Results of Patients (Referenced as [1–12])
	Random forests, decision trees, Naive Bayes,	Palpebral Conjunctiva Images (Referenced as [8, 13])
	Classification and Regression Trees (CART), Linear Discriminant Analysis (LDA),	Pictures of Blood Samples for Sickle Cell Anemia Detection (Referenced as [14–16])
Deep learning algorithms	GA-CNN, GA-SAE, ANN, AlexNet	Eyes-defy-anemia dataset (images of palpebral conjunctiva from 218 anemic and non-anemic individuals [17, 18, 3])

3 Methodologies

"Eyes-defy-anemia" is the dataset used for this investigation. The dataset includes 218 photos of the palpebral conjunctivas of the eyes of both anemic and non-anemic individuals, as well as an Excel sheet with each patient's hemoglobin level. As the data is already clean, there is no need to perform any pre-processing (Fig. 1).

The pre-processed data is pushed onto DL models. In this research, CNN and pre-trained DL models were used (such as VGG16, VGG19, ResNet50, and MobileNet). The layers of the CNN model are rearranged as needed, but the original model's layers remain unchanged. After the data has been flattened, it is sent to the fully connected layer for prediction.

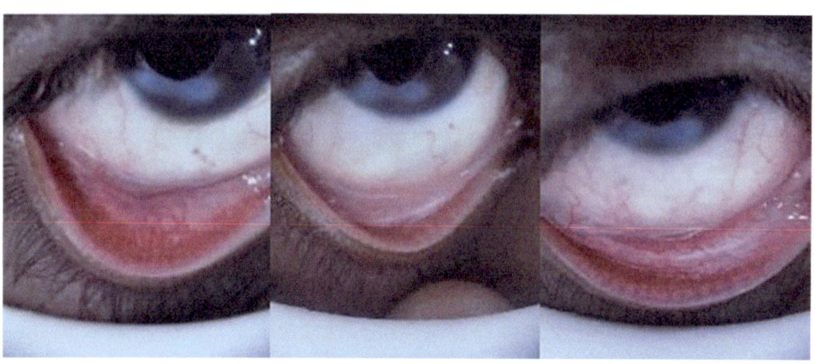

Fig. 1 Sample image present in the dataset

3.1 Convolution Neural Network

When it comes to feature extraction and classification, deep neural networks like convolutional neural networks (CNNs) are unparalleled. Input data, typically an image or image sequence, will be sent to the first convolution layer. The convolutional layers apply convolutional operations to the input data by applying a collection of filters to extract features. The output is a collection of feature maps that emphasize different aspects of the input. The activation layers apply activation functions, such as ReLU, to introduce non-linearity. Pooling Layers minimize the spatial dimensionality and enhance efficiency by down-sampling feature maps. After the output of the preceding layer has been flattened, it is sent to the Fully Connected Layers, where it is multiplied by a matrix with a set of weights to get the final output, which is often passed to a sigmoid activation function to generate a binary classification. Finally, the output layer produces the final output, which is usually a class label or a set of probabilities [25].

A convolutional neural network (CNN) can be expressed mathematically as follows:

Let X represent the image that is being processed, and Y represent the probabilities of the classes that are being generated. Convolutional layers, pooling layers, and fully linked layers are just a few of the several layer types that go into constructing a CNN.

The convolutional layer's filters are applied after the source image X. Let W be the set of learnable filters, and b be the bias term.

The output feature map Z is given by:

$$Zi = f\,(\text{sum}i\,(Wj * X\{i + j\}) + b) \tag{1}$$

when f is an activation function, I is the index into the output feature map, and j is the index into the filter (like ReLU).

Pooling Layer: Spatial dimension reduction is applied to the feature maps that the convolutional layer generates. Maximization pooling, or max pooling, is a common method of selecting the highest possible value from a set of candidates. Let S be the pooling size and stride, the output feature map Y is given by:

$$Yi = \max j\,(Z\{j * S + i\}) \tag{2}$$

where j is the index of the pooling window, and I is the index of the output feature map.

Fully Connected Layer: After the previous layer's output feature map has been flattened, the fully linked layer is responsible for producing the final output Y. Let W be the set of learnable weights, and b be the bias term.

The output Y is given by:

$$Y = \text{Sigmoid}\,(W * X + b) \tag{3}$$

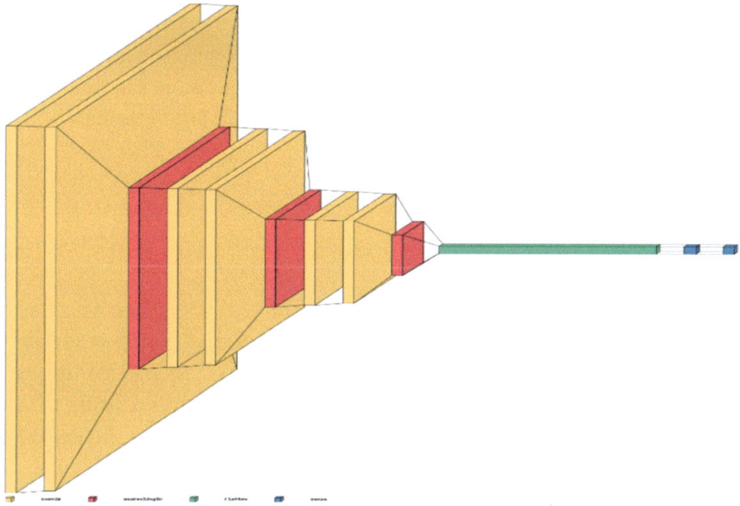

Fig. 2 Convolution neural network architecture. As can be seen, the yellow layer represents the convolution layer, the red layer represents the pooling layer, the green layer represents the flattening layer, and the blue layer represents the dense layer

To produce a binary class from the fully linked layer's output, an activation function called sigmoid is being used.

Each layer of the CNN contributes to the final output Y, which is a function of the input picture X. The network's parameters, including the learnable filters and weights, are optimised to decrease the discrepancy between the anticipated output and the ground truth label during training (Fig. 2).

Depending on the task and dataset, the number and arrangement of these layers may vary. Some architectures, such as VGG, have many layers with small filters, while others, such as ResNet, have fewer layers with larger filters. The choice of architecture depends on the task's complexity and the size of the dataset.

3.2 VGG16

The Visual Geometry Group (VGG) at Oxford University proposed the VGG16 architecture for convolutional neural networks (CNNs). Out of a total of 16, 13 are convolutional layers, and 3 are fully linked ones. A 224 $\times$ 224 RGB picture is sent into the first layer [26]. The ReLU activation function comes after the convolutional layers, which use a 3 $\times$ 3 filter with a stride of 1 and padding of 1. There are 64 filters in the first two convolutional layers, 128 in the next two, and 256 in the last two. The final three convolutional layers all consist of 512 filter nodes. Max-pooling layers with a 2 $\times$ 2 window and a stride of 2 are placed after every pair of convolutional layers. Three fully linked layers are used to flatten the output of the final convolutional layer. The first two fully connected layers each include 4096 neurons, while the third layer has 1000 neurons to represent the various ImageNet class representations. All fully connected layers use the ReLU activation function, except the final one, which uses the SoftMax activation function to provide the final output probabilities.

3.3 VGG19

VGG19 employs a deep neural network architecture with 19 layers, 16 of which are convolutional, 5 of which are max-pooling layers, and 3 of which are fully linked. For more precise detail recognition, the network employs 3×3 convolutional filters with a stride of 1 pixel, followed by a max-pooling layer with a stride of 2 pixels. As the spatial scale decreases, the total number of convolutional layers employed increases from 64 to 512 filters. Each of the fully linked layers in the network has 4096 units, while the final output layer contains 1000 units to account for the 1000 item categories in the ImageNet dataset.

Over 1.2 million pictures from the ImageNet dataset were used to teach VGG19. Overfitting was prevented during training by using weight decay regularization and dropout strategies in addition to a stochastic gradient descent optimizer with momentum. The network was trained over 370,000 rounds using a mini-batch size of 256. The max-pooling layers were used to decrease the output volume's spatial dimensions while retaining crucial information. Furthermore, VGG19 employs transfer learning to capitalize on the features learned from the ImageNet dataset and fine-tune the model for other tasks. To prevent overfitting and improve the network's robustness, data augmentation techniques were used. However, deploying the VGG19 model on low-resource devices can be challenging due to its large size and computational complexity.

3.4 ResNet50

ResNet50 is a 50-layer architecture with 49 convolutional layers and one fully linked layer large units called neurons in a deep convolutional network. The use of residual connections, which help to alleviate the vanishing gradient problem that occurs when training very deep neural networks, is the main novelty of ResNet50. In conventional deep neural networks, each layer learns how to map the input to the output. However, training the network becomes more challenging as its depth increases since the gradient may become very small. To solve this problem, residual connections include skip connections that let the gradient go immediately from one layer to another. The vanishing gradient issue is alleviated, and very deep neural networks may be trained with the help of this technique. There are a total of four nodes in this network, and each one is comprised of many convolutional layers and a skip connection. The first block has 64 filters, while the subsequent blocks include 128 filters, 256 filters, and 512 filters. Following the convolutional layers in each block is a batch normalization layer and a ReLU activation function. One of the primary advantages of ResNet50 over other designs is its higher performance on very deep networks. In a few contexts, including those requiring object identification, detection, segmentation, and image captioning, it has been shown to outperform state-of-the-art algorithms. Another feature of ResNet50 is its capacity to adapt to diverse input sizes. The architecture can take inputs of various sizes, making it useful for activities that need input pictures of varying resolutions. This adaptability is especially important in applications like object identification and segmentation, where the input image size might change greatly. To summarize, ResNet50 is a powerful deep neural network architecture that has significantly enhanced deep learning model performance on a variety of tasks. The vanishing gradient issue can be solved, and very deep networks can be trained due to the use of residual connections. Furthermore, its adaptability to different input sizes and utility as a pre-trained model for transfer learning make it a popular choice in many applications.

3.5 *MobileNet*

The MobileNet network family consists of lightweight convolutional neural networks (CNNs) designed with mobile and embedded devices in mind. Due to their capacity to achieve high accuracy with significantly fewer parameters than conventional CNNs, Google researchers first introduced these networks in 2017 and they have since gained popularity. The mix of depthwise separable convolutions and pointwise convolutions gives MobileNet its remarkable accuracy and speed. Normal convolutions can be broken down into a depth convolution and a point convolution using depthwise separable convolutions. Depthwise convolution, as opposed to pointwise convolution, combines the results by conducting a 1×1 convolution on each input channel. Using these two strategies, MobileNet can reduce the number of parameters and calculations required while maintaining accuracy.

MobileNet's major goal was to create a CNN architecture that could perform well on mobile and embedded devices with limited processing power and memory. Traditional CNNs are usually too huge and computationally expensive for these devices to run, resulting in long inference times and poor performance. In addition to depthwise separable convolutions, MobileNet employs numerous ways to reduce network size and complexity. For example, channel shuffle is a type of channel pruning in which the feature maps between depthwise and pointwise convolutions are swapped to reduce correlations between them. This helps to reduce network parameters while maintaining accuracy.

MobileNet is a family of lightweight CNNs optimized for mobile and embedded devices. It achieves high accuracy with fewer parameters and computations by reducing the network's size and complexity with depthwise separable convolutions and other techniques. It is a popular choice for many computer vision applications on mobile and embedded devices due to its efficiency and effectiveness (Fig. 3).

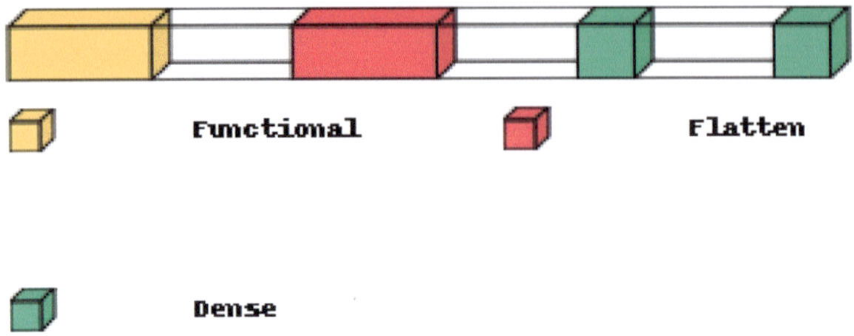

Fig. 3 Pre-trained model architecture: the functional layer depicts the structure of the pre-trained model

4 Result and Discussion

This table displays the performance metrics of deep learning models over several epochs of training. Epochs indicate the number of iterations the model has gone through during the training process. The proportion of cases in the test dataset that were properly categorized is known as accuracy. Loss indicates the value of the loss function (usually cross-entropy) used to optimize the model during training. Precision is defined as the fraction of projected positive cases that were positive. The average squared discrepancy between the predicted and actual values is known as the mean squared error (MSE) (Tables 2, 3, 4, 5 and 6).

The CNN methodology had the best performance measure according to the table, with the highest accuracy of 88.02% in 60 epochs. As the layers, optimization, loss function, and activation function are chosen specifically for the dataset it gives much better results than the pre-trained models (Table 7).

Table 2 Accuracy of CNN model and its model description

Epochs	Accuracy (%)	Layer (type)	Output shape	Parameter
10	79.72	Conv2d	(None, 254, 254, 32)	896
15	79.72	Conv2d_1	(None, 252, 252, 32)	9248
20	80.18	Max_pooling2d	(None, 126, 126, 32)	0
25	82.95	Conv2d_2	(None, 124, 124, 32)	9248
30	83.41	Conv2d_3	(None, 122, 122, 32)	9248
35	84.79	Max_pooling2d_1	(None, 61, 61, 32)	0
40	81.11	Conv2d_4	(None, 59, 59, 32)	9248
45	84.33	Conv2d_5	(None, 57, 57, 32)	9248
50	82.49	Max_pooling2d_2	(None, 28, 28, 32)	0
55	83.41	Flatten	(None, 25,088)	0
60	88.02	Dense	(None, 228)	5,720,292
		Dense_1	(None, 1)	229

Table 3 Accuracy of VGG16 model and its model description

Epochs	Accuracy (%)	Layer (type)	Output shape	Parameter
10	66.36	VGG16 (functional)	(None, 512)	14,714,688
15	71.89			
20	75.12	Flatten (flatten)	(None, 512)	0
25	76.5			
30	72.81			
35	75.58	Dense (dense)	(None, 128)	65,664
40	82.03			
45	81.11			
50	81.11	Dense_1 (dense)	(None, 1)	129
55	82.49			
60	82.03			

Table 4 Accuracy of VGG19 model and its model description

Epochs	Accuracy (%)	Layer (type)	Output shape	Parameter
10	64.52	VGG19 (functional)	(None, 512)	20,024,384
15	70.97			
20	70.51	Flatten (flatten)	(None, 512)	0
25	75.58			
30	75.12			
35	72.81	Dense (dense)	(None, 128)	65,664
40	75.12			
45	76.5			
50	76.5	Dense_1 (dense)	(None, 1)	129
55	79.72			
60	79.89			

Table 5 Accuracy of ResNet50 model and its model description

Epochs	Accuracy (%)	Layer (type)	Output shape	Parameter
10	61.29	ResNet50 (functional)	(None, 2048)	23,587,712
15	69.12			
20	74.19	Flatten (flatten)	(None, 2048)	0
25	73.27			
30	76.5			
35	76.5	Dense (dense)	(None, 128)	262,272
40	75.58			
45	78.8			
50	76.04	Dense_1 (dense)	(None, 1)	129
55	77.42			
60	76.04			

Table 6 Accuracy of MobileNet model and its model description

Epochs	Accuracy (%)	Layer (type)	Output shape	Parameter
10	68.2	MobileNet_1.00_224 (function)	(None, 1024)	3,228,864
15	69.59			
20	70.97			
25	70.97	Flatten (flatten)	(None, 1024)	0
30	75.12			
35	75.12			
40	76.50	Dense (dense)	(None, 128)	131,200
45	76.50			
50	76.5			
55	76.5	Dense_1 (dense)	(None, 1)	129
60	76.96			

Table 7 Comparison of loss, precision, and MSE of all models

	CNN (%)	VGG16 (%)	VGG19 (%)	ResNet50 (%)	MobileNet (%)
Loss	32.71	47.21	47.79	49.14	47.93
Precision	89.92	91.0	81.7	73.1	76.69
MSE	9.21	14.91	15.53	16.24	15.63

5 Conclusion

This study evaluates the performance of multiple deep learning algorithms for predicting the possibility of anemia using images of the conjunctivae around the eyes. The study's findings show the possibility of using deep learning in healthcare. The results reveal that the proposed CNN model predicts anemia with an accuracy of 88.02% using palpebral conjunctiva images. Furthermore, because palpebral and fornical conjunctiva imaging are non-invasive, patients can be screened without the requirement for blood testing. However, there are several drawbacks to this strategy, such as the requirement for high-quality photographs and the possibility of image quality variance between imaging systems. Additional study is required to create and evaluate these models for usage in medical facilities. Lastly, research into the palpebral conjunctiva of the eye as a technique for developing deep learning models for the early diagnosis of anemia seems encouraging. While there are obstacles to overcome, the application of deep learning models in this context offers an interesting breakthrough in the field of healthcare.

References

1. Khan JR, Chowdhury S, Islam H, Raheem E (2019) Machine learning algorithms to predict childhood anemia in Bangladesh. J Data Sci 17(1):195–218
2. Mitani A, Huang A, Venugopalan S, Corrado GS, Peng L, Webster DR, Varadarajan AV (2020) Detection of anemia from retinal fundus images via deep learning. Nat Biomed Eng 4(1):18–27
3. Pingel MS (2021) Leveraging machine learning and process mining to predict anemia with the help of biomarker data. Master's thesis, University of Twente
4. Aliyu HA, Razak MAA, Sudirman R, Ramli N (2020) A deep learning AlexNet model for classification of red blood cells in sickle cell anemia. Int J Artif Intell 9(2):221–228
5. Dukhi N, Sewpaul R, Sekgala MD, Awe OO (2021) Artificial intelligence approach for analyzing anemia prevalence in children and adolescents in BRICS countries: a review. Curr Res Nutr Food Science J 9(1):01–10
6. An R, Man Y, Iram S, Kucukal E, Hasan MN, Solis-Fuentes A, Gurkan UA (2020) Computer vision and deep learning assisted microchip electrophoresis for integrated anemia and sickle cell disease screening. Blood, 136:46–47
7. Acharya S, Swaminathan D, Das S, Kansara K, Chakraborty S, Kumar D, Aatre KR (2019) Non-invasive estimation of hemoglobin using a multi-model stacking regressor. IEEE J Biomed Health Inform 24(6):1717–1726
8. Noor NB, Anwar MS, Dey M (2019) Comparative Study Between Decision Tree, SVM and KNN to Predict Anaemic Condition. In: IEEE international conference on biomedical engineering, computer and information technology for health, November 2019. IEEE, pp 24–28
9. Vohra R, Hussain A, Dudyala AK, Pahareeya J, Khan W (2022) Multi-class classification algorithms for the diagnosis of anemia in an outpatient clinical setting. Plos One 17(7):e0269685
10. Çil B, Ayyıldız H, Tuncer T (2020) Discrimination of β-thalassemia and iron deficiency anemia through extreme learning machine and regularized extreme learning machine based decision support system. Med Hypotheses 138:109611
11. Anand P, Gupta R, Sharma A (2020) Prediction of anemia among children using machine learning algorithms, pp 469–480

12. Laengsri V, Shoombuatong W, Adirojananon W, Nantasenamat C, Prachayasittikul V, Nuchnoi P (2019) ThalPred: a web-based prediction tool for discriminating thalassemia trait and iron deficiency anemia. BMC Med Inform Decis Mak 19:1–14
13. Dimauro G, Camporeale MG, Dipalma A, Guarini A, Maglietta R (2023) Anemia detection based on sclera and blood vessel colour estimation. Biomed Signal Process Control 81:104489
14. Meena K, Tayal DK, Gupta V, Fatima A (2019) Using classification techniques for statistical analysis of anemia. Artif Intell Med 94:138–152
15. Alzubaidi L, Fadhel MA, Al-Shamma O, Zhang J, Duan Y (2020) Deep learning models for classification of red blood cells in microscopy images to aid in sickle cell anemia diagnosis. Electronics 9(3):427
16. Alzubaidi L, Al-Shamma O, Fadhel MA, Farhan L, Zhang J (2020) Classification of red blood cells in sickle cell anemia using deep convolutional neural network. In: Intelligent systems design and applications: 18th international conference on intelligent systems design and applications (ISDA 2018), vol 1. Springer International Publishing, pp 550–559
17. Ayyıldız H, Tuncer SA (2020) Determination of the effect of red blood cell parameters in the discrimination of iron deficiency anemia and beta thalassemia via neighborhood component analysis feature selection-based machine learning. Chemom Intell Lab Syst 196:103886
18. Memmolo P, Aprea G, Bianco V, Russo R, Andolfo I, Mugnano M, Merola F, Miccio L, Iolascon A, Ferraro P (2022) Differential diagnosis of hereditary anemias from a fraction of blood drop by digital holography and hierarchical machine learning. Biosens Bioelectron 201:113945
19. Kilicarslan S, Celik M, Sahin Ş (2021) Hybrid models are based on genetic algorithm and deep learning algorithms for nutritional anemia disease classification. Biomed Signal Process Control 63:102231
20. Sari M, Ahmad AA (2019) Anemia modelling using multiple regression analysis. Int J Anal Appl 17(5):838–849
21. Jaiswal M, Srivastava A, Siddiqui TJ (2019) Machine learning algorithms for anemia disease prediction. In: Recent trends in communication, computing, and electronics: select proceedings of IC3E 2018. Springer Singapore, pp 463–469
22. Kasiviswanathan S, Bai Vijayan T, Simone L, Dimauro G (2020) Semantic segmentation of conjunctiva region for non-invasive anemia detection applications. Electronics 9(8):1309
23. El-kenawy ESMT (2019) A machine learning model for hemoglobin estimation and anemia classification. Int J Comput Sci Inf Secur 17(2):100–108
24. Yıldız TK, Yurtay N, Öneç B (2021) Classifying anemia types using artificial learning methods. Eng Sci Technol Int J 24(1):50–70
25. An R, Huang Y, Man Y, Valentine RW, Kucukal E, Goreke U, Gurkan UA (2021) Emerging point-of-care technologies for anemia detection. Lab Chip 21(10):1843–1865
26. Lakshmi M, Manimegalai P, Bhavani S (2020) Non-invasive haemoglobin measurement among pregnant women using photoplethysmography and machine learning. J Phys: Conf Ser 1432(1):012089. IOP Publishing
27. Karami M, Chaleshgar M, Salari N, Akbari H, Mohammadi M (2022) Global prevalence of anemia in pregnant women: a comprehensive systematic review and meta-analysis. Matern Child Health J 26(7):1473–1487

A Machine Learning-Based Novel Cyber-Intrusion Detection System for Network Administration

E. M. Priyesh, S. Nagaraj, and Ramkumar Krishnamoorthy

Abstract In the present growth of the Internet, people are adjusting to early life of communication that save time and enable speedy completion of tasks, which is their main necessity. Here, the Internet has become a crucial component of today's generation, influencing all facets of daily life, including online commerce, online schooling, and many other crucial services. The topic of user data or personnel information protection, security, and safety issues will result from the enormous rise of Internet technology. A computer system used for data transfer must meet all these requirements and be equipped with the necessary security measures to prevent hostile attacks and unauthorized access from which the user's data can easily be compromised and the system infected. All these cyber-related dangers will significantly harm companies' or individuals' data, but they shouldn't reach an unrecoverable level. Therefore, the creation of a Cyber-Intrusion Detection System (CIDS) using machine learning techniques is the best option for the early detection of cyber-attacks. The suggested work is more successful at handling the many kinds of cyber-attacks that happen for specific data packets and notifying users or system engineers of the attack as soon as possible via social media tools like e-mail or telegram. As a result, we wanted to create the cyber-intrusion system by modifying ML algorithms that deal with cyber-attack issues dynamically and improve early detection rates, decrease false alarm rates, and minimize communication costs.

Keywords Security · Cyber-attacks · Decision tree · Machine learning · IDS and CIDS

E. M. Priyesh
EY Global India, Bangalore, India

S. Nagaraj
Department of CSE, Alliance University, Bangalore, India

R. Krishnamoorthy (✉)
School of CS & IT, JAIN (Deemed-to-be University), Bangalore, India
e-mail: kramdharma@gmail.com

A. P. Agrawal et al. (eds.), *Advances of Artificial Intelligence and Machine Learning in Image Processing, Security and Data Analytics*, Lecture Notes in Electrical Engineering 1464, https://doi.org/10.1007/978-981-95-0148-9_13

1 Introduction

Intrusion detection system is a methodology where the network traffic has been monitored in a search of any suspicious activity or exploitations which damages the CIA of the organization. The application or the software scans for the environmentally unfriendly actions which may be policy breaches, private data exploitation, remote backdoors, spywares, etc. The malicious activity has been passed on to the administrator with the help of SIEM systems. The output of the scans can be fetched through an output window within the system. CIDS (cyber-intrusion detection system) is an application categorized under intrusion detection system. CIDS helps the user with detection and sorting of the attacks or the intrusions over a network more accurately with the help of basic machine learning algorithms. To overcome the traditional output or result-notifying process to the end user, CIDS uses the socio-applications and a web-based login portal to report the results of the scan which majorly helps the improvement of detection rate matrix and the prevention wing which helps in immediate attack fixing and notifying to the administrators or the end users.

CIDS is designed and developed using Python language, Jupiter IDE, and deployed with machine learning algorithms to train and detect the type of cyber-attack which occurred in a specific network. The role of ML is to sort the type of attack (DOS, phishing, etc.) and categorize them using any classification algorithm available and to train, detect the attack occurred while data transmission during the communication that takes place between the host(s) effectively which improve the intrusion detection rates and keep the offensive knowledge databases up to date and more comprehensive. Here, once the attack detected in a network point, immediately the user or system engineers will receive the notification through socio-web apps like e-mail and telegram. CIDS security—early detection will catch the entered attacks and notify problems, which can be categorized as prior attacks detected or accidently prevented attacks. Datasets or outsourced data has been protected from eavesdropping attacks during upload, download, update, and retrieval of the user data. Cyber-Intrusion Detection System (CIDS) is more efficient in terms of effective attack detection with very less time than the existing systems, where it usually takes maximum days or months in identifying the attack caused and by whom and also a cost-efficient software which saves you minimal and huge amount of money.

Figure 1 shows about the part of proposed methodology along with the overall network. In this, CIDS assisted the system administrator to avoid the attacker get into the system.

2 Review of Literature

Intrusion detection system (IDS) is the most significant and unique method which is an essential technology in the field of internetworking and communication systems with an objective to recognize the abnormal system access or common attacks on

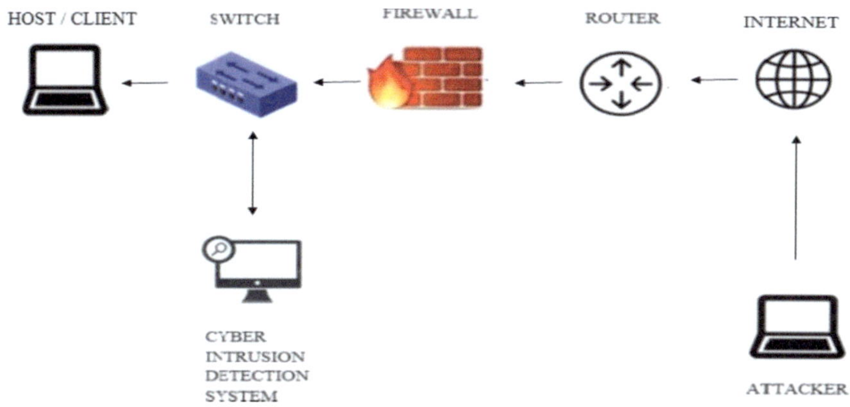

Fig. 1 Role of CIDS in the sample configured network

security systems of internal networks [1]. The author in [2] has implemented IDS using the two techniques, namely, misuse based and anomaly-based. In the first, the cyber-attacks were detected for specific patterns or signatures in data traffic but not worthy against zero-day attacks. In the secondary technique, say, anomaly based is used to identify abnormalities from the normal attacks, i.e., they detect unknown attacks and difficulty to distinguish by the attackers as they are tailored to each particular system, application, and network. Therefore, [1, 3] A Javaid et al. have designed IDS based on Deep Learning (DL) algorithm with set of high-level features for network security and to create an efficient cyber-defense system. The conceptual deep learning (DL) framework model for cyber-security designed and implemented in real time by the authors has been shown in Fig. 2 [4], in which a clear step-by-step process is explained within the architecture itself.

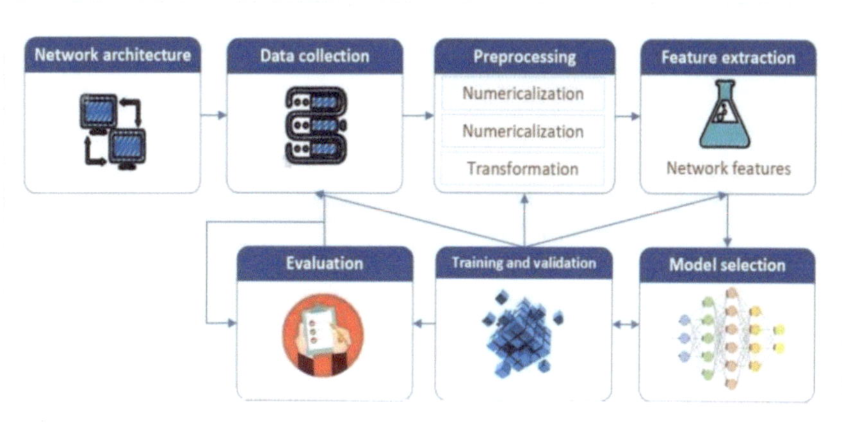

Fig. 2 Model cyber-security framework

Initially, a huge amount of qualitative and quantitative data is aggregated from the particular network organization and preprocessed to extract the network features by cleaning and mapping the data sources with each other. Later, the collected data is made bought for training and validation using the specific DL models.

Madhusudhan and Nair [5] have surveyed and documented cyber-security threats and attacks on multidisciplinary ML approaches with several ML models. The authors focused on, how ML and DL are the best solutions in resolving the cyber-attacks that are causing the huge damage in the fields of network security and to have secured communications. Further their work also extended to list the defensive strategies in order to build the reliable AI/ML applications to enrich and complement the proposed model of their work.

Multidisciplinary ML is used for developing applications to identify the list of attacks in various field of healthcare, transportation, military, robotics, surveillances, intrusion and malware detection attacks, and fraud detection [6, 7]. Therefore, the data processing and computation in these applications using ML are processed very fast in a very less time, and also data interpretation is very effective for classification and faster decision-making based on available ML algorithms.

Sen Tan et al. come up with a new work based on attack detection mechanism for cyber-physical systems (CPS) [8], an integrated computer science and communication network domain in which the systems are controlled and monitored by the algorithms of advanced computer science. A CPS is composed of intelligent transportation networks, advanced technology of cellular networks, robotics, smart grids, etc. as these are communication-based systems interacting with physical devices which makes them more vulnerable to cyber-attacks like denial of service (DOS) and deception attacks that are injected into systems at the physical and cyber-layer. This leads to the irreparable and a potential damage on the physical systems.

Rishabh et al. have proposed E-Shield [9, 10] for website attacks, in which the designed system provides security against maximum type of attacks (Internet worms, Denial of Service (DOS), spam control) in order to protect the user and their personnel data from the attackers. The authors primarily focused on preventing this attack through monitoring both user and system activities by tracking and blocking the malware initially. To achieve this objective, the Intrusion Prevention System (IPS) techniques are the stringent solution and substantially reduce the range of cyber-crimes and further to facilitate the security measures to the organizations and small-scale firms or entrepreneurs.

Kamrul et al. [11, 12] designed and implemented AI-based Advanced Persistent Threat (APT) which is used to check the different patterns of threats that occur and to intimate as earliest and to understand the attack. To achieve the said objective, authors have used the foremost IoC methods, cyber-software tools, malwares and tactics and procedures. Therefore, the authors conclude the exhibiting of their proposed Artificial Intelligence (AI)-based system will analyze the security events by interfacing the existing Security Information and Event Management (SIEM)-based analyzed data with secured cloud. Elangovan et al. [13] discussed about the features of cryptographic applications in general usage along with some algorithms.

This article is prepared as follows: Section 1 places the Introduction, Sect. 2 lists the related articles, and Sect. 3 and Sect. 4 show the Proposed Methodology and Sect. 5 displays the Results and Discussion part. Conclusion is discussed in Sect. 6. Finally references got placed at Sect. 7.

3 Proposed Methodology

3.1 Proposed Methodology

The cyber-intrusion detection and notifying systems using Machine Learning (ML) approaches will detect the attacks like DOS, Imap, Rootkit, Nmap, etc. which are associated to a particular network of the user, and further notifies. Here we incorporate several ML classification algorithms, such as Bayesian network, naive Bayes classifier, decision tree, random decision forest, random tree, decision table, and artificial neural networks, to notice the attack type occurred to the network and the number of packets affected in a previous and present context. It also provides the intelligent services with practical experiments to test the effectiveness of abundant datasets with many forms of cyber-attack types and assess the system performance measure, potency, and correctness by plotting a resultant graph of cyber-attack statistics with number of packets attacked versus type of attack made, presented in a web page with a login portal designed with the help of HTML.

This proposed method adapts the decision tree algorithm which combines the features of selecting the attacks of the packets from the sender and the receiver point. It deeply investigates in all format of networks in identifying the path as well as the availability of the flows. Sample representation of decision tree is shown in Fig. 3 [14].

Here, root node is taken as decision-maker or solution, child node defines the in-depth investigation of packets, leaf node as the outcome of the decision, and class variables considered for prediction.

The work flow of CIDS is as follows:

Step 1: Selecting the path of the dataset file which is present in the local system. Dataset file consists of the data transmission or the network traffic of the system, i.e., reading of data packets has been done here.

Step 2: Specifying the desired number of packets to get scanned from the large dataset. The number of data packets can be described at the code which helps in reducing the work of system engineer or the user.

Step 3: After the further processing of the code, the said ML algorithms will be applied for the dataset specified. Now the cyber-attack statistics will be generated in the form of graph, with attack types versus attack packets. We can see it in the below graph.

Step 4: Once the log graph is represented, immediately the attack details are sent to client mail ID and telegram account as mentioned in the system.

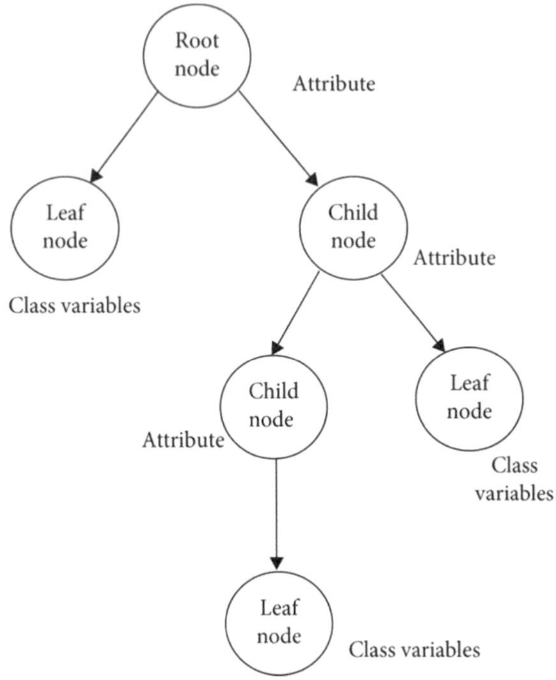

Fig. 3 Decision tree representation

Step 5: The client receives a notification/alert message to his/her mail IDs and accounts. The message contains the attack information as mentioned in Step 6.
Step 6: The alert message contains the following details:
Cyber-attacks have been detected, please check your mail.
Total attack packets: 1826.
Attack category: DOS => suspected packets count: 139.
Attack category: r2l => suspected packets count: 11.

3.2 System Architecture

CIDS is designed and modeled with the basic ML decision tree techniques which gives an advanced progression result to the client or the user. The complete system has been divided into three modules, namely, input module, processing module, and output module which have been diagrammatically represented in Fig. 4.

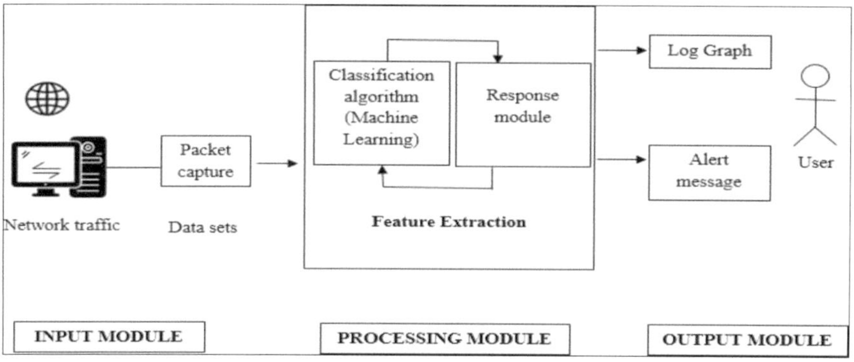

Fig. 4 System framework of CIDS

3.3 *Input Module*

The data transmission over any network has been represented as network traffic, which will be generally represented as data packets. The network traffic of an organization helps in securing and evaluating the quality of service. Packet capturing is a term where a data packet has been intercepted which is traversing at a point in a network.

3.4 *Processing Module*

Classification is a kind of supervised machine learning technique. Here, the label is utilized by every point of data to classify them into their respective classes. The dataset is available in abundance. Incident response is a term utilized to denote the process by which an organization reacts to an incident of data breach or cyber-attack if any, including the way in which it attempts to handle the consequences of the attack or breach incident.

3.5 *Output Module*

This is the module where a user can access the scan results in the means of log graph and alert messages over social media platforms. Using the SMTP protocols, we have implemented the auto-delivering of the mail which contains the results of the scan and with the help of bot system facility on telegram, we have driven the result delivering over there. Upon all of these, CIDS comes up with a web login portal which gives an entire clear view of the present scan and also containing the data or result of previous scan.

3.6 *Some Discussions of Machine Learning Algorithm*

The proposed model CIDS gives factual results of scan with the combinational use of scripting language (Python) and basic ML algorithms also discussed.

3.6.1 Decision Tree Algorithm

CIDS algorithm is being implemented in the DT algorithm which outperforms well in detecting as well as taking the excellent classification of the attacked packets. It helps the researchers to bring some useful conclusion and make the system more secure enough and avoids the future damages of the system. Decision tree algorithm has generally been used for the classification problems. The dataset has been split into multiple homogenous sets which depends on the variables, elements, for the sake of processing into groups. Chi-square, entropy, information gain, and gini are the other techniques which are used for classification.

3.6.2 Random Forest Algorithm

Random forests are a group of decision trees together. When a new object has been classified based on the attributes, it is hence called as votes for the said class. Based on these votes a forest chooses the classification.

3.6.3 Logistic Regression Algorithm

Logistic regression is used to approximate the discrete values, these are also called as binary values based on the provided variables. The prediction is based on the probability of the occurrence of an event by representing the data to a function. As the probability is predicted, the values of output lie between 1 and 0.

3.6.4 Limitations of the Proposed CIDS Algorithm

This research fails to address for the real-time traffic applications in low mobility and high mobility scenarios, also the quality of service not discussed much. In future work, these may be addressed with bringing some possible use case scenarios and implementations.

4 Results and Discussions

The following section illustrates the results and discussion part carried out for the proposed research. Initially, the proposed method has analyzed the attacked packets in the transfer. The following section illustrates the observed results which is shown in Fig. 5.

The sample analyzed results of a taken cyber-attack of e-mail communication is displayed in Fig. 6.

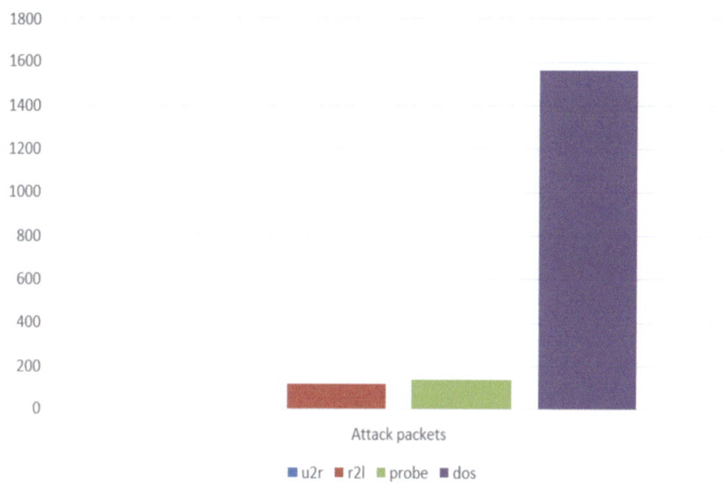

Fig. 5 Attack levels in the packets

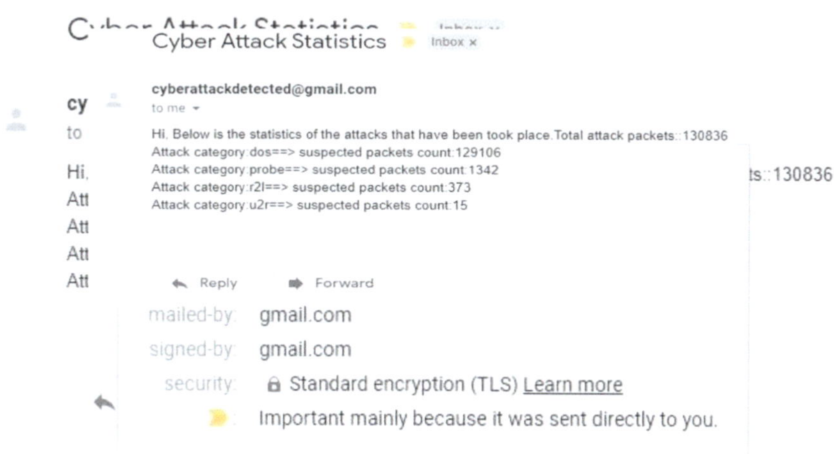

Fig. 6 Cyber-attack statistics in e-mail

5 Conclusion

CIDS helps the user/system admin or made their work easier, while effectively the system detects and diagnoses the attacks which have been occurred in a network by sending the scan results on e-mail and other socio-sites like Telegram. The process report can be fetched through the web page which consists of log graphs and the attack detected information of the scan. Web page also consists of a pre-fetched data available of the previous scan with the present scan too. This functionality makes the admin or the user to compare the data for the further processing. The use of domain knowledge of machine learning algorithms has resulted in higher cyber-detection rates and lowered false alarm rates, reasonable computation and communication costs. All of these factors are because of the ML techniques and algorithms that the system can adopt itself automatically for changing malicious behaviors. Therefore, CIDS provides effective solutions for the attack that occurs during data transmission in communication system, which is more secured and has high-performance metrics too.

References

1. Mayra M, Chunming W (2020) Review: deep learning methods for cybersecurity and intrusion detection systems. In: IEEE Conference Proceedings, vol 20, pp 1–6
2. Axelsson S (2010) Intrusion detection systems: a survey and taxonomy, vol 99. Technical report
3. Javaid A (2016) A deep learning approach for network intrusion detection system. In: Proceedings of 9th BIONETICS & ICST, pp 21–26
4. Mahdavifar S, Ghorbani AA (2019) Application of deep learning to cyber security: a survey. J Neurocomput 347:149–176
5. Sheema M, Nair RR (2019) Converging security threats and attacks insinuation in multidisciplinary machine learning applications: a survey. In: ISRITI IEEE Conference Proceedings, pp 217–222
6. Moosavi S, Fawzi A, Frossard P (2015) Deepfool: a simple and accurate method to fool deep neural networks. CORR, vol 15
7. Carlini. N, Wagner DA (2016) Towards evaluating the robustness of neural networks. In: IEEE conference proceedings, vol 16
8. Sen T, Joesp MG, Peilin X, Renke H, Juan CV (2020) Brief survey on attack detection method for cyber-physical systems. IEEE Syst J 1–11
9. Rishabh J, Aditi S, Savy M, Yash B, Prathamesh C (2017) E-shield: detection and prevention of website attacks. In: RTEICT IEEE conference, vol17, pp 706–710
10. Naived GE, Raghavendra RA, Debabrata S, Nismon R, Ramkumar K, Lokesh GH (2021) Security aspects for mutation testing in mobile applications. In: Cyber intelligence and information retrieval: proceedings of CIIR 2021, Springer, pp 17–27
11. Kamrul H, Sachin S, Sharif U (2019) Artificial intelligence empowered cyber threat detection and protection for power utilities. IEEE CIC 19:354–359
12. Nagaraj S (2022) Manageable e-governance architecture in health care system using cloud storage. Neuroquantology 20(14)

13. Elangovan K, Ramkumar K (2013) A novel algorithm to perform cryptographic operations using Diffie Hellman approach and its applications. J Theor Phys Cryptogr 3(1):11–14
14. Brijain RP, Kushik KR (2014) A survey on decision tree algorithm for classification. Int J Eng Dev Res 2(1):1–5

Protecting Systems from Cyber-Threats Using Multi-layer Security Technology

Syed Ahmeduddin, Gagan Deep Arora, Manish Chhabra, Kuldeep Chouhan, and K. L. Raghavender Reddy

Abstract In the aftermath of the COVID-19 pandemic, the use of the internet has become exceedingly common and widespread. However, this surge in internet usage has also led to a rapid escalation in the misuse of technology within cyberspace, both on a national and international scale. Over the past 2 years, there has been a noticeable surge in cyber-attacks targeting various entities, including businesses, governments, healthcare institutions, and social media platforms. These attacks often involve the dissemination of false information and the creation of deceptive videos on websites, with topics ranging from the virus to unemployment, soliciting fraudulent donations, price manipulation, national emergencies, and financial instability. As a result, e-crimes pose significant threats to nations, organizations, and individuals across the globe, impacting millions of people daily. Tackling e-crime is a formidable task, and relying solely on standard protocols is insufficient to completely eradicate this issue. However, to effectively address the serious challenges posed by e-crimes, it is essential to establish a collective understanding of such criminal activities at an international level. By fostering this shared understanding, it becomes possible to enhance collaboration and develop more effective strategies to combat cyber-threats. Equipping ourselves with a collective knowledge of e-crimes and their global implications will enable us to confront these challenges more successfully and reduce the risks posed by cybercriminals. With the intention of regulating criminal activities in the cyber-world and protecting the technological advancement system, we have defined three layers: biometric identification with face recognition technology, robot test, PIN generations. The government can implement the model in websites and

S. Ahmeduddin (✉) · G. D. Arora · M. Chhabra
Department of Artificial Intelligence and Machine Learning, Vardhaman College of Engineering, Hyderabad, India
e-mail: ahmeduddin8@gmail.com

K. Chouhan
Department of Computer Science and Applications, Sharda School of Engineering and Technology, Sharda University, Greater Noida, India

K. L. Raghavender Reddy
Department of Computer Science and Engineering, Vardhaman College of Engineering, Hyderabad, India

A. P. Agrawal et al. (eds.), *Advances of Artificial Intelligence and Machine Learning in Image Processing, Security and Data Analytics*, Lecture Notes in Electrical Engineering 1464, https://doi.org/10.1007/978-981-95-0148-9_14

social media platforms for the security and privacy policies to identify cyber-security threats. E-crime is the dark side of our society that's not visible but may be reducible by implementations.

Keywords Robot test · Biometric identification · PIN generation · Cyber-security · E-crime · Face reorganization technology

1 Introduction

In contemporary society, technology has assumed a crucial role, becoming an indispensable part of our lives. The global usage of the Internet has reached an astounding figure of over 5 billion users [1]. Throughout the computer age, researchers and scientists have been dedicated to exploring software developments to uncover potential vulnerabilities.

Recently, many small and medium companies and government organizations around the world are facing problems of cyber-security issues [5, 6].

The cybercrime can broadly be classified into four types [2].

1.1 E-Crime Against People

The tactics used by hackers are indeed concerning and pose significant risks to individuals' privacy and safety. Subsequently, hackers can exploit this information for various harmful purposes, including online harassment, cyberbullying, and unwelcome engagement in conversations. In a few cases, they may even deceive victims into meeting them in person, leading to serious crimes such as murder, theft, or rape [7–9, 15].

1.2 Cybercrime Against Property

Hackers utilize malicious software through various channels such as websites, emails, and personal chats to compromise devices and steal information [9]. This malware can cause device destruction, steal information, or enact ransomware attacks [10].

1.3 Cybercrime Against Businesses

The hackers breach the systems of various companies, gaining unauthorized access to sensitive data stored on servers. Moreover, in situations involving financial transactions, they can transfer funds from the company's account to another person's account. As a result of such incidents, customers lose trust in the organization, leading to a considerable decline in the company's customer base [3, 9, 15, 16].

1.4 Cybercrime Against Government

Cybercriminals have the ability to launch attacks on government organizations, with the purpose of filtrating secure databases and manipulating sensitive information. This includes stealing or altering data such as Aadhaar Information (UIDAI), project details, government funds, schemes, national security data, and future plans or policies [3, 9, 15].

Hence, the consequences of cyber-warfare can include the following attacks [12]:

- Catastrophic threat to national security.
- Damage to the national economy.
- Disrupting the air traffic control system.

The research article concludes that cyber-attacks across the globe have an average success rate of 75%, as illustrated in Fig. 1 [5, 15].

Figure 2 highlights the most significant information sought after by cybercriminals, along with the corresponding percentages.

To address this issue, this research paper proposes a practical model to identify and reduce cybercrime incidents.

2 Model Description

The "Cybercrime Reduction Model" comprises four essential layers or steps that users can implement while accessing the Internet. By following these steps, users can significantly decrease the risk of falling victim to cybercrime.

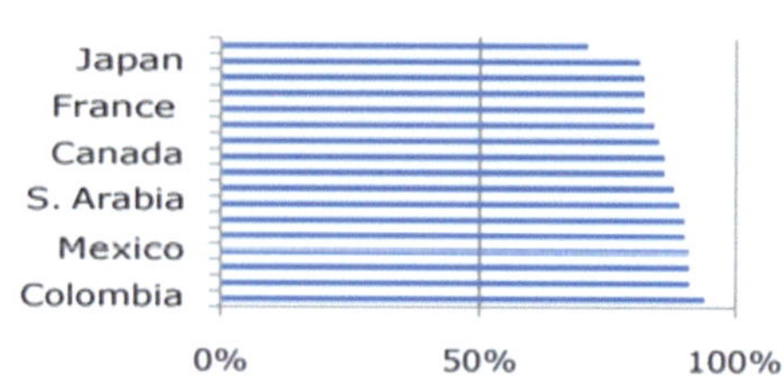

Fig. 1 Cyber-successful attacks last 12 months

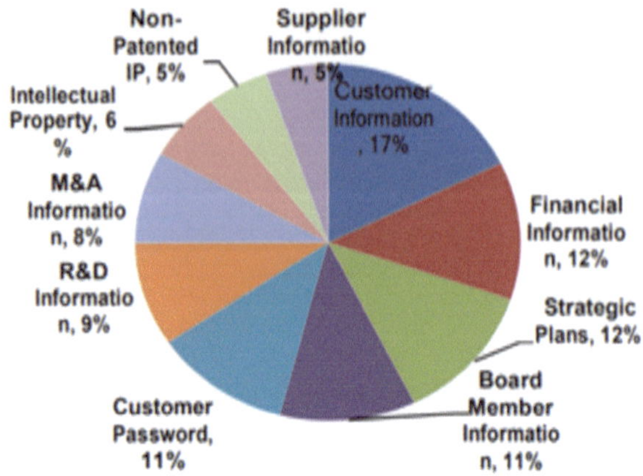

Fig. 2 Valuable information to cybercriminals

- **Step 1**: Secure Password Practices—Users are encouraged to create strong and unique passwords for their online accounts.
- **Step 2**: Awareness of Phishing and Social Engineering—Educating users about the risks of phishing and social engineering is paramount.
- **Step 3**: Regular Software Updates and Security Measures—Staying up to date with software patches and security measures is essential to protect against vulnerabilities that cybercriminals may exploit.
- **Step 4**: Encrypted Communications and Data Protection—The decisive step in the model focuses on ensuring encrypted communications for sensitive information exchange.

Our aim is to develop a machine learning-based real-time framework to detect human faces and keep them in database reference to recognize the human faces for verification in browsers or other applications to control E-crime all over the world [21]. The system utilizes a wide range of algorithms, each serving distinct functions and tasks. These methodologies encompass K-nearest neighbor (K-NN) [24], decision tree, random forest, convolutional neural network (CNN) [25], geometric methods for face detection, template based face detection, simple templates, face recognition algorithms, iterative closest point-based alignment, simulated annealing-based alignment, average-based face model, cat swarm optimization (CSO), particle swarm optimization (PSO), genetic algorithm (GA), face recognition using deep learning [22, 23], and a face recognition using machine learning algorithms.

2.1 *Biometric Face Recognition Based on Real-Time Framework (Step 1)*

Each algorithm serves a specific purpose, contributing to the system's capability to process and recognize diverse patterns and features in the data. The combined use of these algorithms enhances the system's versatility, enabling it to tackle complex tasks and yield precise results. By integrating this array of algorithms, the system becomes a robust and effective solution, proficiently managing a wide range of tasks in various contexts. Our proposed method (step 1) consists of many significant steps shown in Fig. 3.

Now, we use Canny Edge Detection (it detects a wide range of edges in an image) to enhance the edge of the image to more accurately achieve exact facial expressions.

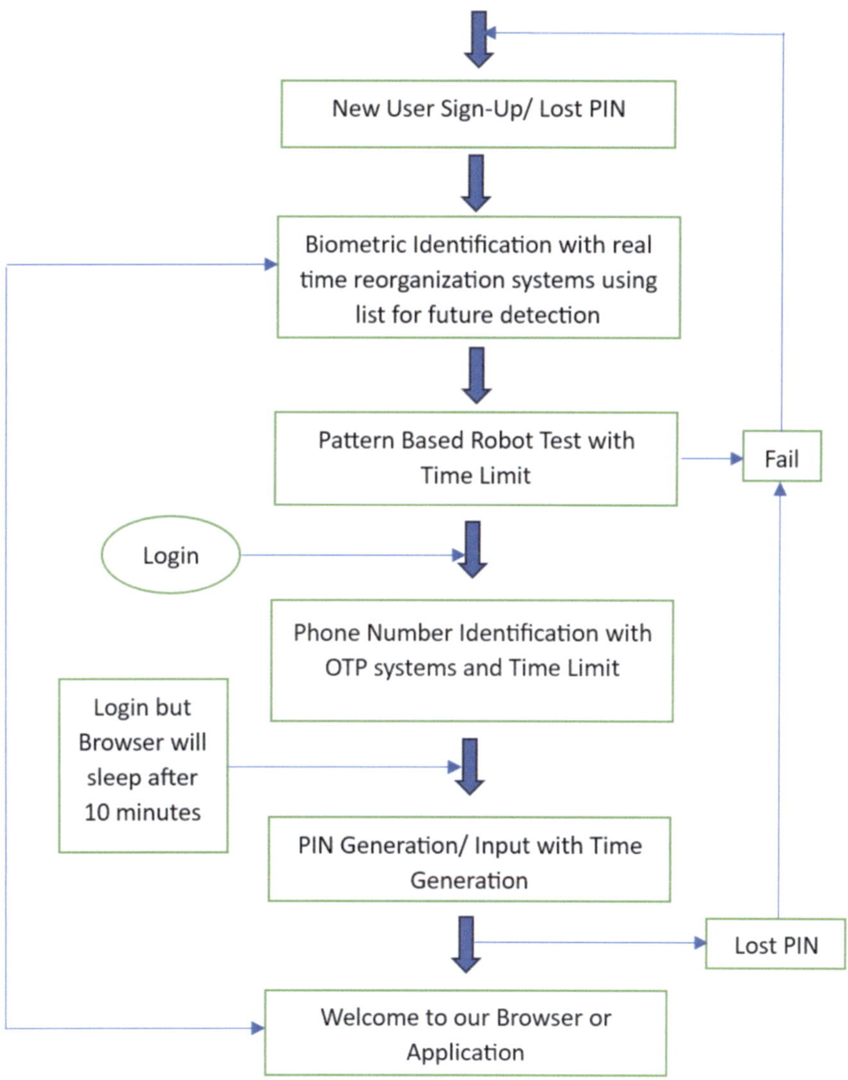

Fig. 3 Cybercrime reduction model and framework for face detection

At first, a Gaussian filter with $ = 1 is applied to smooth it, the equation for the size (2*k* + 1) * (2*k* + 1) is

$$H_{ij} = \frac{1}{2\pi S^2}\left(-\frac{(i-(k+1))^2+(j-(k+1))^2}{2S^2}\right) \tag{1}$$

where 1 <= i, j <= (2 k + 1).

Now by Eq. 1, we input the outcomes in a matrix H to filter, and the result is Gout shown in Figs. 4 and 5a.

```
G=fspecial('gaussian', [3 3], 1); % as $=1;
H= [2 3 6; 6 7 9; 5 8 9];
Gout=imfilter(G,H,'replicate');
```

Now second step Sobel filter applied to detect the intensity in both directions (horizontal x and vertical y) is

$$(G_x = -101 - 202 - 101)\, and\, \left(G_y = 121000 - 1 - 2 - 1\right) \tag{2}$$

Now, when the image is smoothed as shown in Fig. 5b, the derivatives *Ix* and *Iy* w.r.t *x* and *y* are calculated. It may be implemented by convolving I with Sobel filters *Gx* and *Gy* to find the size and angle (θ).

$$|G| = \sqrt{I^2 + I^2, \theta = -1 I_y \tan(I_x)} \tag{3}$$

Fig. 4 Image captured through camera

Now, we applied the double threshold in Fig. 5b and the resulting image in Fig. 6a and b.

Then the last step of the Canny edge detector, the hysteresis threshold is applied to the image in Fig. 6a and the final image in Fig. 6b, the threshold is expressed by

$$H = 1/(1 + e^{-x}) \quad (4)$$

At last, the edges are added to the original image to get the final enhanced image.

Equations 3 and 4 shall be useful if critical issues are created while face recognition technology is used during step 2.

Now next step is using the Viola–Jones algorithm to detect the face from the image then the image is extracted and resized as a c * c image (c is constant), at last technique is to extract features of the faces by Principal Component Analysis (PCA), which reduces the dimension database.

(a). Filtered Gray Scale Conversion

(b). Filtered Image by Image Processing Technology

Fig. 5 Image based on biometric real-time face recognition technology

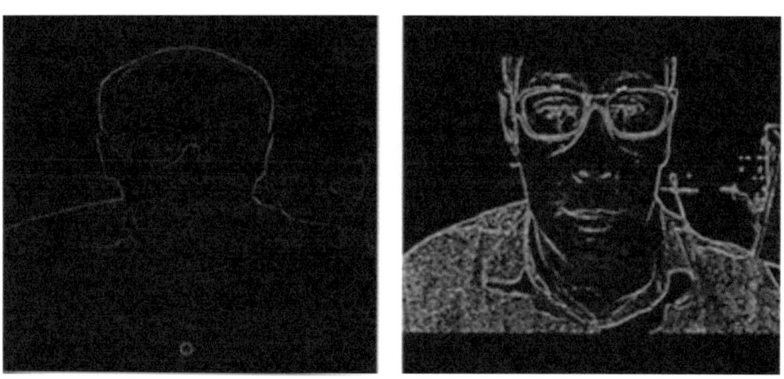

Fig. 6 **a** and **b** Using double threshold: applied hysteresis threshold

Now here we will use the collections of the images (multiple frames per second) that have been uploaded by users then it will be helpful to find the covariance of the matrices. Eigenvalues are the key features of an extracted face; the values will be used for actual face recognition in our system.

The final image is also stored in the database for future reference. We know an image is nothing, but a matrix of pixel values let the matrix for the final image (after using step 1 on the original image Fig. 5a and the resulting image added to the original image) be

$$G = [2003 \quad 10 \quad 022001 \quad 30 \quad 04] \sim [G_1 G_2 G_3 G_4] \tag{5}$$

where the eigenvalues $\gamma 1 = 5, \gamma 2 = 4, \gamma 3 = 2, \gamma 4 = 1$ with corresponding eigenvectors:

$$X_1 = \left(1^{0}_{01}\right), X_2 = \left(0^{1/2}_{11}\right)$$

$$X_3 = \left(-2^{0}_{01}\right), X_4 = \left(0^{-1}_{10}\right) \tag{6}$$

2.2 *Robot Test or ReCAPTCHA (Step 2)*

The algorithm uses a JavaScript code to track the movements of the user mouse, speed of typing, and tracks the information from cookies. Hence "robots" is a software program for verification of the request generated from human or artificial intelligence done by image processing technology.

But there are many cases where this will not work, like:

1. If the user deactivates the tracking scripts, or a programmer creates a program (code) for generating fake mouse clicks and movements then reCAPTCHA will not be sure if the user is human or fake user.
2. There is also another verification method, which asks you to "select all images containing a bus/cycle/traffic light, etc." for such visual tasks the advanced machine learning techniques cannot perform with 100% accuracy for such cases the programmer can write such programs using image processing technologies to submit fake response into the visual task submission.

Now to solve the problems, we have already taken the image of the user by biometric real-time face reorganization technology and applied our framework shown in Fig. 4.

The biggest vulnerability in the server/system is the data/file already exists so the hackers can tamper or depolarize it hence we have found a loophole. If the user corrects his/her image for reCAPTCHA then it will be hard for crooks to exploit.

Now after using the framework (shown in Fig. 4) in the image of the user then we applied a pattern like four submatrices of the given matrix G shown in Eq. 3 with eigenvalue and vector shown in Eq. 6.

$$NewG = [2003 \quad 30 \quad 041002 \quad 20 \quad 01] \sim [G_1 G_4 G_2 G_3] \tag{7}$$

Hence, the new matrix will be after changing the position of the sub-matrix in Eq. 7, the resulting matrix is Eq. 5 and the visual display for the user is shown in Fig. 6b.

We must note that the human face is symmetric but according to the camera, the face may never be symmetric like more hair in the left eyebrow than the right, etc. because the camera captures each frame hence it will count how many pixels in each sub-matrix, where, the eigenvalues,

$$\gamma_1 = 5, \gamma_2 = -1, \gamma_3 = 2 - \sqrt{3}, \gamma_4 = \sqrt{3} + 2$$

with corresponding eigenvectors:

$$X_1 = \left(2^{0}_{01}\right), X_2 = \left(-1^{0}_{01}\right)$$

$$X_3 = \left(0^{-\sqrt{3}}_{01}\right), X_4 = \left(0^{\sqrt{3}}_{01}\right) \tag{8}$$

Now, if the hacker uploads a matched face Fig. 5a, then the matrix will be different because eigenvalues will not match. Suppose, the user selects the different unmatched image in reCAPTCHA then:

$$G = [1002 \quad 20 \quad 033004 \quad 20 \quad 01] \sim [G_2 G_1 G_4 G_3] \tag{9}$$

where, the eigenvalues $\gamma 1 = 5$, $\gamma 2 = 4$, $\gamma 3 = -1$, $\gamma 4 = -2$ with corresponding eigenvectors:

$$X_1 = (0101), X_2 = \left(\frac{2}{3}010\right), X_3 = \left(0^{-1}10\right), X_4 = \left(-3/4^{0}_{01}\right) \tag{10}$$

We can see if we change the sub-matrix (Eq. 8 in place of Eq. 7) then eigenvalues are not matching (Eqs. 8 and 10 are different). Hence, the consequence is registration failed but here there will be fewer chances (0.01%) that the two similar matrices have the same eigenvalues because the user has only one pattern-matching facility already updated in the system with a time limit of 2 min.

K-Nearest Neighbor Algorithm

K-NN algorithm [24] is used to make classifications or predictions about the grouping of an individual pixel/data. Hence, for different eigenvectors, the new face image will pass through all the steps (the hacker may tamper with the system) in the database and the nearest one will be the results. We used Manhattan distance formula to calculate the distance metric for more accuracy in our framework (shown in Eq. 11).

This method is applied to existing users who want to change their PIN or forget the four-digit PIN.

$$d|x, y\Big| = (\sum m \Big|xi - yi|) \tag{11}$$

where x is the image that passed step 1 (defined in Fig. 4) and y is the test image (defined in Fig. 6); hence, we obtained by simulating different values of k, the results with maximum accuracy of 95%.

Convolutional Neural Network

It is a deep learning algorithm [22, 23, 25, 26] that takes an input image to various objects then able to differentiate two images without losing features, now at last we obtained a maximum of 95.67% accuracy.

Our research proposal is a CNN model for face recognition which is better for more accuracy while applying in step 2.

2.3 *Phone Number Identification (Step 3)*

Now in the third step we proposed a secure one-time password (OTP) system to provide more security against e-crime [3].

To generate OTP by mobile number the user must pass through these two steps:

1. Current location.
2. Mobile number.

If the location does not match (the browser/application will access the location of the user in real time) then verification fails.

If the user already registered, then he has to login every time from the third step.

2.4 *PIN Generation (Step 4)*

Now the new user has completed the three steps, he/she will create a PIN (four digit) with confirmation in the fourth step.

If there are no movements by users in the last 10 min, then the browser/application will sleep, hence in such cases the user has to input the PIN every time in the system. The user has a time limit and there is no hint to remember the PIN so if the user fails then he/she must follow all four steps (shown in Fig. 3) to access the Internet facility.

3 Potential Challenges

3.1 Challenges Related to Google ReCAPTCHA and Image Recognitions

1. Script Disabling and CAPTCHA Bypass:
   a. Users may disable tracking scripts, compromising the effectiveness of Google reCAPTCHA.
   b. Advanced image recognition models may pose a challenge in bypassing CAPTCHA systems that rely on image-based challenges.

3.2 Challenges Related to the "Cybercrime Reduction Model" Layers

1. Biometric Face Recognition:
   a. The accuracy and reliability of the chosen face recognition algorithms may vary in different scenarios and lighting conditions.
2. PIN Generation and Security:
   a. Balancing user convenience with security in the PIN generation process may be challenging.
   b. Enforcing a strong PIN without compromising user experience is crucial.
   c. Mechanisms to recover or reset PINs securely need to be set up to prevent potential lockouts.

3.3 General Challenges

1. User Education and Adoption:
   a. Ensuring that users understand and adopt the recommended security practices may require effective education and training programs.
   b. Users may resist adopting new security measures, especially if they perceive them as cumbersome or inconvenient.
2. System Integration and Compatibility:
   a. Integrating the proposed model into existing systems and platforms may be complex and may require compatibility adjustments.
   b. Ensuring the model works seamlessly across different browsers and applications could be a challenge.

3. Regulatory Compliance:
    a. Adhering to data protection laws and regulations on the collection and processing of biometric data is essential.
4. System Performance:
    a. Ensuring the real-time processing capability of the system may be challenging, especially when dealing with a variety of algorithms.
    b. Resource-intensive algorithms may impact system performance, requiring optimization for efficiency.

4 Conclusion

The paper concludes by underscoring the paramount importance of computer security in our increasingly interconnected world. As global networks become more intertwined, ensuring a secure approach to conducting essential transactions in the cyber-realm becomes imperative. However, the ever-evolving landscape of cybercrime introduces novel challenges, as new tools and attack vectors emerge relentlessly, posing threats to both organizations and individuals. Safeguarding infrastructures and data in such a dynamic environment demand not only innovative technologies but also a strategic approach and intelligence. The paper presents a novel architecture model, offering significantly improved security compared to existing models. Despite this progress, there remains an opportunity for researchers to further refine the model's accuracy, leaving the door open for future advancements. In essence, the paper underscores the fact that there is no panacea for completely eradicating cybersecurity threats. Nevertheless, the focus should be on persistent efforts to mitigate such risks and set up a secure environment for the future of cyberspace. By investing in robust cyber-security practices and supporting vigilance against emerging threats, we can pave the way toward a safer and more secure digital landscape.

References

1. Homepage Statista. http://www.statista.com/statistics/617136/digital-population-worldwide/
2. Samuel, C, Han, Yu L, Ayush, K, David, Y (2022) Hacker types, motivations and strategies: a comprehensive framework. Elsevier 5:1–8
3. Yuchong L, Qinghui L (2021) A Comprehensive review study of cyber-attacks and cyber security: Emerging trends and recent developments. Science Direct Elsevier 7:8176–8186
4. Ho HT, Luong HT (2022) Research trends in cyber crime victimization during 2010–2020: a bibliometric analysis. SN Soc Sci Springer 2(4)
5. Cristos V (2022) Cyber crime and artificial intelligence. In: An overview of the work of international organizations on criminal justice and the international applicable instruments, vol 23. Springer Link, pp 109–126
6. Rohit C, Jaya LT, Lohith RRK, Kalavathi A (2022) A study of cyber security issues and challenges. In: IEEE Bombay section signature conference, Gwalior, India

7. Sara Giro C (2021) Patterns of online repeat victimization and implications for crime prevention. IEEE Xplore 20987453:27
8. Tamilarasi P, Uma Rani R (2020) Diagnosis of crime rate against women using K-fold cross validation through machine learning. In: 4th international conference on computing methodologies and communication. IEEE Xplore, Erode
9. Himanshu A, Tanuj M, Geetanjli B, Shikha C (2022) Cyber security challenges and trends on recent technologies. In: 6th international conference on computing methodologies and communication, vol 21687999. IEEE Xplore, Erode
10. Frank C, Barry S, Michael F, Arash NK, Martin M, Finbarr M, Stefan M (2022) Cyber risk and cyber security: a systematic review of data availability. Springer Nature 47(3):698–736
11. Sagiroglu S, Arslan B (2019) Fighting with cyber terror and terrorism: threats and precautions. In: 4th international conference on computer science and engineering, Samsun, Turkey
12. Gunasekaran M, Daphne L (2017) A survey of big data architectures and machine learning algorithms in healthcare. Int J Biomed Eng Technol Inderscience 25(2/3/4)
13. Mohammadi H, Rezaee R, Nawaz, Farhadi H (2015) Review on cyber security risk assessment in cloud computing. J Cloud Comput: Adv Syst Appl Elsevier 4(1):1–15
14. Cong W, Qian W, Kui R, Wenjing L (2009) Ensuring data storage security in cloud computing. In: Proceedings of the 17th international workshop on quality of service, vol 9689759, pp 1–9
15. Wenli D, Mengchu Z, Abdullah AS (2022) A survey of cyber attacks on cyber physical systems: recent advances and challenges. IEEE/CAA J Autom Sin 9(5):784–800
16. Homepage Finances online. https://Financesonline.com/cybercrime-trends/
17. Binny N, Ashir M, Hiteshri Y, Manan S (2022) The impacts of artificial intelligence technologies in augmentation of cyber security: a comprehensive review. Complex Intell Syst Springer 8:1763–1780
18. Alashhab ZR, Anbar M, Singh MM, Leau YB, Al-Sai ZA, Alhayja'a SA (2021) Impact of coronavirus pandemic crisis on technologies and cloud computing applications. Science Direct 19:100059
19. Serghei O, Grigori B (2022) Estimation of the key segments of the cyber-crime economics. In: IEEE 8th international conference on problem of info communication, science and technology, vol 21730993
20. Cybersecurityventures.com/hackerpocalypsecybercrime-report-2016/
21. Iqbal HS (2021) Machine learning: algorithms, real-world applications and research directions. SN Comput Sci Springer Nature 2:160
22. William T, Moustafa MS, Shonan H (2021) Real-time deep learning- based object detection framework. In: IEEE symposium series on computational intelligence, vol 20341101, Australia
23. Waleed BS, Baber A, Haider A, Saad BK, Hammad A (2022) An enhanced deep learning-based framework for web attacks detection, mitigation and attacker profiling. Science Direct 198:103270
24. Pengbo W, Yongqiang Z, Wenting J (2021) Application of K-nearest neighbor algorithm for human action recognition. In: 4th advanced information management, communicates electronics and automation control conference. IEEE Xplore, China
25. Rahul C, Kamal KG, Joshi RC (2019) Convolutional neural network for image detection and recognition. In: 1st International conference on secure cyber computing and communication. IEEE Xplore, India
26. Dharminder Y, Avinash S, Sharik A, Umesh C (2022) Political sentiment analysis, case study of Haryana using machine learning. In: Mobile radio communications and 5G networks. Lecture notes in networks and systems, vol 339. Springer Nature, Singapore

Determining Construction Objects and Equipment on Site by Analyzing Construction Activities and Using Deep Learning Networks

Swati Sah, Shweta Kumawat, and Kofi Immanuel Jones

Abstract Reliable equipment that automatically and instantly detects activities on building sites helps to lower expenses, improve operational costs, and eliminate pollution. To track resource progress and productivity, a comprehensive perspective and resource allocation of different construction-related tasks are very important. Despite the task's importance, all of the work must be done by hand. Nonetheless, research studies have made significant progress in examining computer vision techniques. Nevertheless, it is focused on a small range of object kinds and activity classes. Furthermore, the emphasis is on identifying individual activities when an image only contains one or more objects performing that activity. This study square approach uses deep convolution neural networks to identify 22 groups of items in photos taken during various construction-level activities. The chance of things coexisting will be represented by semantic relevance, and the proximity at the positions of the image and patterns of activity will be represented by spatial relevance. It is advantageous to have the ability to automatically recognize structure activity. So, this is a potential method to manage important time when gathering information to identify construction-related activities.

Keywords Activity recognition · Semantic relevance · Spatial relevance · Recurrent faster convolutional networks · Relevance networks

---

S. Sah (✉)
Sharda University, Greater Noida, Uttar Pradesh, India
e-mail: swati.sah@sharda.ac.in

S. Kumawat
Tokyo, Japan

K. I. Jones
Jain University, Bengaluru, Karnataka, India

A. P. Agrawal et al. (eds.), *Advances of Artificial Intelligence and Machine Learning in Image Processing, Security and Data Analytics*, Lecture Notes in Electrical Engineering 1464, https://doi.org/10.1007/978-981-95-0148-9_15

1 Introduction

Huge scales and numerous simultaneous operations are seen at construction sites. Primary resources consist of two limited classes of items, such as the construction work done by excavators, which has two limited categories. Almost two other groups of objects and operations can be estimated using this method. There could be a way to create sensors that can be trained using a smaller amount of constrained object classes using handcrafted features [1]. Increasing the amount of object categories is a difficult task despite the fact that these methods have led to complacency in anticipated scenarios [2]. Instead, attention is being paid to activity recognition as well as detection when an image contains multiple objects performing a single action [3]. Models which are composed of various preprocessing layers to train data representation with different abstracted levels. The method described in the study uses a quicker recurrent convolution neural network and CNN to identify activities associated with building on-site photos. Faster object detection with the network ResNet-50 CNN as a network for regional suggestions, for identifying and classifying the observed items, and for creating relevant networks.+

2 Background and Related Work

2.1 *Object and Human Activity Recognition Based on Computer Vision with Visual Object Detection*

Researchers made more rapid CNN which uses CNN to classify proposed objects, which speeds up training and testing. Swifter R-CNN permits free-cost region proposals and communicates a full picture of convolutional properties with the discovered network [4]. Earlier construction-related detection research has focused on detecting personnel, equipment, and various object classifications, but the resulting detections are sparse. The recovery of images using fundamental features is the topic of research that includes materials relevant to building [5]. Shape based, stochastic, rule based, and space-time are the four categories that make up human activity recognition. Human activities can be recognized using pose classification and shape-based algorithms that characterize human operations in three-dimensional position models [6]. By treating human actions as activity entities that are randomly predicted in a sequential state, stochastic approaches can model human behavior. Space-time techniques examine actions based on pixel-based traits occurring throughout the frames. Rule-based strategies examine the activity using probabilistic matching patterns by framing the activity as activity patterns characterizing atomic operations [7].

2.2 *Construction Activity Recognition and Analyzing the Productivity*

Based on their complexity, gestures, interactions between people, interactions between objects, interactions between atoms, and events, human activities are divided into different levels. Rule-based approach examination of the operation of pouring concrete with buckets and a tower crane. This approach separates construction activities into a range of work tasks, such as semantic context, and illustrates how this working task develops in a predetermined state (spatial context) [8]. Moreover, sequences are like temporary contexts. Real-time video is manually coupled with the spatial context by selecting particular locations in video scenes and designating a functioning state for each region separately [9]. The initialization of the rule-based activity detection approach involves calculating the productivity of excavators and garbage trucks when it comes to earthwork [10]. While excavator and dump trucks are visible nearby, the activity state—static, moving, absent, and filling—is observed. Photogrammetry is used to measure the amount of Earth that has been removed and the productivity of the earthwork. The emphasis is on identifying individual activities, and the graphic shows a few things performing individual activities. We need a method that can evaluate numerous items on on-site photos and identify construction-related activities because of the overall awareness of the challenges in construction project resource leveling performance monitoring and productivity analysis, and concurrent activity [11].

3 Implementation and Methodology

The issue of interactions between construction-related activities on the job site is addressed in this study. In order to detect objects' on-site photos, R-CNN and ResNet-50 are deployed. This study identifies construction-related activities by analyzing network and activity patterns. The two categories of semantic relevance and geographical relevance are used to initialize the relevance network. Whereas spatial relevance describes a 2D pixel distance-related image, semantic relevance defines the chance of coexistence between two things in the construction sector. In order to detect 17 different types of construction-related on-site activities, this research builds a two-dimensional activity pattern depending on an object with respect to a relevance network [12]. The overview of the R-CNN training and testing procedures is shown in Fig. 1.

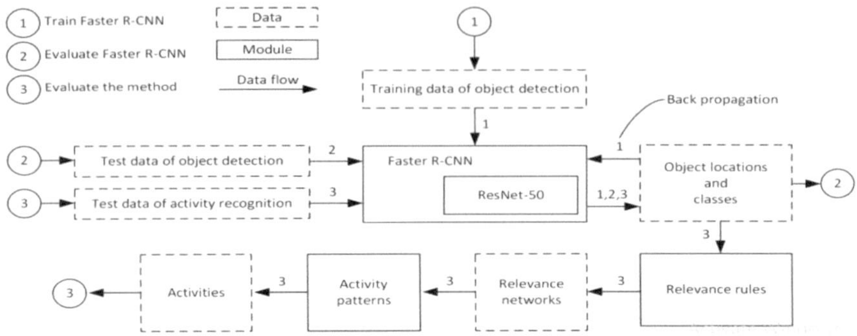

Fig. 1 The architecture of data and training algorithms

3.1 Dataset

Deep neural convolution networks' training and testing datasets are separated out. The study focuses on photographs taken during the planning and building phases, and it covers a total of 22 different object types. 80% of the photos were used for training, and 20% of the randomly chosen trained data was tested. The first things are divided into four categories: personnel, goods, general transportation, and equipment. Tasks connected to formwork, concrete, rebar, and scaffolding are included in the second category's classes inside subcategories [13].

3.2 Semantic and Spatial Relevance

The study uses Rse, or semantic relevance, to characterize the likelihood that two objects would cohabit near a construction site. Rse computes among classes using a 5-point Likert scale with intervals of 0.25. In contrast to Rse 0, which displays a result with no semantic relevance, Rse 1.0 depicts the same class significance of two objects. The figure illustrates the inclusion of these scores. Rse 0.75 displays the relation across classes across particular subcategories. A strong relationship of unique subcategories between classes is presented by Rse 0.5. It is possible that there are two classes of materials: one is treated, and the other is dependent on treatment. An example of this would be the placement of concrete and the formwork for beams or slabs, which might be directly related to handling and being handled by personnel or materials [14] (Figs. 2 and 3).

The physical distances describing the complicated proximity in the region would be best determined at a vertical angle. The mission of special relevance was developed by the investigators because there are three different ways to illustrate the significance of connecting the bounding boxes in the context of items A and B. Predicting the bounding box and the bounding box with actual data is what is meant by intersection

Fig. 2 Representation of different objects and classes in input data

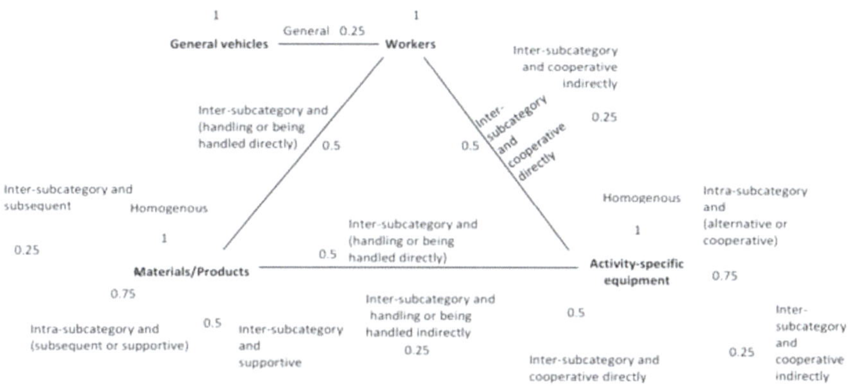

Fig. 3 Special relevance for bounding box

over the union. In the first case, the RSP has achieved a maximum of 1.0 when one bounding box contains another box. This establishes that the workers identified here are all engaged in the same activity (WKR-3, WKR-4, WKR-5, WKR-6, WKR-10, and WKR-20), and (M-RSB-11) must contain the same Rsp value, which will be 1, even though the bounding frames of some workers (WKR-10 and WKR-6), for example, are smaller than those of others (WKR-10 and WKR-6, respectively) [15]. A bounding frame crosses another in the second place. At the time of joining and the smaller bounding frame area, M-RSB-11 and WKR-0's spatial importance is analyzed in red (WKR-0). The first two situations investigate the function area that determines the state of an enclosing frame. This intersecting significance is set inside 120.5. The bounding boxes do not intersect at the third step. Using lengths and distances first, followed by areas, the spatial importance between the two distinct things is stated. The minimal range between M-RSB-11 and WKR-1 is indicated in the figure with a yellow mark, while that between M-RSB-11 and M-FMW is indicated with a blue mark (Fig. 4).

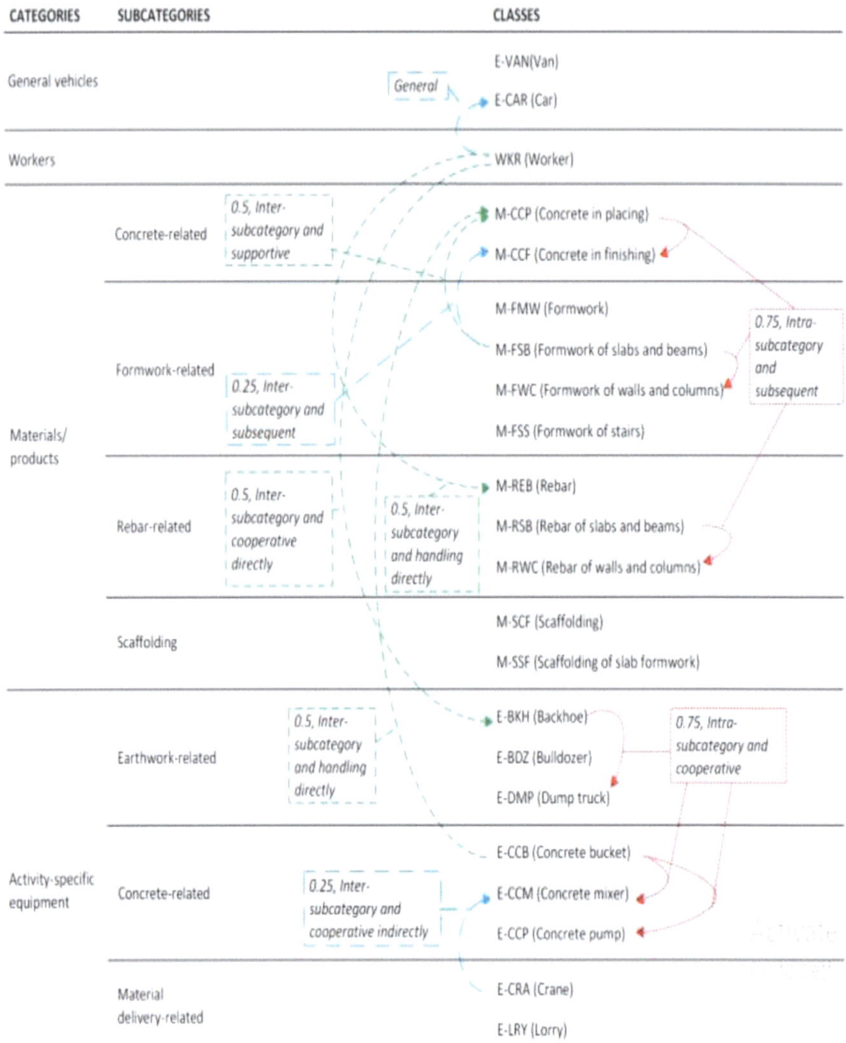

Fig. 4 Flow chart indicating boundaries

3.3 *Relevance Networks and Activity Recognition Patterns*

A circle that has been spotted in an image and whose center conforms to the bounding box in the picture is described by addresses and notes in the relevance network. The relevance score measures the edge width when two relevant objects are connected. Due to the direct correlation between the capacity of workers with decreased activity, this separates worker and non-worker nodes when creating relevant networks. As a result, non-worker objects are first put together in the network and their relevance scores are totaled throughout the object's non-worker object scale and radius. Moreover, worker nodes are gathered into the earlier constructed network [16]. This focuses on the visual identification of construction-related activities, classifying the 20 patterns into 4 divisions based on the combination mode of object patterns' ability to indicate activity. To signal activity directly on equipment right away, the initial activity pattern requires specialized construction equipment. They include concrete mixtures (E-CCM) and concrete pump trucks (E-CCP) that slide and carry information about the progress of placing concrete. In order to demonstrate land leveling, we experimented with using a (E-BDZ) bulldozer and dump truck (E-DMP) combination [17]. In the third section, it is said that concrete is being poured and that, if it is detected, it will alert workers to the impending arrival of concrete laying equipment (M-CCP). The third segment discusses how employees and equipment influence construction-related activities. Next, an excavator will be used to lift the foundation, including pipes and precast slabs, and workers will work as a team to install and put the components. The presence of employees and materials to reflect ongoing activity is included in the fourth component [18] (Fig. 5).

Fig. 5 Combination of ongoing activities identification and detection

4 Experiments and Results

4.1 Detection of Objects Using Evaluation Matrices

Using a precise–recall curve, the experiment gauges object performance. True positives represent the number of successful detections, whereas false positives represent the number of unsuccessful detections. Missed items are depicted as false negatives with the decision estimate of the initial metric having a ratio of TP to TP + FP and the subsequent recall having a ratio of TP to TP+FN. Precision–recall curve r is calculated by setting the recall precision to the maximum precision guaranteed for each recall r0 > r. The experimentation can be carried out by categorizing all of the detections in accordance with the probability value in descending order [19].

4.2 Training Faster R-CNN Sample Heading (Third Level)

The trained dataset, which contains a mini-batch size of 64, a momentum of 0.9, and an initialized learning rate of 0.001, is utilized on the pre-trained ResNet-50 to fine-tune 100,000 iterations. This sorting is reduced to 0.1 times the initial after every 25,000 iterations. The quicker R-CNN model has two groups: one gives the precise location of the bounding box from the original box and the other predicts specific class scores. Figure 6 shows a process that is numerically convergent but has noise from missing annotations and inconsistent training data. Little objects are prone to missing annotation and the uncertain edge of the object is sensitive to inconsistent annotation.

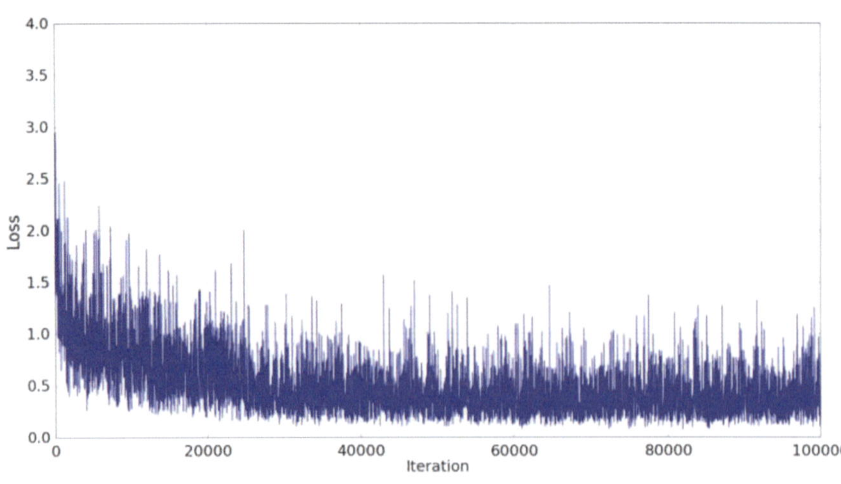

Fig. 6 Training prediction, loss, and score

4.3 Activity Recognition with Evaluation Matrices

Three instances are diagnosed using various approaches to evaluate the efficacy of activity recognition. It displays the discovered items as labels with the classifications of class code, ID, confidence, and relevance networks with construction activity patterns. The class code, ID, and total relevance score are used to categorize non-worker points in relevance networks, with the total relevance score section missing from worker nodes. An initial relevance threshold of 0.25 is used to partition the network. When the value is higher than the relevance, it says that two entities are relevant. In line with the relationship between the non-worker points, it produces the partition into sub-networks and aids in identifying team activities (Figs. 7 and 8).

Part 1. In this section, three activities are named and acknowledged. As the first activity, WKR-1 + M-RWC-1 in Pattern WM7 and WKR-3 + M-RWC-1 in Pattern WM7 are used to produce rebar for walls and columns. The following task is building scaffolding systems at Pattern WM8 under the control of WKR-0 and M-SCF-9. Building formwork for walls and columns using WKR-2 + M-FWC-0 as Pattern WM3 is the next activity. The first and last activities involved teams

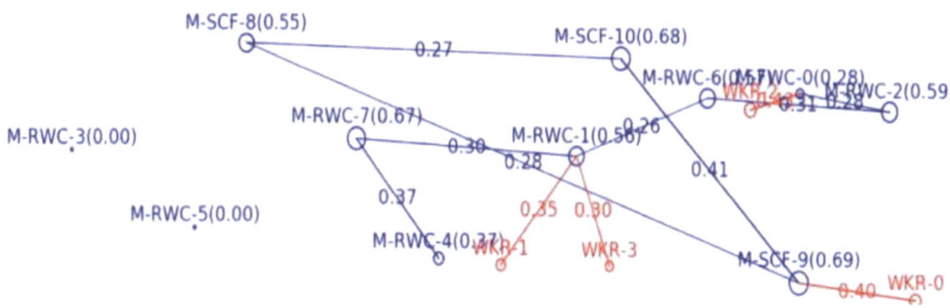

Fig. 7 Activity recognition with evaluation matrices

Fig. 8 Construction activity recognition and class identification with accuracy

building formwork. Together, they make up the M-RWC-10, M-RWC-6, M-RWC-2, and M-RWC-0 non-worker points (Fig. 9).

Part 2. Four employees are discovered by the model, i.e., WKR-0, WKR-1, WKR-2, and WKR-3. All three demonstrate M-FWC-8 relevance. In the end, three activity entities—WKR-1 + FWC-8, WKR-2 + FWC-8, and WKR-3 + FWC-8—are identified. The first activity entails building formwork for walls and columns. In close proximity to M-SCF-11, WKR-0 frequently builds scaffolding structures in accordance with the WM8 Pattern. E-CCM-1, a concrete mixer, was identified. DE3 Pattern matching indicated that concrete placement was about to begin, and erroneous object detection occurred in this section. Container M-FMW-5 was mistakenly discovered since it is a raw formwork product (Fig. 10).

Part 3. Three activities are identified here. Concrete placement is the first activity, and it is accomplished with WKR-0 + E-CCB-0 in Pattern WE2. Due to the lack of activity, this constitutes a false positive detection. Temporarily positioned E-CCB-0

Fig. 9 Determining the worker's activity and framing with accuracy

Fig. 10 Detection of constructing formwork of slabs and beams

is near WKR-0. The following activity is built using a formwork of columns and walls and consists of WKR-4 + M-FWC-5. Building a formwork of slabs and beams is the final activity, which entails three activity entities shared by three workers (WKR-1, WKR-2, and WKR-3) and the output M-FSB-4. Four individuals WKR-1, WKR-2, WKR-3, and WKR-4 could construct formwork as a group in the subnetwork by moving from M-FSB-4 to M-FSB-5.

4.4 Results

Individual actions are sorted into distinct activity kinds based on the findings of the experiment, and operations that were aggregated are disregarded. Further strategies were introduced and tested with IoU for the positive detection and recognition of all activities at 0.5. In the end, the performance of the methods with recall and precision was evaluated. The experiment's results, which comprise 152 true positive detections and 90 false positive detections, including 22 false negative detections, indicate that the proposed approaches have the capacity to identify and detect building activities with 63% accuracy and 87% recall.

4.5 Conclusions and Limitations

In situations where image classification methods are ineffective, this work develops novel techniques for identifying and detecting numerous construction activities carried out by numerous objects in site photos. The methods involve two steps: identifying the activity and detecting the construction-related objects. Modern object detection tools like Faster R-CNN and ResNet-50 are trained to create the object detection problem. This methodology performs well in terms of mAP and the improved AP when compared to the state of the art for object detection, but it exhibits a comparatively high AP variance. Lower APs are ultimately caused by free forms with blurred edges and poorer resolutions. Semantic and geographical relevance were tried for relevance network formation. Semantic relevance displays the likelihood that two things are involved in the same building process. The image points' spatial relevance and 2D pixel distances state the observed potential that is concurrent with the construction activity. Also on basis of relevance networks, a collection of activity patterns is discovered. According to the experiments, a collection of activity pattern was identified. The method's distance limitations result from two-dimensional photographs on a single camera, which might be represented along with 3D physical distances obtained by rotating different camera perspectives. Another drawback is the inability to distinguish between challenging tasks and transitional states when there is a sudden shortage of transitory data within site images. Future analysis of surveillance footage from building sites can be used to create dynamic relevance networks by correlating related things and spotting them throughout the frame.

References

1. He, K, Zhang, X, Ren, S, Sun, J (2016) Deep residual learning for image recognition. http://image-net.org/challenges/LSVRC/2015/. Accessed 08 Mar 2023
2. D'Innocente A, Carlucci FM, Colosi M, Caputo B (2017) Bridging between computer and robot vision through data augmentation: a case study on object recognition. Lecture Notes in Computer Science (including subseries Lecture notes in artificial intelligence and Lecture notes in bioinformatics), vol 10528 LNCS, pp 384–393
3. Bao R, Sadeghi MA, Golparvar-Fard M (2016) Characterizing construction equipment activities in long video sequences of earthmoving operations via kinematic features. In: Construction research congress 2016: old and new construction technologies converge in Historic San Juan - proceedings of the 2016 construction research congress, CRC 2016, pp 849–858
4. Inoue M, Inoue S, Nishida T (2018) Deep recurrent neural network for mobile human activity recognition with high throughput. Artif Life Robot 23(2):173–185
5. Liang M, Hu X (2015) Recurrent convolutional neural network for object recognition, pp 3367–3375
6. Gong J, Caldas CH, Gordon C (2011) Learning and classifying actions of construction workers and equipment using bag-of-video-feature-words and Bayesian network models. Adv Eng Inform 25(4):771–782
7. Seo J, Han S, Lee S, Kim H (2015) Computer vision techniques for construction safety and health monitoring. Adv Eng Inform 29(2):239–251
8. El-Omari S, Moselhi O (2011) Integrating automated data acquisition technologies for progress reporting of construction projects. Autom Constr 20(6):699–705
9. Cheng T, Teizer J (2013) Real-time resource location data collection and visualization technology for construction safety and activity monitoring applications. Autom Constr 34:3–15
10. Yang JB, Nguyen MN, San PP, Li XL, Krishnaswamy S (2015) Deep convolutional neural networks on multichannel time series for human activity recognition. In: Proceedings of the twenty-fourth international joint conference on artificial intelligence
11. Rashid KM, Behzadan AH (2017) Risk behavior-based trajectory prediction for construction site safety monitoring. J Constr Eng Manag 144(2):04017106
12. Ergen E, Akinci B, East B, Kirby J (2007) Tracking components and maintenance history within a facility utilizing radio frequency identification technology. J Comput Civ Eng 21(1):11–20
13. Song L, Eldin NN (2012) Adaptive real-time tracking and simulation of heavy construction operations for look-ahead scheduling. Autom Constr 27:32–39
14. White DJ, Alhasan A, Vennapusa P (2023) Conference on autonomous and robotic construction of infrastructure.www.intrans.iastate.edu. Accessed 08 Mar 2023
15. Vepakomma P, De D, Das SK, Bhansali S (2015) A-wristocracy: deep learning on wrist-worn sensing for recognition of user complex activities. In: 2015 IEEE 12th international conference on wearable and implantable body sensor networks, BSN 2015
16. Walse KH, Dharaskar RV, Thakare VM (2016) PCA based optimal ANN classifiers for human activity recognition using mobile sensors data. In: Smart innovation, systems and technologies, vol 50, pp 429–436
17. Hammerla NY, Halloran S, Plötz T (2016) Deep, convolutional, and recurrent models for human activity recognition using wearables. In: ijcai international joint conference on artificial intelligence, vol 2016, pp 1533–1540
18. Bengio Y (2013) Deep learning of representations: looking forward. Lecture notes in computer science (including subseries Lecture notes in artificial intelligence and Lecture notes in bioinformatics), vol 7978 LNAI, pp 1–37
19. Zeng M et al (2015) Convolutional neural networks for human activity recognition using mobile sensors. In: Proceedings of the 2014 6th international conference on mobile computing, applications and services, MobiCASE 2014, pp 197–205
20. Kim J, Chi S, Seo J (2018) Interaction analysis for vision-based activity identification of earthmoving excavators and dump trucks. Autom Constr 87:297–308

21. Chari V, Sturm P (2009) Multiple-view geometry of the refractive plane. In: British machine vision conference, BMVC 2009 - proceedings, pp 1–11

Novel Approach for Medical Image Security Using Image Encryption and Watermarking

S. Aruna Deepthi, K. Deepthi, C. H. Neetu, K. R. Deepthi, and Y. Harshith

Abstract Medical images are a very vital aspect of medical practice and are required for the detection of several ailments and conditions that could not be diagnosed with the naked eye. Every pixel in the image can be required for the diagnosis procedure, and any distortion could lead to a faulty diagnosis. The healthcare industry has made significant advancements, allowing medical images to be shared and transferred across various regions. As medical images are transferred over public networks, there is a possibility of various security issues arising. These images are often stored in third-party cloud servers, and health information systems contain sensitive patient information. In this paper, we have combined the techniques of image cryptography and watermarking techniques to build a secure model for protecting medical images with the goal of achieving full recoverability and process reversibility. In the cryptography part, two methods for image encryption are proposed which use XOR operation for both encryption and decryption processes. Both methods use a common concept of using an image of the same size as a key for encrypting the original image and these methods are implemented in MATLAB.

Keywords PBE · AES · DES · Encryption · Watermarking · Bitplane encryption · XOR method

S. Aruna Deepthi (✉) · K. Deepthi · C. H. Neetu · K. R. Deepthi · Y. Harshith
Department of ECE, Vasavi College of Engineering, Hyderabad, Telangana, India
e-mail: sadeepthi@staff.vce.ac.in

K. Deepthi
e-mail: deepti@staff.vce.ac.in

C. H. Neetu
e-mail: ch.neetu@staff.vce.ac.in

A. P. Agrawal et al. (eds.), *Advances of Artificial Intelligence and Machine Learning in Image Processing, Security and Data Analytics*, Lecture Notes in Electrical Engineering 1464, https://doi.org/10.1007/978-981-95-0148-9_16

1 Introduction

In today's digital world, we heavily rely on technology for storing and transmitting data. This data can be personal, financial, or sensitive in nature. As a result of the rapid development of communication networks, additional security and privacy issues have recently emerged. The need for digital security arises from the fact that as technology advances, so do the threats and vulnerabilities that can be exploited to obtain access to this data. Cyberattacks, data breaches, [1] and identity theft are just a few examples of the consequences of weak digital security. It is critical that image and video material transmission be safe and reliable, not to mention any other pertinent issues that should be considered. Images are currently regarded as the most significant source of information. They can contain sensitive information about individuals, organizations, or even nations. If these images [2] are not secured, it can lead to severe consequences like identity theft, blackmailing, leaking of confidential information, cyberbullying, and even national security breaches.

Every pixel in the image can be required for the diagnosis procedure, and any distortion could lead to a faulty diagnosis. Medical images are a very vital aspect of medical practice, and are required for the detection of several ailments and conditions that could not be diagnosed with the naked eye. This makes them valuable targets for cyber-attackers, who can use this information for identity theft, financial fraud, or even extortion. Cryptographic techniques can provide a strong security system for medical images by ensuring the confidentiality, integrity, and authenticity of the data. By encrypting the images using a secure key, access to the data can be restricted only to authorized users. Additionally, the use of watermarking technique can ensure that the data has not been altered.

A secure cryptographic system can prevent any potential damage to the medical system, which may have serious consequences for the patients. Cryptography can also provide a secure and efficient method of sharing [3] medical images among healthcare professionals without compromising patient privacy. Therefore, cryptographic techniques are an essential tool for ensuring the safety and privacy of medical images and maintaining the integrity of the medical system.

As a result, it is critical to use strong cryptographic techniques that can provide effective security measures for medical images, while also ensuring their accessibility and usability for medical professionals [4] who need to view and analyze them. Real-time and security requirements for the encryption technique will be more[5, 6] demanding for this particular picture[7] encryption. Also, image and video data require a separate encryption approach.

2 Literature Survey

In the past few years, the digital transformation of data has seen a remarkable growth across diverse domains, such as medicine, driven by technological advancements and the convenience of storing, exchanging, and analyzing digital information. And post-COVID, the healthcare systems have also been digitalized to a huge extent [8] with medical images being transferred over the Internet. The need for medical image encryption arises due to the sensitive nature of medical images and the need to protect patient privacy. In this paper, we are using a symmetric [9] approach to secure the medical images. This is done using a key image. We also used two methods, one of which is the pixel-based approach. Another method uses the concept of bit planes to encrypt the image. A random [10] shuffling process is done to get a scrambled image. These techniques are used to provide a more randomized image that is difficult to crack. Also, the concept of watermarking based on DCT is used to add authentication to the encryption process.

A survey and analysis of the image encryption methods [11] have been implemented in which various basic ciphers like symmetric encryption techniques like RC4 cipher, Stream Cipher, Block Cipher, Triple-DES, etc. and also asymmetric techniques like ECC and also chaotic systems were discussed.

Steganography technique [12] that uses AES and IDEA algorithms to convert hidden messages into ciphertext and embed them into a stego image is discussed. The encrypted stego image is then encrypted again using the same algorithms, providing two layers of protection against attacks. Image evaluation metrics such [13] as PSNR and SSIM are also used to evaluate the stego images.

The requirement of image encryption in the E-health services [14] and most common attacks are observed. A deep insight on different spatial domain-based encryption techniques like using chaotic maps, ECC, and DNA provides comparisons between these approaches. A Chaotic ARX Block Cipher [8] explores the creation of a new lightweight picture encryption system that employs the Lorenz 3D superchaotic map and an addition-rotation-XOR block cipher. The chaotic addition-rotation-XOR block image cipher proposed is found to be effective in terms of speed and randomness.

The information on three different types of encryption techniques [15, 16] which are Block Cipher Image Encryption using AES method, encryption algorithm based on chaotic map using the concept of Arnold map, and a hybrid chaos approach using logistic maps to enhance the chaotic system's security.

An encryption model was [17] proposed for medical images using chaotic logistic and tent maps. The security was achieved through diffusion and confusion processes based on the C-function. Adaptive grasshopper optimization algorithm was utilized to select key to the system. The encryption was evaluated using PSNR and correlation coefficient fitness function.

The novel approaches for image encryption [18] using a key image and provides analysis on the approaches. The paper contains three different methods which are Key Bitplane Encryption, Key RGB Displacement Encryption, and Key Scan approach, shows that using a key image for encryption can give good results and the different approaches using key image.

This work presents a selective bitplane [19] encryption approach for securing image data transfer in mobile environments. The encryption scheme's security is assessed by examining several bitplane features and comparing the encryption of normal and histogram equalized enhanced images.

The paper discusses that [20] watermarking can give high security to an image also providing authentication. This is also reliable since it results in a completely reversible encryption which is apt for medical [21] images due to the highly sensitive data contained in medical images. A reliable and [3] blind watermarking method is discussed [4] to secure medical images used in telemedicine. The method uses three transformations to extract the frequency content of the image and Schur decomposition to modify the upper triangular matrix values.

3 Proposed Method

Access to health information system data can pose a serious threat to patient privacy and security, particularly in the case of medical images. Unauthorized access to these images can undermine the medical system's integrity and put patients at risk. As a result, several inquiries are being conducted to strengthen the security of medical images and guarantee their confidentiality. Medical images have unique properties that require a different kind of encryption than other digital images, making it necessary to develop specialized encryption techniques for this purpose. Therefore, traditional encryption techniques like Triple-DES, Data Encryption Standard (DES), International Data Encryption Algorithm (IDEA), and Advanced Encryption Standard (AES) are not appropriate for securing medical images.

In this context, the current work proposes an optimal encryption technique [12] based on the usage of the image as a key. This approach is particularly relevant for medical images, as patients often need to share these images online with [22] only a select group of people, and need to protect them from various kinds of attacks. This approach can help to improve the overall security of the medical system and protect patient privacy.

All the proposed methods use an image as a key for encrypting a given input image. The image used as [23] key may or may not be of same size as the plain image. The key image is reshaped according to the dimensions of the plain image provided as the input. These encryption methods are symmetric encryption algorithms, and hence the same key image is used for the decryption process as well. The XOR operation plays a major role in both encryption and decryption processes. Along with encryption, watermarking approach is also introduced in the proposed methods. The watermarking process is performed in spatial domain and its main objective is to provide [24] authentication.

The entire process of the proposed approach is described briefly in the block diagram as shown in Fig. 1.

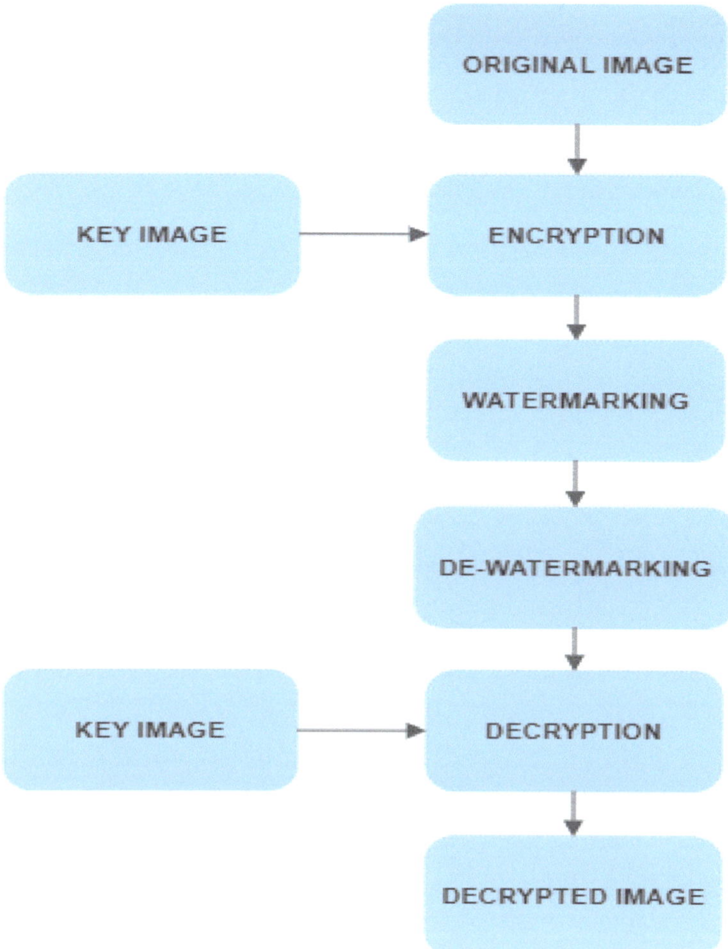

Fig. 1 Proposed block diagram

The encryption is symmetric and the watermarking process is applied on the encrypted image produced after the encryption process. Then, the watermarked image undergoes de-watermarking. The decryption using the same key image provided during the encryption process is done on the de-watermarked image to get back the decrypted image. The proposed system aims to address the security concerns related to the storage and sharing of medical images in third-party cloud-based systems. By providing effective confidentiality and authentication services, the system ensures that patient data is secure and protected from unauthorized access or malicious attacks. This is particularly important in the healthcare industry, where patient privacy and confidentiality are paramount.

4 Results

The proposed methods are implemented in MATLAB R2022a with I5 processor and 8 GB RAM. The test images are taken from National Library of Medicine (NIH) database and are of different sizes. They include XRAY, CT scans, etc. For all the methods, same key image is used as given in Fig. 2.

4.1 Method 1: Pixel XOR Method

Figure 3 shows the implementation of Method 1: Pixel XOR method. In that figure, (a) is the original input image, (b) corresponds to the encrypted image, (c) corresponds to the watermarked image, and finally (d) is the decrypted image.

Fig. 2 Key image

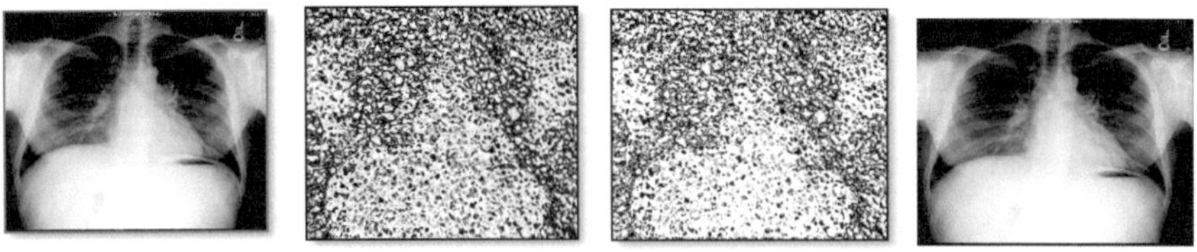

a)Original Image b)Encrypted Image c)Watermarked Image Decrypted Image

Fig. 3 Method 1 implementation

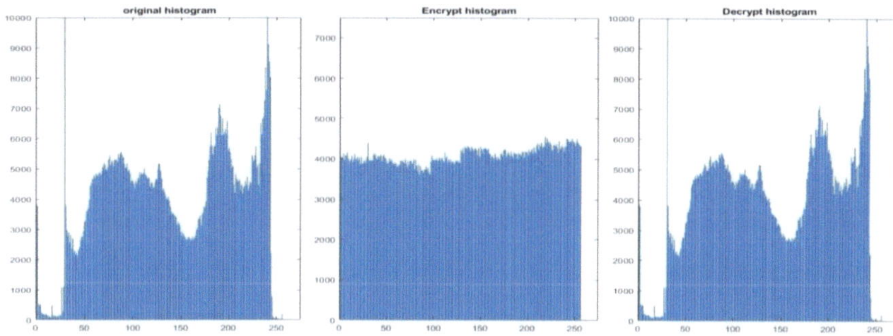

a)Original histogram b)Encrypted histogram c)Decrypted histogram

Fig. 4 Histogram analysis of method 1

Figure 4 shows the histograms of the (a) original input image, (b) encrypted image, and (c) decrypted image. As seen in the figure, the encrypted image's histogram is uniform, indicating that it does not provide any statistical information to attackers. Additionally, the histograms of the original and the decrypted photos coincide, demonstrating that there was no loss in the decoded image.

4.2 Method 2: Bitplane XOR Method

Bitplane XOR method is subdivided into two methods: (a) partial bitplane encryption (PBE) and (b) full bitplane encryption (FBE). In partial bitplane encryption (PBE) method, implementation of (2-b) PBE method is shown in Fig. 5, where (a) is the original input image, (b) corresponds to the encrypted image, (c) corresponds to the watermarked image, and finally (d) is the decrypted image. In this method, only four bitplanes of the input image are selected for encryption, and the rest of the bit planes are left untouched.

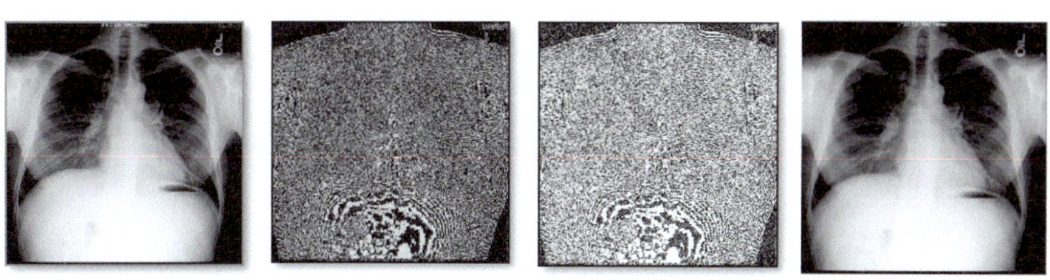

Original image Encrypted image Watermarked image Decrypted image

Fig. 5 PBE method implementation

Original image Encrypted image Watermarked image Decrypted image

Fig. 6 FBE method implementation

Full bitplane encryption (FBE) method implementation of (2-b) FBE method is shown in Fig. 6. Here, all the bitplanes of the input image are selected for encryption process.

4.3 *Performance Parameters*

Various measures such as peak signal-to-noise ratio, entropy, histogram analysis, number of pixels change rate (NPCR), and structural similarity index measure (SSIM) are used to evaluate the efficiency of the proposed medical image encryption method. We have taken four test images to compare the parameters between various methods proposed (Fig. 7).

Peak Signal-to-Noise Ratio (PSNR)

It is a commonly used metric in image encryption to evaluate the quality of the decrypted image. It calculates the ratio of a signal's maximal achievable power to the power of corrupting noise that influences the fidelity of its representation. The PSNR between encrypted and actual input images should be as low as possible. Table 1 compares the performance of the proposed methods in terms of PSNR between encrypted and real images, as shown in Tables 2 and 3.

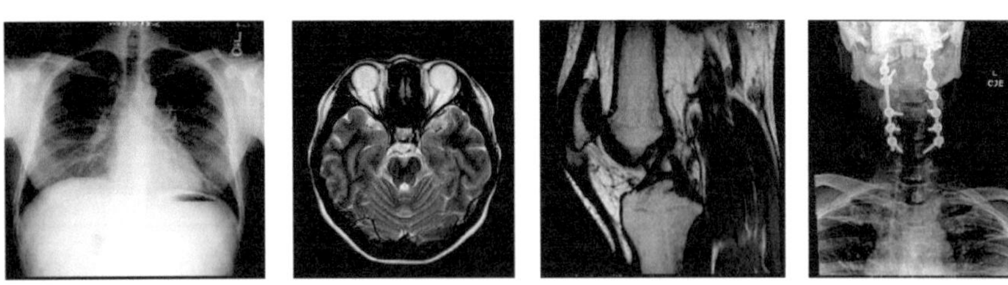

Img-1 Img-2 Img-3 Img-4

Fig. 7 Test images

Table 1 PSNR values between encrypted and actual images

Test image name	Method 1: pixel XOR method (dB)	Method 2: bitplane XOR method	
		(a) PBE (dB)	(b) FBE (dB)
Img-1	8.15	7.91	7.86
Img-2	7.47	7.17	7.08
Img-3	7.1	6.97	6.76
Img-4	7.24	7.09	7.07

Table 2 Entropy values of encrypted image

Test image name	Method 1: pixel XOR method	Method 2: bitplane XOR method	
		(a) PBE	(b) FBE
Img-1	7.86	7.78	7.85
Img-2	7.54	7.62	7.71
Img-3	7.28	7.51	7.58
Img-4	7.39	7.57	7.69

Table 3 SSIM values between original and encrypted images

Test image name	Method 1: pixel XOR method	Method 2: bitplane XOR method	
		(a) PBE	(b) FBE
Img-1	0.0534	0.0182	0.0143
Img-2	0.0421	0.00987	0.00902
Img-3	0.0335	0.0114	0.00921
Img-4	0.0475	0.0108	0.00899

Structural Similarity Index Measure (SSIM)

It is a widely used metric for measuring the similarity between two images. The SSIM index compares the structural information of the two images being compared. In particular, it measures the similarity between the luminance, contrast, and structure of the images. The index ranges from -1 to 1, with a value of 1 indicating that the two images are identical.

Number of Pixels Change Rate (NPCR)

The NPCR measures the extent to which an encrypted image changes when a single pixel in the original image is changed. A high NPCR value indicates that even a small change in the original image leads to a significant change in the encrypted image, which is desirable for cryptographic security. Table 4 compares the NPCR values for the proposed methods.

Pixel XOR method is implemented on a test image and the encrypted, watermarked, and decrypted images are observed. Histograms are plotted for the input

Table 4 NPCR values between original and encrypted images

Test image name	Method 1: pixel XOR method	Method 2: bitplane XOR method	
		(a) PBE	(b) FBE
Img-1	99.127	99.274	99.325
Img-2	98.453	99.481	99.548
Img-3	99.076	98.827	99.137
Img-4	97.983	98.952	98.989

original image, encrypted image, decrypted image, and are analyzed based on the distribution of various grayscale levels. Then using the same key image, bitplane XOR methods, i.e., (a) Partial Bitplane Encryption (PBE) and (b) Full Bitplane Encryption (FBE) methods are implemented on the test image in same manner. In PBE method, only four bit planes are selected for encryption while, in FBE method, all the bit planes are selected for encryption process. Here also, histograms are taken for original, encrypted, and decrypted images and are analyzed. Second part of the paper is about observing the performance parameters for all the methods. Various statistical parameters are calculated for a set of test images for all the proposed methods. The observed parameters for all the methods are then compared with each other. On the basis of performance metrics, Full Bitplane Encryption (FBE) method has got the best results. Also, Partial Bitplane Encryption (PBE) approach produced equivalent results to FBE method while requiring less computing.

5 Conclusion

In this paper, a robust security model has been proposed to safeguard medical images which include two encryption algorithms based on the concept of key image and a reversible watermarking approach. Image encryption will ensure that the medical image remains safe from unauthorized access and data breaches. The two encryption methods namely, pixel XOR method and bitplane XOR method are implemented in MATLAB platform and good results were obtained. Various performance parameters are evaluated and analyzed for both the methods. The performance metrics evaluated for two methods are compared and the findings revealed that (Method 2b) Full Bitplane Encryption (FBE) yielded the best results in terms of security and (Method 2a) Partial Bitplane Encryption (PBE) offered a comparable level of security to FBE while maintaining a balance between encryption strength and computational efficiency. In addition to image encryption, watermarking is employed to embed a unique watermark into the encrypted image. The watermarking process makes it possible to verify the authenticity of the image and protect it against any unauthorized tampering. The proposed model has successfully attained full recoverability and reversibility, ensuring that the original medical image can be reconstructed from the encrypted and watermarked image with no loss of essential information.

6 Future Scope

Image encryption algorithms using key images have a lot of potential for future development and improvement. We can explore the integration of bitplane-based image encryption algorithms with artificial intelligence (AI) and machine learning (ML) techniques to enhance the security and efficiency of the encryption process.

We can explore the development of hybrid encryption schemes that combine bitplane-based image encryption algorithms using key images with other encryption techniques such as Elliptic Curve Cryptography (ECC) for enhanced security and efficiency for various applications. Key image-based encryption algorithms can be integrated with block chain technology to enhance the security and traceability of the encryption process. This can enable secure and tamper-proof storage and sharing of encrypted images on a decentralized network.

Acknowledgements The authors would like to thank the authorities of Vasavi College of Engineering for allowing access to resources to complete this paper in college and declare there is no conflict of interest and no funding is received.

References

1. Chavali S, Kandavalli C, Sugash T, Gopalakrishnan P (2022) Comparative study of image encryption and image steganography using cryptographic algorithms and image evaluation metrics. https://doi.org/10.1007/978-981-19-7126-6_22
2. Priyanka, Singh A (2022) A survey of image encryption for healthcare applications. Evol Intell. https://doi.org/10.1007/s12065-021-00683-x
3. Gosul M, Gandhewar N (2022) A reversible digital image watermarking technique for integrity and authenticity of MR images. Int J Eng Res Comput Sci Eng 9:41–46. https://doi.org/10.36647/IJERCSE/09.09.Art012
4. Ajili S, Hajjaji MA, Mtibaa A (2021) Crypto-watermarking algorithm using weber's law and AES: a view to transfer safe medical image. Sci Program 2021:5559191. https://doi.org/10.1155/2021/5559191
5. Chaudhary N, Shahi T, Neupane A (2022) Secure image encryption using chaotic, hybrid chaotic and block cipher approach. J Imaging 8:167. https://doi.org/10.3390/jimaging8060167
6. Ratan R, Yadav A (2021) Security analysis of bit plane level image encryption schemes. Defense Sci J 71. https://doi.org/10.14429/dsj.71.15643
7. Mohananthini N, Mohamed Parvees MY, Abdul Samath J (2020) Lightweight image encryption: a chaotic ARX block cipher. J Circuits Syst Comput 30:2150026
8. Shankar K, Elhoseny M, Chelvi E, Lakshmanaprabu SK, Wu W (2018) An efficient optimal key based chaos function for medical image security. IEEE Access 1. https://doi.org/10.1109/ACCESS2018.2874026
9. Mishra S, Dastidar A (2018) Hybrid image encryption and decryption using cryptography and watermarking technique for high security applications. In: International conference on current trends towards converging technologies (ICCTCT), Coimbatore, India, March 2018, pp 1–5. https://doi.org/10.1109/ICCTCT.2018.8551103
10. Mohammad OF, Shafry M, Rahim M, Rafeeq S, Zeebaree M, Ahmed F (2017) A survey and analysis of the image encryption methods. Int J Appl Eng Res 12:13265–13280

11. Kankonkar JTG, Naik N (2017) Image security using image encryption and image stitching. In: International conference on computing methodologies and communication (ICCMC), Erode, India, July 2017, pp 151–154, https://doi.org/10.1109/ICCMC.2017.8282663
12. Kester Q-A, Nana L, Pascu A, Gire S, Eghan MJ, Quaynor N (2015) A cryptographic technique for security of medical images in health information systems. Procedia Comput Sci. Elsevier. https://doi.org/10.1016/j.procs.2015.08.070
13. Madhu S, Hussain MA (2014) Securing medical images by image encryption using key image. Int J Comput Appl
14. Zhou Y, Cao W, Chen C (2014) Image encryption using binary bitplane, Signal Process 100:197–207. https://doi.org/10.1016/j.sigpro.2014.01.020
15. Mahmood A, Hamed T, Dony R (2013) Improving the security of the medical images. Int J Adv Comput Sci Appl 4. https://doi.org/10.14569/IJACSA.2013.040922
16. Magdy M, Hosny KM, Ghali NI, Ghoniemy S (2022) Security of medical images for telemedicine: a systematic review. Multimed Tools Appl 81:25101–25145. https://doi.org/10.1007/s11042-022-11956-7
17. De' R, Pandey N, Pal A (2020) Impact of digital surge during covid-19 pandemic: a viewpoint on research and practice. Int J Inf Manag 55:102171. https://doi.org/10.1016/j.ijinfomgt.2020.102171
18. Xue X, Jin H, Zhou D, Zhou C (2021) Medical image protection algorithm based on deoxyribonucleic acid chain of dynamic length. Front Genet 12:654663. https://doi.org/10.3389/fgene.2021.654663
19. Kasim Ö (2022) Secure medical image encryption with Walsh-Hadamard transform and lightweight cryptography algorithm. Med Biol Eng Comput 60(6):1585–1594. https://doi.org/10.1007/s11517-022-02565-5
20. Fang P, Liu H, Wu C, Liu M (2022) A survey of image encryption algorithms based on chaotic system. Vis Comput 39(11):1975–2003. https://doi.org/10.1007/s00371-022-02459-5
21. Magdy M, Hosny KM, Ghali NI, Ghoniemy S (2022) Security of medical images for telemedicine: a systematic review. Multimed Tools Appl 81:25101–25145
22. Li D, Li J, Bhatti UA, Nawaz SA, Liu J, Chen Y-W, Cao L (2023) Hybrid encrypted watermarking algorithm for medical images based on DCT and improved DarkNet53. Electronics 12:1554. https://doi.org/10.3390/electronics12071554
23. Raghunandan KR, Dodmane R, Bhavya K, Rao NSK, Sahu AK (2023) Chaotic-map based encryption for 3D point and 3D mesh fog data in edge computing. IEEE Access 11:3545–3554
24. Chikyal N, Veera Swamy K (2019) Performance assessment of various thyroid image segmentation techniques with consistency verification. J Adv Res Dyn Control Syst 11(02)

Advances in Fall Detection Using Artificial Intelligence

Tuhina Raj, Pallavi Gupta, and Manisha Rajoria

Abstract Falls have been known to cause serious concerns among the human beings, especially when it comes to elderly, children, and people working on construction sites or living in high-rise buildings. Falls can lead to serious injuries, long hospital stays, rehabilitation, mental trauma as well as death. With the advancements in the fall detection technology through sensors, there has been an increase in the safety as well as in detection of these falls within an appropriate time to reduce the complexities. This paper elaborates the role of Artificial Intelligence (AI) in enhancing the fall detection system more accurate as well as effective by reducing the False Alarm Rate (FAR), missed events, latency, etc. The paper also reviews the common machine learning algorithms used for the detection of fall events as well as compare the parameters for the system with or without these algorithms.

Keywords Fall detection · Sensors · Artificial intelligence · False alarm rate · Machine learning algorithms

1 Introduction

A fall is characterized as an occurrence that causes a person to inadvertently come to rest on the ground, a floor, or a lower level [1]. Traumatic brain injuries as well as fractures of the wrist, arm, ankle, and hip are fall-related injuries. Falls are a significant public health issue, particularly for the elderly. They may culminate in catastrophic injuries like hip fractures, head traumatic events, and loss of mobility.

In fact, according to data from the World Health Organization, between 28 and 35% of adults over 65 fall each year. Even worse, for persons aged 70 and older, these

T. Raj · P. Gupta (✉) · M. Rajoria
Department of Electrical Electronics and Communication Engineering, Sharda University, Greater Noida, UP, India
e-mail: pallavi.gupta2@sharda.ac.in

M. Rajoria
e-mail: manisha.rajoriya@sharda.ac.in

A. P. Agrawal et al. (eds.), *Advances of Artificial Intelligence and Machine Learning in Image Processing, Security and Data Analytics*, Lecture Notes in Electrical Engineering 1464, https://doi.org/10.1007/978-981-95-0148-9_17

percentages rise to as high as 32–42%. These findings additionally show that when age and fragility increase, the frequency of falls increases. It is widely believed that the overall number of injuries brought on by falls would double by 2030 if no preventive strategies are investigated due to the steadily rising average life expectancy. Because of all of this, automatic fall detection is a vital topic for research [2].

There are many intrinsic risk factors for falls, including neurological deterioration, the presence of chronic diseases, problems with balance, and compromised sensory capacities like vision and hearing as well as depression [3, 4]. The extrinsic factors such as poor lightings and uneven surfaces also account for large number of trigger events for falls [5, 6].

The presence of someone else for support and personal care is one of the alternatives. However, it is physically impossible for someone to constantly be there with an elderly person [7]. Apart from elderly continuous news and articles have been flooded with daily reports of children falling from high-rise buildings, or falls due to extrinsic or intrinsic factors. With the advancements in the technology, sensors are being used to develop a system where the fall is detected and an appropriate measure is taken to alert the fall to the care-takers. Depending on the data acquisition the sensors could be used for: wearables, ambient and vision-based approaches [8].

According to these three criteria, the performance of the systems is sometimes measured in terms of accuracy (AC), specificity (SP) and sensitivity (SE), or precision (PR) and recall (RE) [9]. The selection of the placement of sensors varies as per the choices taking in account the factors such as the environmental factors, user's choice, cost dependency, nature of output, size, and weight factors. [10].

Recently the data from the sensors are being integrated with different machine learning algorithms to produce a much better result.

Artificial intelligence (AI) with sensor integration has several benefits, including as more substantial accuracy and dependability, reduced responsibilities for human operators, increased productivity and efficiency, and enhanced decision-making [11]. For instance, AI-powered sensors are used in healthcare to track patients' vital signs, in manufacturing to spot disease early on, and in transportation to enhance traffic flow.

Machine learning algorithms can reliably differentiate between seasonal events and routine activities. These models learn how to recognize distinct patterns in sensor data associated with falls by training on an assortment of datasets that include various falling scenarios [12, 13]. As a result, there are fewer false alarms, resulting in very accurate detection of actual fall accidents.

This ongoing analysis ensures that falls are identified as they happen, enabling rapid reactions that are essential in real-time scenarios allows for merging of many sensor modalities [14].

The study examines how artificial intelligence (AI) might help fall detection systems become more accurate and efficient by tackling problems with latency, missed events, and false alarm rate (FAR). For fall event detection, it examines many machine learning techniques, such as support vector machines (SVMs), recurrent neural networks (RNNs), decision trees (DTs), and convolutional neural networks (CNNs). An in-depth examination of the literature is part of the technique and databases like IEEE Xplore, ScienceDirect, Scopus, and Google Scholar. Peer-reviewed English publications on AI-assisted fall detection are the main emphasis of the inclusion criteria, while duplicates, reviews, surveys, and unrelated topics are eliminated by the exclusion criteria. The study uses data from a subset of papers, including its primary findings and restrictions as well as the title, authors, publication year, sensor kinds, AI approaches, and performance parameters. The data is synthesized using a thematic approach, which highlights patterns and trends. It covers the benefits and drawbacks of various sensors and AI methods for fall detection, as well as highlights difficulties and proposing areas for further investigation.

2 Algorithms Used in Camera-Based Sensors

Systems for detecting falls using vision rely significantly on machine learning methods. These systems are made to identify and monitor incidents when people fall, especially in places like residences, healthcare facilities, or assisted living facilities [15]. Deep learning's recent impact on computer vision has impacted the field, boosting performance in several pertinent tasks like object detection, segmentation, and picture captioning. Few of the algorithms have been discussed below.

2.1 Convolutional Neural Network (CNN)

A potential approach for creating non-invasive and contactless fall detection devices is vision-based fall detection utilizing CNN. A deep learning algorithm known as a CNN can be trained to extract useful information from pictures and videos. With an emphasis on detecting human individuals [16], these networks are trained to recognize and distinguish between distinct objects and features inside the video frames. The system needs to establish a baseline for human presence and movement in the environment, therefore this first step is essential. High accuracy has been established for CNN-vision-based fall detection systems in real-world situations [17, 18]. Figure 1 depicts the algorithm for CNN in vision-based systems.

There are several limitations in developing CNN-based vision-based fall detection systems. One significant barrier is obstruction, which makes it difficult for CNN to correctly identify the subject of surveillance if they are partially or entirely hidden by nearby objects [19]. Additionally, because CNNs heavily rely on training data, if the dataset has an orientation toward types of falls or scenarios, the system may find it difficult to generalize to a variety of unexpected situations, decreasing its overall accuracy as well as efficacy in real-world applications.

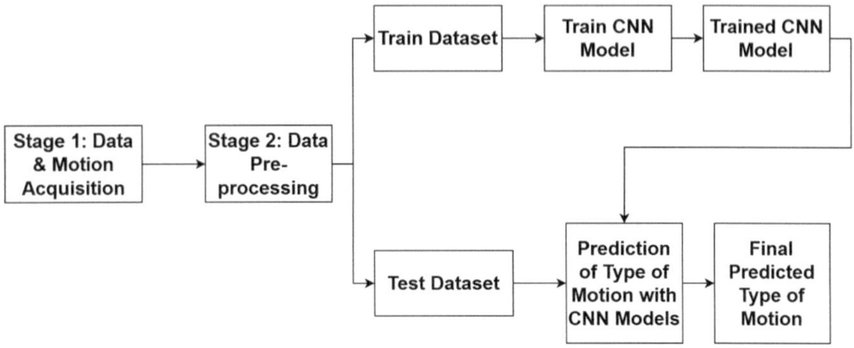

Fig. 1 Process of CNN in vision-based system

2.2 Recurrent Neural Networks (RNNs)

Recurrent neural networks (RNNs) are an innovative method for vision-based fall detection that makes use of sequential data from video footage to rapidly detect and deal with fall accidents. Using cameras or depth sensors, this technique entails recording video data, which is then processed using an RNN framework like Long Short-Term Memory (LSTM) [19] or Gated Recurrent Unit (GRU). The RNN can recognize abrupt and abnormal changes in an individual's stance or movements indicative of a fall occurrence by constantly analyzing the video stream frame by frame [20].

Although recurrent neural networks (RNNs) have shown potential in vision-based fall detection, they have some drawbacks that must be taken into account. The first issue is that because of their computational complexity, training and deployment can be resource intensive, especially when dealing with enormous and complex datasets [21]. Second, RNNs' resilience may be impacted by noisy input data, such as poor-quality video feeds or circumstances in which the person's fall is slightly hidden.

2.3 You Only Look Once (YOLO)

The innovative object identification algorithm YOLO (you only look once) is ideal for vision-based fall detection. Since YOLO is a single-stage detector, it can identify objects in a picture in just one pass. Because of this, YOLO is extremely quick and effective, which is crucial for real-time fall detection systems. An image is divided into a grid of cells in YOLO [22]. By training YOLO on a collection of photographs and videos of people falling, it can be utilized for vision-based fall detection with better accuracy and robustness [23].

The YOLO model may fail to effectively detect falls, potentially leading to false negatives [24] or decreased dependability in real-world applications. This is because the training dataset may not fully represent the range of fall occurrences and scenarios that are frequently encountered in real-world settings.

2.4 Random Forests and Decision Trees

Random forests and decision trees can be used to categorize video frames as either including a fall or not in vision-based fall detection [25, 26]. For the algorithm to accomplish this, a dataset of video frames that have been labeled as either including a fall or not is used to train the algorithms. The algorithms can be used to identify new video frames and detect falls in real time after being trained. Because decision trees and random forests are generally comprehensible machine learning algorithms, it is possible for one to understand how the algorithms anticipate the future [27]. This can help with fall detection system troubleshooting and performance optimization.

Random forest classifier relies on Gini index to select its attributes. Equation of Gini index can be written as follows:

$$\mathrm{Gini}(p) = 1 - \sum\nolimits_{k=1}^{c} p(k|p)^2 \tag{1}$$

where p(k|p) is the probability that the selected cases related to the class *k* [25].

While useful in many machine learning applications, random forests and decision trees do have certain drawbacks when used for vision-based fall detection. In the case when working with large and complicated datasets, their computational complexity can become a problem. Real-time fall detection may not always be practical due to the high computational resource requirements of the training process [28].

Table 1 demonstrates the comparison for the above-mentioned algorithms.

Table 1 Comparison chart for the above algorithms based on vision sensors

Algorithm	False alarm rate (%)	Accuracy (%)	Sensitivity (%)	Latency (ms)
CNN	1–5	95–99	95–99	10–100
RNN	5–10	90–95	90–95	100–200
YOLO	10–15	90–95	90–95	5–10
Random forest	5–10	95–99	95–99	100–200

3 Algorithms Used in RADAR-Based Sensors

Real-time fall detection can be achieved with contactless, non-invasive technology such as radar-based sensors. They function by sending out electromagnetic waves and measuring the echoes that are returned from the surrounding objects. Compared to other fall detection methods like wearable technology and pressure mats, radar-based sensors have several advantages. They are less intrusive and simpler to install because they can be mounted in a room's walls or ceiling. Additionally, they are less likely to be set off by unfounded alarms like moving furniture or pets.

The machine learning algorithms helps in achieving better performance, few of the algorithms have been discussed below:

3.1 Support Vector Machines (SVM)

Support Vector Machines (SVMs) are a subclass of supervised machine learning method that may be applied to several tasks, including regression and classification. SVMs split the data points into two classes with the greatest possible margin by locating a hyperplane in the feature space [29]. Radar-based fall detection systems track a person's motions and look for signs that they have fallen using radar sensors. The height, speed, and acceleration of the individual can be extracted from the radar data together with other parameters that are important for fall detection [30, 31].

Radar data that has been labeled with fall and non-fall occurrences can be used to train SVM classifiers. The SVM classifier can be used to categorize fresh radar data points as fall or non-fall events once it has been trained [32].

Radar fall detection systems based on SVM have several intrinsic limitations. First and foremost, they are prone to false positives—predicting a fall event when none has occurred—which are frequently due to noise in the radar data, shifting ambient circumstances, or the presence of nearby people or objects. Secondly, due to factors like the fall not matching the capabilities of the radar sensor or the person being too far from the sensors, these systems may also cause false negatives, failing to identify a true fall [33].

3.2 Hidden Markov Model (HMM)

Sequential data can be modeled using hidden Markov models (HMMs), an efficient statistical tool. As a fall may be characterized as a series of multiple states, such as standing, falling, and resting on the ground, this makes them well suited for fall detection. By identifying the most likely state sequence for the new data, the HMM may be used to detect falls in new radar data once it has been trained. The Viterbi algorithm can be used to do this [34].

The HMM-based fall detection method has a few drawbacks. It requires large resources to train and instantaneous deployment because of its computational complexity. It also causes false alerts since it is susceptible to noise and clutter in radar data [35]. The accuracy significantly depends on the design of the HMM and the quality of the training data; poorly configured HMMs or bad data can produce inaccurate results. Additionally, the algorithm's efficacy varies depending on the type of fall, making falls from sitting or kneeling, for instance, harder to detect than falls from a standing position [34].

3.3 Extreme Gradient Boosting (XGBoost)

Radar-based fall detection is an appropriate fit for the machine learning technique known as Extreme Gradient Boosting (XGBoost). This uses a gradient boosting approach; a model is built by sequentially including weak learners in it [36]. Until the target level of performance is attained, each weak learner is educated using the residuals of the prior learner. Radar-based fall detection systems [37] can benefit from XGBoost's resilience and ability to handle complicated patterns in data. XGBoost is a scalable as well as a flexible algorithm, and it can be used to train models on large datasets with smaller errors [38].

The use of XGBoost for radar-based fall detection provides benefits but also has certain drawbacks. The crucial process of feature selection is made more difficult by the size of radar datasets. The process of fine-tuning model parameters takes time. Due to its sensitivity to noisy radar data, XGBoost must be handled carefully in real-world settings.

3.4 Naive Bayes

Naive Bayes learning algorithm uses Bayes method for classification. It is a straightforward method built on the Bayes theorem. The Bayes theorem is a mathematical formula that can be used to determine the likelihood that an event will occur given the likelihood that other occurrences will also occur. Naive Bayes classifiers may be trained on a range of characteristics, making them a good choice for radar-based fall detection [39]. The algorithm would utilize the probability distributions which it acquired to determine the likelihood [40] that a new radar data point would belong to each class. The class of the data point would be projected to be the one with the highest probability. It is practical for a variety of categorization issues and is relatively easy to install and train. When used for radar-based fall detection, naive Bayes has some drawbacks. It is sensitive to the quality of the training data because it relies on labeled data, which could lead to detection errors. In complicated radar data, the premise of feature independence might not hold, which would have an influence on accuracy. Because of its simplicity, it cannot handle complex feature interactions,

Table 2 Comparison chart for the radar-based algorithms

Algorithm	False alarm rate (%)	Accuracy (%)	Sensitivity (%)	Latency (ms)
SVM	5–10	90–95	90–95	10–100
HMM	10–15	85–90	85–90	100–200
XGBoost	5–10	95–99	95–99	100–200
Naïve Bayes	10–15	85–90	85–90	10–100

which is a problem for fall detection systems. Finally, naive Bayes cannot guarantee perfect predictions even with high-quality data because it is a probabilistic model [41].

Table 2 represents the comparison factor for the above-mentioned algorithms.

4 Algorithms Used in Wearable-Based Sensors

In order to distinguish falls from other daily activities (ADLs), wearable-based sensors for fall detection use machine learning techniques. These algorithms identify new sensor data as either a fall or a non-fall using knowledge, they have gained from training on a collection containing sensor data from both falls and non-falls.

The common as well as most efficient algorithms have been identified in this paper.

4.1 Linear Support Vector Machine

A supervised machine learning approach called the Linear Support Vector Machine (SVM) is useful for both classification and regression problems. When dealing with high-dimensional data, such as that obtained from wearable sensors, SVMs are especially effective. The classification of sensor data into falls and non-falls using linear SVMs can be implemented to detect falls [42]. The SVM can be used to categorize fresh sensor data after being trained. Due to their ease of implementation and training, linear SVMs are frequently used for fall detection algorithms [43]. They can also perform fall detection jobs with a high degree of accuracy. They can handle data that is not completely linearly separable, are noise resistant, and offer efficient computational capabilities.

While promising as a machine learning paradigm for wearable-based fall detection, linear support vector machines (SVMs) do have several drawbacks.

When dealing with complex datasets, linear SVMs may not be as precise as their non-linear versions [44]. The linear SVM's reliance on precise feature extraction and selection is another problem.

Outliers in the data might negatively affect linear SVMs. The SVM may acquire a decision boundary that does not accurately reflect the true connection in cases where the training dataset contains anomalies.

4.2 *Logistic Regression*

A fundamental machine learning algorithm called logistic regression is commonly used in wearable devices to detect falls. This technique was chosen because it may differentiate between fall events and non-fall activities simply and effectively in binary classification tasks [45]. Furthermore, logistic regression can be quite simple to read, which is useful for figuring out the elements that cause falls.

The sigmoid function, also referred to as the logistic function, is the foundation of logistic regression. With the use of this function, any real-valued number is converted into a probability represented by a value between 0 and 1.

The sigmoid function may transform the input information to a probability score because of its S-shaped curve [46].

Figure 2 represents the graphical form of the typical sigmoid function.

These algorithms do have certain limitations which needs to be addressed. Since it has its fundamental linearity, it is limited in how well it can handle complicated datasets, which may contain non-linear connections between input features and fall events [47]. Advanced techniques, such as deep learning models or non-linear support vector machines (SVMs), are favored in these complex situations due to their ability to catch complicated patterns and attain higher accuracy [48].

Figure 3 represents the overall algorithm for the logistic regression in wearables for the fall detection

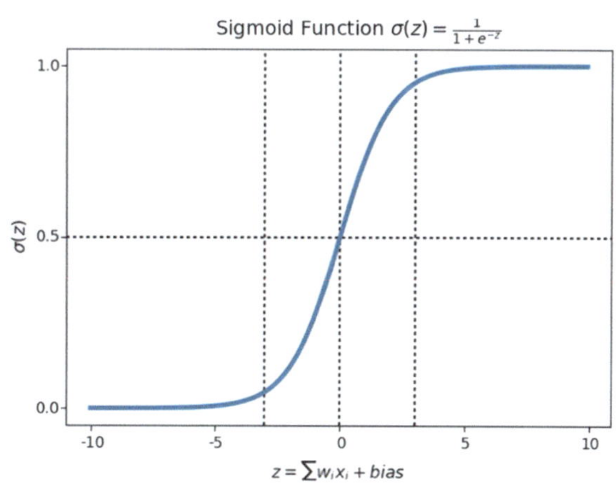

Fig. 2 Sigmoid function [46]

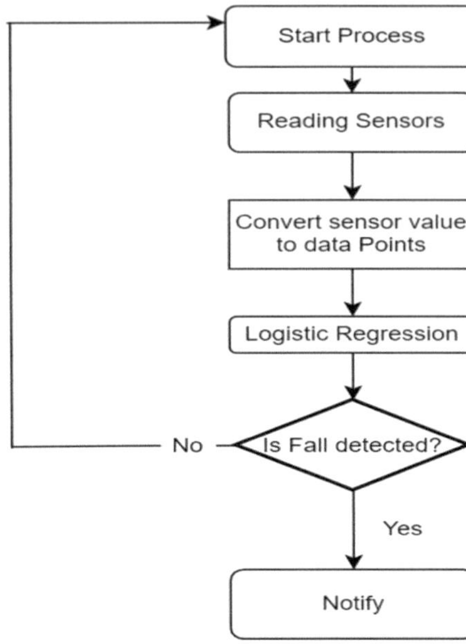

Fig. 3 Algorithm for system

Furthermore, selecting and extracting features for logistic regression require attention to detail. The selection of pertinent features that are both reliable wearable measurable and related to fall detection is crucial to the model's effectiveness.

4.3 AdaBoost (Meta-Learning)

The machine learning technique known as AdaBoost, or adaptive boosting, can be used to enhance the performance of other machine learning algorithms. Being a meta-learning algorithm, it understands how to discover [49]. In order to create a strong learner, AdaBoost first trains a few weak learners by using basic machine learning algorithms and then combines their predictions. Simple algorithms like decision trees or even linear models might serve as these weak classifiers. The goal is to train these classifiers iteratively by assigning greater weight to instances that were incorrectly labeled in earlier iterations. In fall detection scenarios, AdaBoost has shown outstanding performance, managing the complexity and variety of wearable sensor data with ease[50]. It is robust to overfitting, which is helpful when there is a dearth of data, as is frequently the case in wearable-based fall detection [51].

The AdaBoost algorithm for fall detection in wearables has limitations such as computational complexity, sensitivity to outliers and noise, overfitting, high-quality data requirement, and sensitivity to weak learners [52]. These issues can lead to poor performance on real-world data, unseen data, and require careful selection of weak learners for optimal performance.

4.4 Polynomial Kernel Function

To discover non-linear associations between data points, machine learning techniques like support vector machines (SVMs) can be employed in combination with the polynomial kernel function. It is a popular option for fall detection in wearables because it can be utilized to understand the intricate connections between fall occurrences and the characteristics of wearable sensor data, like acceleration and gyroscope data.

By permitting to alter the polynomial's degree, the polynomial kernel provides versatility, one can change the degree to manage the decision boundary's complexity. The polynomial kernel may reveal hidden structures by translating the data into a higher dimensional space [53], which makes it simpler for the SVM or other algorithms to locate optimal decision boundaries. This can be useful when dealing with dimensional sensor data from wearables [54].

Despite being a useful algorithm in the field of wearable-based fall detection, the polynomial kernel function has some drawbacks. First, it involves a heavy computational load, especially when dealing with large datasets or using higher degree polynomials. Considering wearables have limited computational capabilities, real-time processing may be hampered by this computational complexity.

Second, the careful selection of the polynomial degree is crucial since it has a significant impact on how well the polynomial kernel performs. On the other hand, a degree that is too high can make the model vulnerable to overfitting, where it essentially memorizes the training data and finds it difficult to generalize successfully to new, unknown data [55]. In addition, using the polynomial kernel function raises the particular concern of overfitting. Polynomial degrees can introduce higher levels of complexity; hence, it naturally has an overfitting risk.

Table 3 represents the comparison factor for the above-mentioned algorithms.

Table 3 Comparison chart for the wearable-based algorithms

Algorithm	False alarm rate (%)	Accuracy (%)	Sensitivity (%)	Latency (ms)
LSVM	5–10	90–95	90–95	10–100
Logistic Regression	10–15	85–90	85–90	10–100
AdaBoost	5–10	95–99	95–99	100–200
Polynomial kernel function	10–15	90–95	90–95	200–300

Table 4 Comparison between systems using sensors and sensors along with AI

Factors	Sensor-based fall detection	AI-based fall detection
Sensitivity	Moderate-high	May achieve higher levels
Accuracy	Varies depending on sensor quality and placement	It can be optimized through training and data availability
Recall (true positive)	Less	High
Specificity (true negative)	Lower	Higher
False positives	False positives due to environmental factors are more	Can be reduced through fine-tuning and data augmentation
False negatives	May miss some falls	Can be reduced through an improved model training
Ease of installation and maintenance	Relatively easy to install and maintain these systems	May require more complex infrastructure, but has the advantage that it can be automated
Cost	Mostly cost-effective	Its development and deployment can be costly
Real-time detection	Provides real-time detection capabilities	Can be achieved with proper hardware and software
Integration with other technologies	Integration with other technologies could be challenging	Integration with other is relatively easier

5 Evaluation of Systems Using AI Versus Non-AI Systems WRT Fall Detection

The thorough analysis outlined in this paper helps to clarify the roles that algorithms and sensor technologies play to improve key metrics. There is a strong need for the refinement of these criteria given the importance of minimizing falls. The precision of the system, a crucial component in real-time settings, is increased using algorithms, which play a crucial part in increasing the system's effectiveness.

The factors examined in this study are summarized in Table 4, which highlights how the use of algorithms helps the system run seamlessly while also reducing complexity.

6 Future Scope

Future potential for machine learning algorithms and sensor technology integration in fall detection systems is extremely promising.

The adoption of increasingly advanced and precise sensors is one development that we may anticipate in the future. For instance, scientists are creating sensors that can track three-dimensional bodily movements or detect variations in blood pressure and pulse rate. In order to create fall detection systems that are more precise and dependable, this data can be integrated into machine learning algorithms.

The application of machine learning algorithms to create more individualized fall detection systems is another development that we can anticipate. For instance, a person's distinctive gait and movement patterns can be looked at with machine learning algorithms to develop a unique fall detection profile. Then, using this profile, it will be possible to identify falls with more accuracy and reduce the amount of false alarms.

The application areas could also further work on the creation of advanced, contextually aware fall detection systems. For instance, a smart home system may have a fall detection system that operates autonomously to turn on the lights or call for help if a fall is detected.

Merging information from several sensor types such as wearables, ambient sensors, radars, and cameras to produce fall detection that is more reliable and precise. In addition, this could also help to overcome limitations and lessen reliability on a single sensor modality.

7 Conclusion

The usage of machine learning is a future trend that we may expect to achieve. We can create fall detection systems that are more specific, reliable, and individualized by combining sensor technology with machine learning algorithms. The safety and well-being of senior citizens and people with disabilities may be enhanced as a result. This study offers a thorough analysis of current machine learning methods under active research. It carefully analyzes the benefits and drawbacks of these algorithms when used with various sensor types, such as wearables, ambient-based sensors, and vision-based sensors, many of which have already gained market share. The study offers a thorough evaluation of their overall effectiveness, focusing on important factors like accuracy, false alarm rates, and latency in real-world settings. The basic importance of these factors inside life-saving systems is highlighted in this study. As a result, the study provides a solid framework for understanding their crucial role, with significant implications for improving the quality of life, especially for the elderly and people with impairments. The basic importance of these factors inside life-saving systems is highlighted in this study. As a result, the study provides a solid framework for understanding their crucial role, with significant implications for improving the quality of life, especially for the elderly and people with impairments. Several essential aspects could be redefined and advanced by further research efforts, which might cause changes and major breakthroughs in these systems.

References

1. WHO: Falls. https://www.who.int/news-room/fact-sheets/detail/falls. Accessed 24 Nov 2023
2. Sannino G, De Falco I, De Pietro G (2015) A supervised approach to automatically extract a set of rules to support fall detection in an mHealth system. Appl Soft Comput 34:57–69
3. Yeong UY, Tan SY, Yap JF, Choo WY (2016) Prevalence of falls among community-dwelling elderly and its associated factors: a cross-sectional study in Perak, Malaysia. Malays Fam Phys 11(1):7–15
4. Hatahira H, Hasegawa S, Sasaoka S, Kato Y, Abe J, Motooka Y, Fukuda A, Nakao S, Takeuchi H, Nakamura M (2018) Analysis of fall-related adverse events among older adults using the Japanese Adverse Drug Event Report (JADER) database. J Pharm Health Care Sci. 4:8
5. Chen X, Lin Z, Gao R, Yang Y, Li L (2021) Prevalence and associated factors of falls among older adults between urban and rural areas of Shantou City, China. Int J Environ Res Public Health 18(13):7050
6. He C, Liu S, Zhong G, Wu H, Cheng L, Lin J, Huang Q (2023) A non-contact fall detection method for bathroom application based on MEMS infrared sensors. Micromachines 14(1):130
7. Preissner CE, Kaushal N, Charles K, Knäuper B (2023) A protection motivation theory approach to understanding how fear of falling affects physical activity determinants in older adults. J Gerontol B 78(1):30–39
8. Nizam Y, Mohd MNH, Jamil MMA (2016) A study on human fall detection systems: Daily activity classification and sensing techniques. Int J Integr Eng 8(1)
9. Abbate S, Avvenuti M, Corsini P, Light J, Vecchio A (2010) Monitoring of human movements for fall detection and activities recognition in elderly care using wireless sensor network: a survey. Wirel Sens Netw Appl-Centric Des 1
10. Mubashir M, Shao L, Seed L (2013) A survey on fall detection: principles and approaches. Neurocomputing 100:144–152
11. Zhang Z, Conly C, Athitsos V (2015) A survey on vision-based fall detection. In: Proceedings of the 8th ACM international conference on pervasive technologies related to assistive environments, pp 1–7
12. Thakur N, Han CY (2021) A study of fall detection in assisted living: identifying and improving the optimal machine learning method. J Sens Actuator Netw 10(3):39
13. Yuwono M, Moulton BD, Su SW, Celler BG, Nguyen HT (2012) Unsupervised machine-learning method for improving the performance of ambulatory fall-detection systems. Biomed Eng Online 11(1):1–11
14. Ogawa Y, Naito K (2020) Fall detection scheme based on temperature distribution with IR array sensor. In: 2020 IEEE international conference on consumer electronics (ICCE). IEEE, Las Vegas, NV, USA, pp 1–5
15. LeCun Y, Bengio Y, Hinton G (2015) Deep learning. Nature 521(7553):436–444
16. Raj T, Nisar T, Abbas M, Priyadarshini R, Naz S, Tiwari U (2023) Qualitative analysis of techniques for device-free human activity recognition. Rev Intell Artif 37(3)
17. Menacho C, Ordoñez J (2020) Fall detection based on CNN models implemented on a mobile robot. In: 2020 17th international conference on ubiquitous robots (UR). IEEE, Kyoto, Japan, pp 284–289
18. Fakhrulddin, AH, Fei X, Li H (2017) Convolutional neural networks (CNN) based human fall detection on body sensor networks (BSN) sensor data. In: 2017 4th international conference on systems and informatics (ICSAI). IEEE, Hangzhou, China, pp 1461–1465
19. Islam MM, Tayan O, Islam MR, Islam MS, Nooruddin S, Kabir MN, Islam MR (2020) Deep learning based systems developed for fall detection: a review. IEEE Access 8:166117–166137
20. Lee DW, Jun K, Naheem K, Kim MS (2021) Deep neural network–based double-check method for fall detection using IMU-l sensor and RGB camera data. IEEE Access 9:48064–48079
21. Hasan MM, Islam MS, Abdullah S (2019) Robust pose-based human fall detection using recurrent neural network. In: 2019 IEEE international conference on robotics, automation, artificial-intelligence and internet-of-things (RAAICON). IEEE, Dhaka, Bangladesh, pp 48–51

22. Raza A, Yousaf MH, Velastin SA (2022) Human fall detection using YOLO: a real-time and AI-on-the-edge perspective. In: 2022 12th international conference on pattern recognition systems (ICPRS). IEEE, Saint-Etienne, France, pp 1–6
23. Yin Y, Lei L, Liang M, Li X, He Y, Qin L (2021) Research on fall detection algorithm for the elderly living alone based on YOLO. In: 2021 IEEE international conference on emergency science and information technology (ICESIT). IEEE, Chongqing, China, pp 403–408
24. Long KZ, Haron H, Ibrahim M, Eri ZD (2021) An image-based fall detection system using you only look once (YOLO) algorithm to monitor elders' fall events. In: Knowledge management international conference (KMICe) 2021, vol 1
25. Alaoui AY, El Fkihi S, Oulad Haj Thami R (2019) Fall detection for elderly people using the variation of key points of human skeleton. IEEE Access 7:154786–154795
26. Merrouche F, Baha N (2020) Fall detection based on shape deformation. Multimed Tools Appl 79
27. Bhatlawande S, Khapre D, Kinge M, Khairnar T (2022) Vision based assistive system for fall detection. In: 2022 2nd international conference on intelligent technologies (CONIT). IEEE, Hubli, India, pp 1–7
28. Ramachandran A, Adarsh R, Pahwa P, Anupama KR (2018) Machine learning-based fall detection in geriatric healthcare systems. In: 2018 IEEE international conference on advanced networks and telecommunications systems (ANTS). IEEE, Indore, India, pp 1–6
29. Zhao Z, Song Y, Cui F, Zhu J, Song C, Xu Z, Ding K (2020) Point cloud features-based kernel SVM for human-vehicle classification in millimeter wave radar. IEEE Access 8:26012–26021
30. Deliali A, Tainter F, Ai C, Christofa E (2023) A framework for mode classification in multimodal environments using radar-based sensors. J Intell Transp Syst 27(4)
31. Li Y, Xie P, Tang Z, Jiang T, Qi P (2019) SVM-based sea-surface small target detection: a false-alarm-rate-controllable approach. IEEE Geosci Remote Sens Lett 16(8):1225–1229
32. Taylor W, Dashtipour K, Shah SA, Hussain A, Abbasi QH, Imran MA (2021) Radar sensing for activity classification in elderly people exploiting micro-Doppler signatures using machine learning. Sensors 21(11)
33. Jiang H, Rusuli Y, Amuti T, He Q (2019) Quantitative assessment of soil salinity using multi-source remote sensing data based on the support vector machine and artificial neural network. Int J Remote Sens 40(1):284–306
34. Zhou Y, Wei S, Zhang L, Zhang W, Ma Y (2022) Dynamic estimation of spin satellite from the single-station ISAR image sequence with the hidden Markov model. IEEE Trans Aerosp Electron Syst 58(5):4626–4638
35. You H, Byun SH, Choo Y (2022) Underwater acoustic signal detection using calibrated hidden Markov model with multiple measurements. Sensors 22(14):5088
36. Liu Y, Liu L, Yang L, Hao L, Bao Y (2021) Measuring distance using ultra-wideband radio technology enhanced by extreme gradient boosting decision tree (XGBoost). Autom Constr 126:103678
37. Almalki N, Alnfiai M, Al-Wesabi FN, Alduhayyem M, Hilal AM, Hamza MA (2023) Deep transfer learning driven automated fall detection for quality of living of disabled persons. Comput Mater Contin 74(3)
38. Wei CC, Hsu CC (2020) Extreme gradient boosting model for rain retrieval using radar reflectivity from various elevation angles. Remote Sens 12(14):2203
39. Tewari RC, Sharma S, Routray A, Maiti J (2023) Effective fall detection and post-fall breath rate tracking using a low-cost CW Doppler radar sensor. Comput Biol Med 164:107315
40. Nandi P, Anupama KR, Agarwal H, Patel K, Bang V, Bharat M, Guru MV (2023) Analysis of ML algorithm for geriatric fall detection due to the effects of various user characteristics
41. Xefteris VR, Tsanousa A, Meditskos G, Vrochidis S, Kompatsiaris I (2021) Performance, challenges, and limitations in multimodal fall detection systems: a review. IEEE Sens J 21(17):18398–18409
42. Pan D, Liu H, Qu D, Zhang Z (2020) Human falling detection algorithm based on multisensor data fusion with SVM. Mob Inf Syst 2020:1–9

43. Guo J, Tang Y, Fu S, Guo C, Yu C (2020) A fall detection algorithm based on a support vector machine. In: Artificial intelligence and security: 6th international conference, ICAIS 2020, Hohhot, China, 17–20 July 2020, proceedings, Part I 6. Springer Singapore, pp 515–526
44. Rastogi S, Singh J (2021) A systematic review on machine learning for fall detection system. Comput Intell 37(2):951–974
45. Nahar N, Hossain MS, Andersson K (2020) A machine learning based fall detection for elderly people with neurodegenerative disorders. In: International conference on brain informatics. Springer International Publishing, Cham, pp 194–203
46. Haykin S, Network N (2004) A comprehensive foundation. Neural Netw 2(2004):41
47. Le TM, Tran LV, Dao SVT (2021) A feature selection approach for fall detection using various machine learning classifiers. IEEE Access 9:115895–115908
48. Otanasap N, Bangkomkun P, Tanantpapat T (2023) Pre-impact fall detection system using logistic regression model: pre-impact fall detection system using logistic regression model. SAU J Sci Technol 9(1):30–43
49. Cai W, Qiu L, Li W, Yu J, Wang L (2019) Practical fall detection algorithm based on adaboost. In: Proceedings of the 4th international conference on biomedical signal and image processing, pp 117–121
50. Hassan CAU, Karim FK, Abbas A, Iqbal J, Elmannai H, Hussain S, Ullah SS, Khan MS (2023) A cost-effective fall-detection framework for the elderly using sensor-based technologies. Sustainability 15(5):3982
51. Collado-Villaverde A, Cobos M, Muñoz P, Barrero DF (2020) A simulator to support machine learning-based wearable fall detection systems. Electronics 9(11):1831
52. Hellmers S, Krey E, Gashi A, Koschate J, Schmidt L, Stuckenschneider T, Hein A, Zieschang T (2023) Comparison of machine learning approaches for near-fall-detection with motion sensors. Front Digit Health 5:1223845
53. Saleh M, Abbas M, Prud'Homm J, Somme D, Le Bouquin Jeannes R (2021) A reliable fall detection system based on analyzing the physical activities of older adults living in long-term care facilities. IEEE Trans Neural Syst Rehabil Eng 29:2587–2594
54. Hsieh CY, Huang CN, Liu KC, Chu WC, Chan CT (2016) A machine learning approach to fall detection algorithm using wearable sensor. In: 2016 international conference on advanced materials for science and engineering (ICAMSE). IEEE, Tainan, Taiwan, pp 707–710
55. Savas C, Dovis F (2019) The impact of different kernel functions on the performance of scintillation detection based on support vector machines. Sensors 19(23):5219

Face Emotion Detection from Facial Expression Using OpenCV

Sangam Patil and Janhavi Sutar

Abstract The real-time face emotion detection represents a complex and critical frontier in the realm of computer vision and artificial intelligence (AI). This paper delves into the intricate landscape of real-time emotion detection from facial expressions, providing an extensive examination of challenges, applications, and cutting-edge methodologies. The ability to decipher facial emotions is paramount for effective human communication, underscoring the significance of developing automated systems that excel in accurate emotion recognition. This comprehensive review underscores the inherent importance of emotions in human communication, elucidating their pivotal role in understanding social interactions and behavior. Real-time emotion recognition emerges as a valuable tool across diverse domains. Notably, this paper explores various commercial and research implementations, revealing promising outcomes in identifying specific emotions such as happiness, sadness, anger, fear, disgust, and surprise with noteworthy accuracy. While acknowledging the significant strides made in real-time face emotion detection, the paper underscores the ongoing need for research to augment precision, adaptability, and generalization across diverse demographics and cultural contexts. The synthesized insights aim to empower researchers, developers, and practitioners in the domain of real-time face emotion detection, guiding them toward the creation of more accurate, efficient, and ethically responsible emotion recognition systems tailored for a myriad of practical applications.

Keywords Convolution neural networks · VGG-16 · Face emotion detection · Deep learning and transfer learning

S. Patil (✉)
Vishwakarma Institute of Technology, Pune, Maharashtra, India
e-mail: sangampatil184@gmail.com

J. Sutar
Sharad Institute of Technology College of Engineering, Ichalkaranji, Maharashtra, India

A. P. Agrawal et al. (eds.), *Advances of Artificial Intelligence and Machine Learning in Image Processing, Security and Data Analytics*, Lecture Notes in Electrical Engineering 1464, https://doi.org/10.1007/978-981-95-0148-9_18

1 Introduction

Real-time face emotion detection is the process of automatically detecting and classifying the emotions on a person's face in real time. A seamless robot to user interaction is critical to the success of service robotics. As a result, a robot ought to be able to infer a user's gender or determine their emotional state only by looking at them. The great diversity of the samples within each challenge has made it difficult to appropriately interpret any of these variables using machine learning (ML) approaches. As a result, millions of parameters are used to train models on thousands of samples. Moreover, $65 \pm 5\%$ of humans correctly identify a face image representing one of the seven main emotions. Facial emotion detection aims to recognize the emotional state of a person from their facial expressions, which can be captured by images or videos. With the advancement of deep learning algorithms and the availability of large datasets, real-time face emotion detection has become more accurate and efficient, enabling various applications in areas such as human–computer interaction, user experience analysis, and sentiment analysis. Several studies have been conducted to build and design numerous facial emotion detection systems, yet several common issues remain in the emotion recognition process. The characteristics are extremely sensitive to changes in noise, light, and occlusion, which raises two key difficulties. This implies that a minor change in noise, lighting, or occlusion may affect the identification process's accuracy rate, and, second, the immense data dimension affects the performance of such systems.

Facial tissue signals might be used to identify the eye expression. However, a few emotions stand out in a handful of distinct facial highlights. This component is also influenced by a person's difference in the subject, such as degree, frequency, or pace of expression. Based on these findings, we may conclude that it is critical to design a facial expression recognition system that recognizes facial expressions in real time with adequate accuracy. But our goal is to develop and test a successful system for facial expression identification and emotion detection on face eye pictures using deep convolution neural networks. Traditional approaches to emotion detection relied on hand-crafted features and classifiers. It frequently had trouble capturing intricate patterns in face emotions. But the advent of deep learning methods has completely changed this area. Because Convolution Neural Networks (CNNs) and Recurrent Neural Networks (RNNs) can automatically learn hierarchical and temporal features from raw data, they offer promising results for real-time face emotion recognition. The study uses deep convolution neural networks to increase identification and emotion classification accuracy. The development of such systems is crucial to enable machines to understand and respond to human emotions, ultimately leading to more natural and empathetic interactions.

2 Literature Review

In the paper, "Building a Computation Savings Real-Time Face Detection and Recognition System," by Cheong et al. [1], the development of a real-time face detection and recognition system with a focus on computational efficiency is described. The authors propose techniques to optimize the system's components for faster performance, particularly in computationally constrained environments. The results of the experiments demonstrate that the suggested system, including the optimization approaches, can run real time even on low-power systems. The proposed techniques are shown to be effective in enhancing system performance without compromising accuracy, making the system suitable for real-time applications like surveillance and security.

This paper titled "Face Detection Techniques: A Review" [2] provides an overview of various techniques explored for face detection in digital images. The authors discuss the increasing need for automatic understanding and analysis of information because of the vast amount of data in the form of images and videos. The paper covers the challenges faced in face detection, including complex background, face occlusion, lighting variations, odd expressions, and more various face detection techniques are discussed, including feature-based approaches like Active Shape Model (ASM) and low-level analysis using skin color models (RGB, HSV, YCbCr, CIELAB). The authors also mention different standard databases available for face detection and the report concludes by noting possible future areas for research in this topic.

The paper "An Efficient Method to Face and Emotion Detection" [3]. proposes an efficient method for face and Viola–Jones facial identification method and K-NN classifier are used to identify emotions. The authors focus on creating a database for face and emotion features and then use the K-NN classifier for real-time recognition. The proposed method involves two phases: 1. database creation for face and 2. creation of an emotion database. The K-NN classifier matches these features with the database to recognize the person. Experimental results indicate an accuracy of 94.5–97% for face and emotion recognition using the proposed method. The authors also compare their approach with other methods, demonstrating its effectiveness in face identification and emotion recognition challenges.

The paper titled "Gesture-based Facial Emotion using Convolution Neural Network" by Naik and Mehta [4] proposes an improved method for facial emotion recognition that incorporates hand-over-face gestures along with facial expressions to identify a wide range of emotions, including basic emotions (happy, sad, angry, etc.) and extra feelings (confident, making decisions, terrified, embarrassed, okay sign). The proposed method combines a large coding schema with various hand motions with Convolution Neural Network (CNN) for automatic feature extraction and emotion categorization. The CNN architecture makes use of two convolution layers, two max-pooling layers, one fully linked layer, and one Softmax layer. To sum up, the suggested Hand-over-Face Gesture-based Facial Emotion Recognition Method (HFG_FERM) accomplishes the following: it effectively recognizes a wider variety of emotions, including those conveyed through hand gestures.

The paper titled "A Novel Online Product Recommendation System Based on Face Recognition and Emotion Detection" [5] offers a new approach for an online product recommendation for online consumers based on emotion detection and facial picture recognition. The system follows five steps: face detection, face recognition, gender identification, product recommendation, and emotion detection feedback. The face detection algorithm is based on Haar-like features, and face recognition uses the eigenfaces method. The Fisher faces algorithm is employed for gender identification, and collaborative filtering is used for product recommendation based on the user's buying history and gender. Emotion detection is conducted using the Fisher faces algorithm to identify happy and sad emotions. The proposed system aims to improve user interaction and enhance product recommendation accuracy on online shopping websites and other platforms.

The paper titled "Emotion Recognition using Fisher Face-based Viola-Jones Algorithm" [6] suggests use of Fisher faces to modify the Viola–Jones algorithm for facial emotion identification. The paper highlights the limitations of image integral features in emotion recognition and proposes Fisher faces as a more robust feature. The proposed approach involves using Fisher faces, a combination of PCA and LDA, to extract low-dimensional features for emotion detection. The experiments are conducted using a dataset of student facial expressions, with emotions classified as either "interest" or "bored." The results show that the Fisher face-based Viola–Jones algorithm outperforms the original Viola–Jones algorithm in terms of accuracy and precision, achieving an accuracy of 0.82 compared to 0.78 for the original algorithm.

The paper titled "Face Detection Based on Adaboost and Face Contour in E-learning" by Wansen Wang and Yaya Peng [7]. The study addresses the issue of identifying learners' expressions in online instruction. The AdaBoost algorithm and a set of Haar wavelet-like features form the foundation of the authors' face detection method. They improve detection performance by using AdaBoost with the facial shape. The experimental findings reveal that when compared to employing AdaBoost or face contour alone, the combined technique produces greater accuracy and reduced false positive rates. The method can effectively detect faces even in the presence of learners wearing glasses and in complex lighting conditions.

This research paper "Content Based Facial Emotion Recognition Model using Machine Learning Algorithm" [8] discusses the development and implementation of using machine learning methods, a Convolutional Neural Network (CNN) is trained to recognize facial emotions. This paper shows challenges on emotion recognition, including face detection, accurate emotion classification, and appropriate datasets. The authors propose a CNN-based framework that combines face detection and emotion classification tasks using their designed CNN architecture. They validate their model using the FER-2013 dataset, which contains facial expressions for emotion analysis. To summarize, the paper presents a research study on facial emotion recognition using CNNs, discusses the challenges, proposes a CNN-based framework, evaluates different models, and explores future application possibilities.

3 Methodology

The proposed approach in this research builds upon a computer vision-based method for extracting face features and automatically recognizing emotions. The classification performance of the system is greatly enhanced by incorporating six geometrical and two appearance-based parameters without compromising real-time processing requirements. The resultant processing pipeline classifier recognizes seven facial emotions, including the six fundamental Ekman's emotions as well as neutral. Understanding human emotion for computers using Convolution Neural Networks (CNN) involves using deep learning techniques to automatically recognize and classify emotions from facial expressions in images or video frames. CNNs are well suited for this task due to their capacity to immediately learn hierarchical features from raw picture data making them much effective in capturing the complex patterns and structures present in facial expressions (Fig. 3).

Here's how CNNs are typically used for emotion recognition.

3.1 Data Preparation

Getting a sizable dataset of photos with captions for different face expressions is the first stage. Images of different people exhibiting various emotions should be included in this collection, together with labels that indicate the appropriate emotion (e.g., joyful, sad, angry, etc.).

3.2 Data Preprocessing

Before feeding the data into the CNN, some preprocessing steps are usually applied. Common preprocessing steps include face detection to locate and crop the face region, and normalization to ensure consistency in facial appearance across different images.

3.3 CNN Architecture

Creating the CNN architecture is the next stage. Usually made up of several convolutional layers, CNNs extract different information from the input images. Following the use of these layers, the spatial dimensions of the feature maps are reduced using pooling layers, and fully connected layers produce the final output for classification.

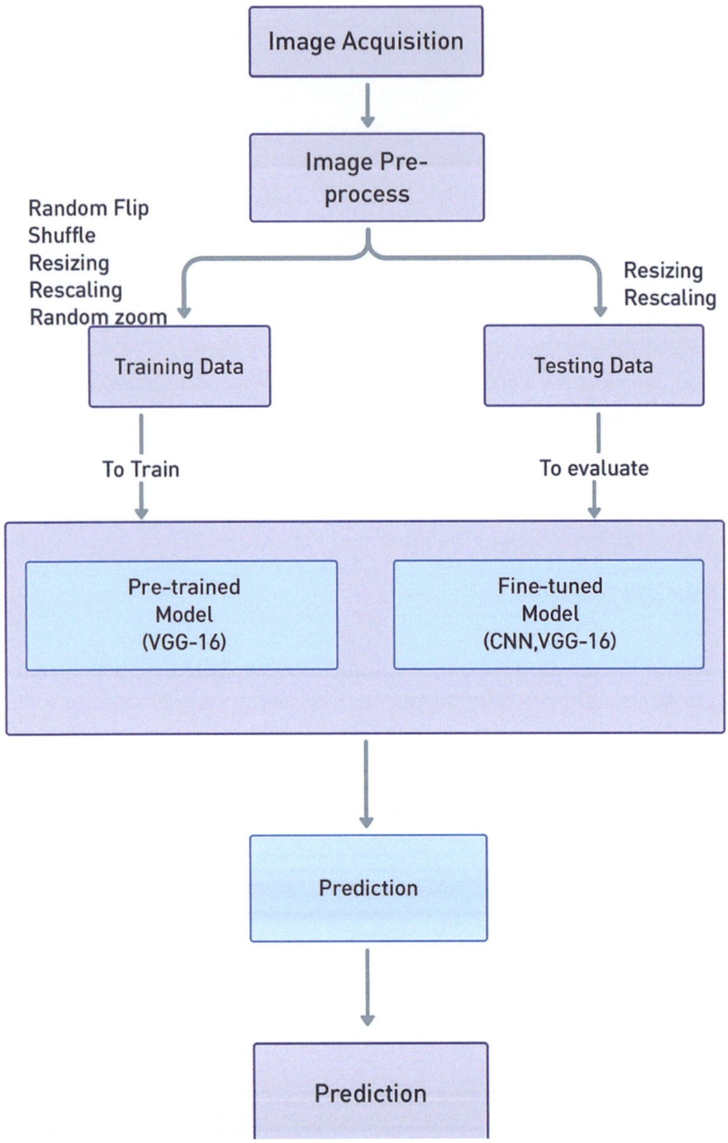

Fig. 1 Proposed methodology

3.4 Training the CNN

In the training phase, the CNN iteratively modifies its weights depending on the labeled data to identify patterns and features from the facial expression images. A loss function, like categorical cross entropy, and an optimization technique, like Adam or stochastic gradient descent (SGD), are used to optimize the model's parameters.

3.5 *Transfer Learning*

Transfer learning involves using a pre-trained CNN, such as VGG, ResNet, or Inception, which has been trained on a large dataset (e.g., ImageNet). The pre-trained CNN is then fine-tuned on the emotion recognition dataset, allowing it to adapt to the new task while still benefiting from the knowledge learned on the large dataset.

3.6 *Evaluation*

A different test dataset is used to evaluate the CNN's performance once it has been trained. Evaluation criteria including accuracy, precision, recall, and F1-score are frequently used.

4 Our Proposed Idea

By leveraging CNNs for facial emotion recognition, computers can be equipped with the ability to perceive and understand human emotions, enabling more natural and intuitive interactions between humans and machines.

4.1 *The VGG-16 Model*

The University of Oxford's Visual Geometry Group (VGG) unveiled the VGG-16 model, a deep convolutional neural network (CNN) architecture. Since its introduction in 2014 as a component of the ImageNet Large Scale Visual Recognition Challenge (ILSVRC), it has gained popularity as a solution for a variety of computer vision applications, including picture categorization. Using the pre-trained VGG-16 architecture—which was initially intended for picture classification—real-time face emotion detection with the VGG-16 model entails modifying it for emotion recognition. The ability of the 16-layer deep convolutional neural network VGG-16 to extract hierarchical information from images is well known (Fig. 2).

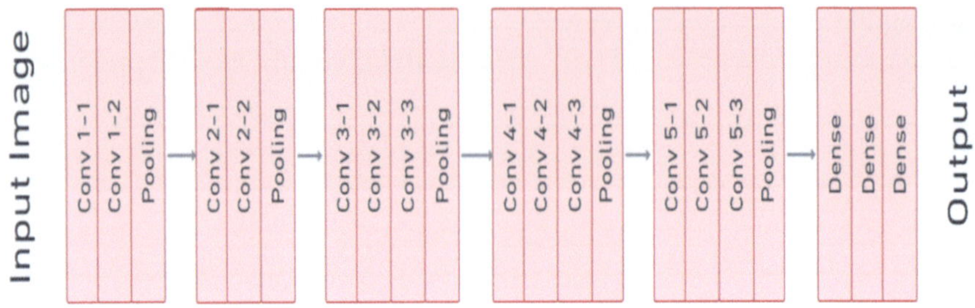

Fig. 2 VGG-16 model

Key Characteristics of VGG-16:

Architecture: VGG-16 follows a straightforward and uniform architecture, where each convolutional block is followed by a max-pooling layer to reduce spatial dimensions. The overall network design is characterized by the repeated pattern of convolutional and pooling layers.

Fully Connected Layers: After the convolution layers, the network contains three completely linked layers that lead to the final output layer that is in charge of categorization.

Classification: VGG-16 was originally designed for image classification tasks. The network is trained to recognize and classify images into one of a predefined set of classes. It has been influential in the development of deep learning for computer vision due to its deep architecture and relatively simple design.

Pre-training: The VGG-16 model was pre-trained on the Image Net dataset, which contains millions of labeled images belonging to 1000 different classes. This pre-training allowed the network to learn generic features from a wide range of images, which can then be fine-tuned for specific tasks using transfer learning.

Transfer Learning with VGG-16: VGG-16 transfer learning involves leveraging pre-trained ImageNet features to adapt the network for new tasks such as object identification, segmentation or facial emotion detection. By replacing fully connected layers with task-specific ones and optimizing on a smaller dataset, this approach is popular for real-world applications, offering efficiency with less labeled data and reduced training time.

4.2 CNN Model

Convolution Neural Networks (CNNs) are a type of deep learning model that is commonly used for image identification, object detection, and segmentation. CNNs are inspired by the structure and function of the human brain's visual cortex. They are meant to learn hierarchical representations of features from input data automatically and adaptively, making them extremely successful in image-related tasks. Real-time facial emotion detection using Convolution Neural Networks (CNNs) involves leveraging the power of CNNs to recognize facial expressions in real-time video streams. This application is particularly useful in various domains, including human–computer interaction, virtual reality, surveillance, and mental health monitoring.

Key Characteristics of CNN

Convolution Layers: The convolution layer is the fundamental component of a CNN. In order to identify local patterns and features, it performs convolution operations on the input data. Convolution is the process of creating a feature map that identifies pertinent patterns in the input by swiping a tiny filter, sometimes referred to as a kernel, over the input data and conducting element-wise multiplication and summing.

Activation Function: An activation function, commonly the Rectified Linear Unit (ReLU), is applied element by element after each convolution operation to introduce nonlinearity into the network. ReLU makes it easier for the network to learn expressive and complex features.

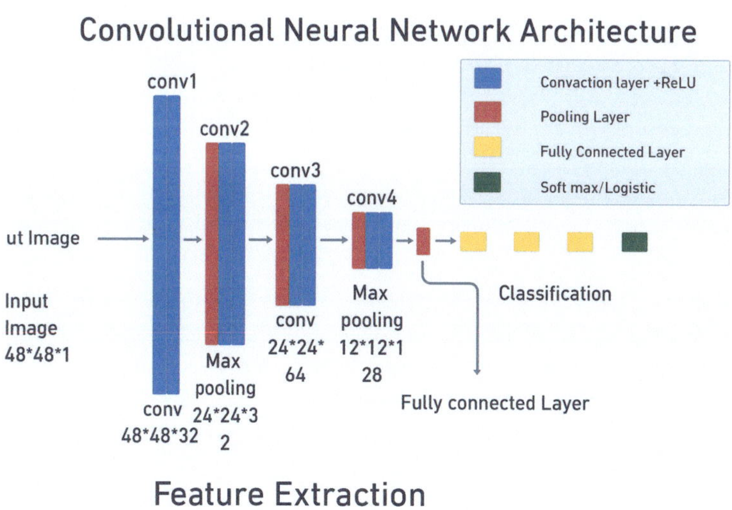

Fig. 3 CNN model

Pooling Layers: Pooling layers, also known as average- or max-pooling, lower the spatial dimensions of feature maps. They achieve this by downsampling the feature maps, helping to reduce computational complexity, control overfitting, and make the network more translation invariant.

Fully Connected Layers: The features are flattened and run through one or more fully connected layers following numerous convolution and pooling layers. These layers are responsible for making the final predictions, such as class labels in image classification.

Stride: The stride is the step size at which the convolutional filter slides over the input data. A larger stride results in downsampled feature maps, while a stride of one preserves the spatial dimensions.

4.3 The Steps for Real-Time Face Emotion Detection

Data Collection: Collecting a dataset of images containing labeled face regions (bounding boxes) and non-face regions.

Preprocess the data: This involves resizing images to a fixed size, normalizing pixel values, and converting labels to one-hot encoding or numerical format.

Model Identification: The first step in real-time face emotion detection is the detection of a face in a picture or video. This can be done using a variety of methods, such as Haar cascades, CNN (convolutional neural networks), or facial landmarks. In our case, we have used CNN and VGG-16.

CNN Model Architecture: Define the architecture of the CNN model that will be used for face detection. Typically, the model will include numerous convolutional layers, then pooling layers for downsampling, and finally fully connected layers for classification. The final layer will usually output the probability of a face being present in a given region of the input image. Experiment with different hyperparameters, model architectures, and optimization techniques to improve the system's accuracy and speed.

Feature extraction: Following the detection of the face, the following stage is to extract characteristics from the face that may be used to identify the emotion. These features include the shape of the eyes, eyebrows, mouth, and nose, as well as the position of these features relative to each other.

Model Training: Following that, the retrieved characteristics are input into a machine learning system that has been trained to categorize emotions. Here we have used Convolution Neural Networks (CNNs) and retrained VGG-16 models. CNNs are well suited for this task because they can figure out the spatial correlations between face characteristics.

Once the machine learning algorithm is trained, it can be used to classify the emotions in real time. This is done by feeding the extracted features from the face into the algorithm, which then outputs the most likely emotion. This is our approach for facial emotion detection.

5 Results and Discussion

We are able to identify the following emotions:

1. Neutral.
2. Happy.
3. Sad.
4. Angry.
5. Fearful.
6. Disgusted.
7. Surprised.

Real-time face emotion detection has a variety of potential applications, including human–computer interaction by allowing computers to understand the user's emotional state. For example, a computer could adjust its behavior based on the user's emotions, such as providing more help or support if the user is feeling sad. Also, real-time face emotion detection can be used to improve the accuracy of virtual assistants. For example, a virtual assistant could use real-time face emotion detection to determine if the user is happy or angry, and then adjust its responses accordingly. Real-time face emotion detection can be used for security purposes, such as detecting people who are feeling angry or frustrated. This could be used to prevent potential violence or other security threats.

We have successfully built this project with 94.03% overall accuracy. Our system is pretty much fast to respond with almost no lag in prediction. In the testing phase, the overall turnaround time (i.e., lag or the time taken for emotion detection after receiving the input) is observed as 0.01 seconds. We have also developed an immersive and appealing user interface. The system shows the intensity of emotion(s). The device can identify several emotions and measure their intensity on a scale of 0–1. Our system is capable of detecting multiple faces also. We have tested our system with 40 individuals, and it detected all the faces and their respective emotions. Here is the confusion matrix of our system (Table 1).

Here are some snapshots of our system (Figs. 4 and 5).

Table 1 Confusion matrix

Expression	Precision %	Recall %	F1-score %
Angry	96.6 (±0.005)	95.3 (±0.005)	95.0 (±0.005)
Disgust	94.2 (±0.005)	96.8 (±0.005)	95.6 (±0.005)
Fear	93.3 (±0.005)	84.8 (±0.005)	94.3 (±0.005)
Happy	97.0 (±0.005)	96.6 (±0.005)	96.1 (±0.005)
Sad	96.2 (±0.005)	96.1 (±0.005)	95.8 (±0.005)
Surprise	95.8 (±0.005)	94.8 (±0.005)	94.4 (±0.005)
Neutral	95.4 (±0.005)	95.9 (±0.005)	96.6 (±0.005)

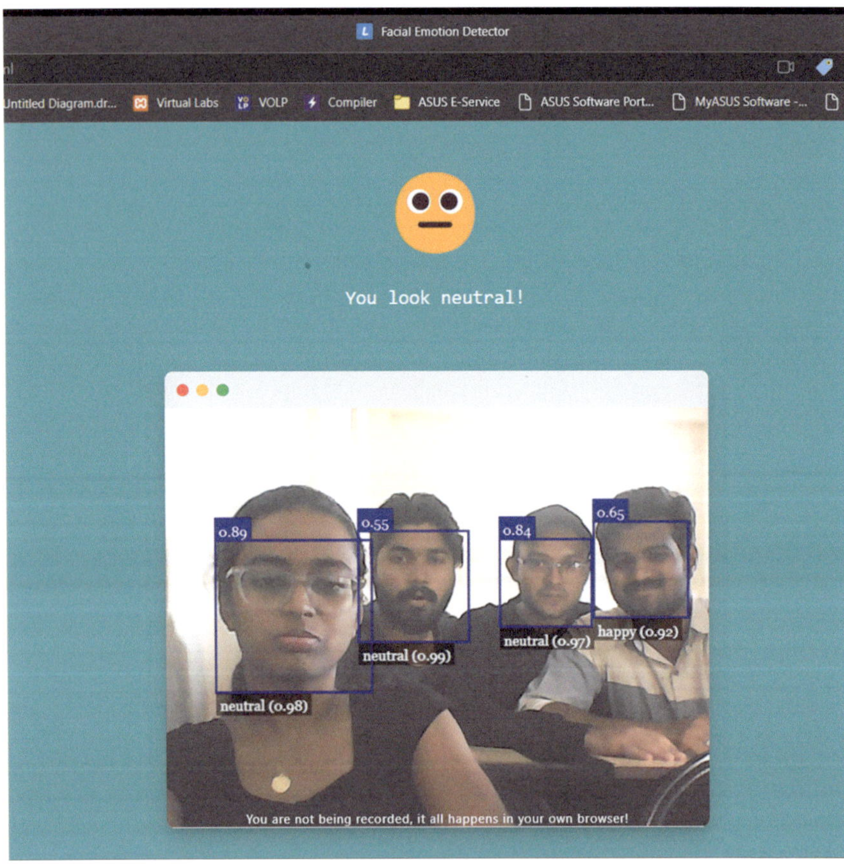

Fig. 4 The model is detecting multiple faces and their respective emotions along with intensity

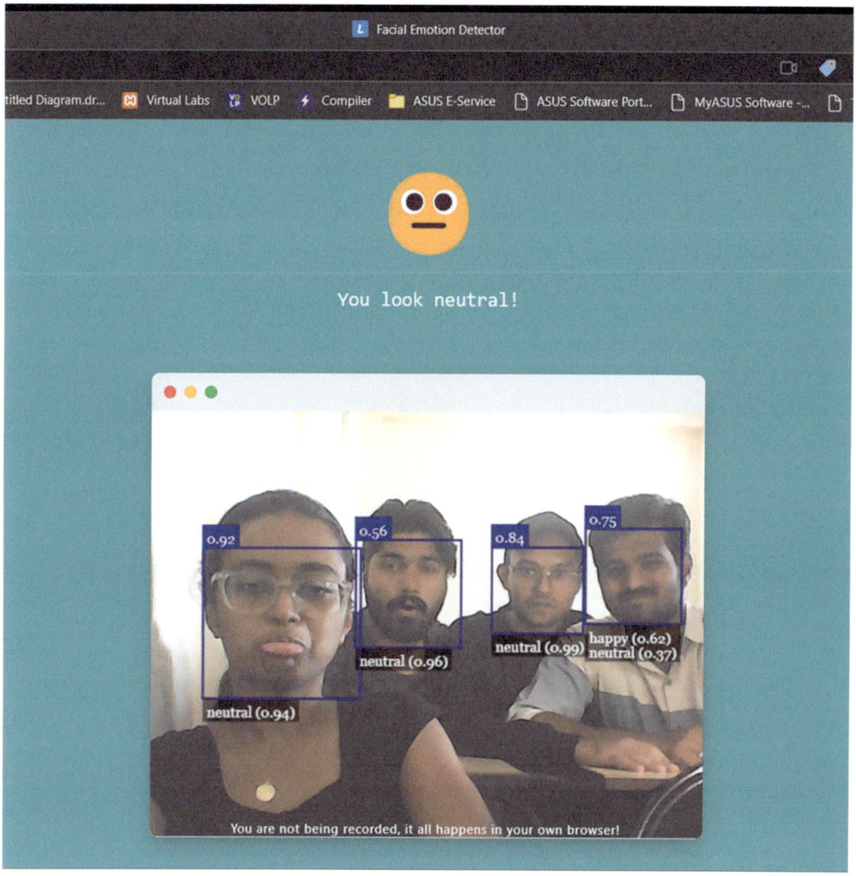

Fig. 5 The model is detecting multiple faces

6 Future Scope

In our future endeavors, we aim to enhance accuracy by introducing the feature of bitmoji—an AI-generated avatar reflecting the user's expressions. Addressing a current limitation, we plan to improve face detection on images displayed on paper or mobile through advanced AI techniques. Presently, the project employs CNN and VGG-16 for visual emotion detection, but expanding its capabilities by incorporating audio (voice analysis) and text (sentiment analysis) can offer a more comprehensive understanding of user emotions. While VGG-16 serves as a solid foundation, experimenting with newer models like ResNet, Inception, and EfficientNet may yield significant improvements in emotion detection. Real-time performance is crucial, and exploring optimization techniques, such as quantization and hardware acceleration, can enhance efficiency in resource-constrained environments. These future steps aim to evolve the project into a sophisticated system with applications in mental healthcare and emotional computing.

7 Conclusion

In conclusion, real-time facial emotion recognition using machine learning has progressed significantly, particularly with the rise of deep learning and the availability of large datasets. Both VGG-16 and CNN models demonstrated effective real-time emotion detection, with the VGG-16 model showing slightly higher accuracy and improved generalization on untested data. The system's overall accuracy surpassed the baseline, showcasing the potential of deep learning in emotion recognition. The system exhibited high-speed real-time performance, enabling seamless emotion detection in live video streams, opening avenues for applications in human–computer interaction, virtual reality, mental health monitoring, and personalized user experiences in entertainment and gaming. Despite achievements, challenges like dataset diversity, hyperparameter optimization, and variations in lighting conditions and facial expressions were encountered. Future enhancements could involve broader and more diverse datasets, fine-tuning hyperparameters, and exploring advanced techniques like transfer learning for improved performance. This project contributes to the field by highlighting the potential of VGG-16 and CNN models in real-time applications, offering new insights into understanding human emotions in real-world scenarios with substantial implications for industries such as healthcare, marketing, and entertainment.

References

1. Cheong WL, Char CM, Lim YC, Lim S, Khor SW (2010) Building a computation savings real-time face detection and recognition system. In: 2010 2nd international conference on signal processing systems, Dalian, China, pp V1-815–V1-819. https://doi.org/10.1109/ICSPS.2010.5555245
2. Kumar A, Kaur A, Kumar M (2019) Face detection techniques: a review. Artif Intell Rev 52:927–948. https://doi.org/10.1007/s10462-018-9650-2
3. Thankachan D, Tripaathi N (2022) An efficient method to face and emotion detection
4. Naik N, Mehta MA (2020) An improved method to recognize hand-over-face gesture based facial emotion using convolutional neural network. In: 2020 IEEE international conference on electronics, computing and communication technologies (CONECCT), Bangalore, India, pp 1–6. https://doi.org/10.1109/CONECCT50063.2020.9198376
5. Pauly L, Sankar D (2015) A novel online product recommendation system based on face recognition and emotion detection. https://doi.org/10.1109/ICCICCT.2015.7475299
6. Kirana KC, Wibawanto S, Herwanto HW (2018) Emotion recognition using fisher face-based viola-jones algorithm. In: 2018 5th international conference on electrical engineering, computer science and informatics (EECSI), Malang, Indonesia, pp 173–177. https://doi.org/10.1109/EECSI.2018.8752783

7. Wang W, Peng Y (2013) Face detection based on adaboost and face contour in e-learning. In: 2013 ninth international conference on natural computation (ICNC), Shenyang, China, pp 1026–1030. https://doi.org/10.1109/ICNC.2013.6818127
8. Jadhav RS, Ghadekar P (2018) Content based facial emotion recognition model using machine learning algorithm. In: 2018 international conference on advanced computation and telecommunication (ICACAT), Bhopal, India, pp 1–5. https://doi.org/10.1109/ICACAT.2018.8933790

Leveraging NLP and Machine Learning to Detect Cyberbullying on Digital Platforms

Prerna Mahajan and Swati Sah

Abstract Everyone nowadays uses social media to comment and express themselves freely. Facebook, Instagram, YouTube, Snapchat, and Twitter are just a few examples of social networking platforms where users may instantly share photos and videos throughout the world. Regrettably, these additionally, as a result of technology advancements, cyberbullying has increased, leading to despair and even suicidal thoughts in children, adults, and both. Recent events have demonstrated the power of a single tweet to sway opinion, the market, and even politics. People now have a great deal of animosity, which they express through verbal abuse and harsh remarks. Finding a method to stop cyberbullying is essential. Through the use of a recent Twitter dataset from Kaggle, this research effort seeks to identify cyberbullying based on various demographic groups. The work also made use of SMOTE for data balancing and TF-IDF for feature extraction to achieve better results than the baseline strategy. Classical supervised machine learning algorithms, BiLSTM and BERT models, were applied on pre-processed dataset. The performance of the various models is evaluated using metrics. SVM and deep learning models fared better than the other models. The study also explains how age, gender, religion, ethnicity, and other forms of cyberbullying and non-bullying interact in prediction models.

Keywords Sentiment analysis · LSTM · BERT

P. Mahajan (✉)
School of CS and IT, Jain (Deemed to be) University, Bangalore, India
e-mail: prerna.m@jainuniversity.ac.in

S. Sah
School of Engineering and Technology, Sharda University, Greater Noida, India
e-mail: swati.sah@sharda.ac.in

A. P. Agrawal et al. (eds.), *Advances of Artificial Intelligence and Machine Learning in Image Processing, Security and Data Analytics*, Lecture Notes in Electrical Engineering 1464, https://doi.org/10.1007/978-981-95-0148-9_19

1 Introduction

The unprecedented technological growth has led to advancement in society at different levels. Internet has become pervasive in our day-to-day life and has offered several social and educational benefits to all. People are becoming more and more dependent on social media for sharing their views, connecting with people, expressing their emotions. However people can fake their identity over social media platforms which can lead to adverse effects across all age groups.

Many cyberbullying trends are observed over social media and a variety of trends have been observed like anonymous bullying, online harassment and trolling, sharing inappropriate content, social manipulation, online memes, impersonification, and fake accounts [1, 2]. With the ability to create anonymous profiles or use fake accounts, cyberbullies often target individuals while hiding their identity. This makes it harder to track down the perpetrators and hold them accountable. Individuals, especially public figures, may face targeted harassment, hate speech, and derogatory comments online. This can include body shaming, racial or ethnic slurs, and personal attacks. Also cyberbullies may share or distribute private or embarrassing photos, videos, or personal information without consent. This practice, known as "revenge porn" or "sexting shaming," can cause significant emotional distress and harm the victims. In certain scenarios, many cyberbullies may engage in social exclusion or manipulation tactics, such as spreading rumors, creating social hierarchies, or intentionally excluding someone from online groups or conversations.

People also cyberbully through memes which are shared widely across social media platforms. Bullies create and circulate memes with the intention of mocking or humiliating specific individuals or groups. It may also be in the form of Doxing, which involves the public release of someone's personal information, such as their address, phone number, or workplace, without their consent. This can lead to offline harassment and safety concerns for the victims. Cyberbullying is prevalent in online gaming communities, where players may engage in verbal abuse, threats, or humiliation toward others during gameplay. They may create fake social media accounts, posing as someone else to harass or intimidate individuals. This tactic can lead to reputation damage and further psychological harm.

It's important to note that cyberbullying trends can vary depending on the specific platform, cultural context, and the demographics of the users involved. Social media platforms continuously work to address these issues by implementing stricter policies, providing reporting mechanisms, and incorporating automated systems to detect and mitigate cyberbullying incidents.

However detecting cyberbullying over social media can be challenging, It requires vigilance, active involvement, and a commitment to creating a safe online environment. Collaboration between individuals, social media platforms, schools, and law enforcement can be essential in addressing cyberbullying effectively. Popular platforms like Facebook, Twitter, Instagram, Snapchat, and TikTok should be monitored regularly to identify potential cyberbullying incidents. People should familiarize themselves with the platform's policies and reporting mechanisms. A healthy environment should be created, where individuals feel comfortable reporting cyberbullying. Digital literacy should be promoted and individuals can be educated about online safety, responsible social media use, and the consequences of cyberbullying. Online tools and machine learning techniques should be employed to scan social media platforms for specific phrases, slurs, or keywords commonly associated with cyberbullying. These tools can help flag potential incidents and provide early detection. Machine learning models can be used to regularly review comments, messages, and posts on social media profiles. It can be used for early detection of online harassment, sudden changes in behavior, withdrawal from social media, or signs of distress from individuals.

Such tools can be used to report and document incidents to platform administrators, school authorities, or law enforcement if necessary. Hence they will be beneficial to offer timely support and reassurance to the victims and if the victim is a minor, involve parents, guardians, or school administrators as appropriate. It is very crucial to help youth from suicidal tendencies and depression due to such cyberbullying incidents. In this research paper, we propose to develop a machine learning model to identify cyberbullying in social media content. The model will eventually help administrators in taking necessary actions such as blocking or reporting the perpetrator can be taken to ensure the victim's safety and prevent further incidents.

2 Literature Review

The section discusses the results obtained by machine learning and deep learning and the challenges posed by them separately as two subsections presented next.

2.1 Machine Learning Methods

Machine learning techniques have been extensively employed to detect and address cyberbullying incidents on different social media platforms. Traditional classifiers such as logistic regression, support vector machines, naïve Bayes, random forest, and gradient boosting have been used and assessed for their performance on popularly used social media platforms like Twitter, Facebook, Instagram, MySpace, etc. Researchers compared linguistically supported word embeddings on the Kaggle cyberbullying dataset, using SVM to assess the embeddings and neural network topologies [3]. Recurrent neural network, multilayer perceptron classifier, convolutional neural network (CNN) with rectified linear units, and long short-term memory (LSTM) were employed as a neuron activation function. It was determined that SVM performs better for pertained word embeddings while CNN performs better for training the language model on target data.

Using SVM classifiers and sentiment analysis with TF-IDF, the author of a recent study [4] offered a solution for cyberbullying detection and prevention.

1000 tweets were manually classified because the dataset was not labeled with performance of 75.1% accuracy and 75% recall, precision, and f1-score. The author of the proposed study noted a shortcoming in the ability to identify sarcastic writing as cyberbullying.

In a further study, the author Rahman et al. [5] offered machine learning techniques as SVM, random forest, logistic regression, decision tree, and MNB to categorize cyberbullying text data. The accuracy and f1-score for random forest with SMOTE were 89%.The author has also noted that Unicode in text was not recognized, which prevented some terms from being removed and was also mentioned as a study constraint. The amount of the dataset and the short training period were further restrictions.

Online tweets were retrieved using Twitter API and machine learning model using TF-IDF was used to identify cyberbullying [6]. Authors classified a tweet as bullying or not using probability predicted by linear kernel-based SVM and naive Bayes method. Additional 10 tweets from user timelines were included based on the decision in order to further identify cyberbullying. For SVM and naive Bayes, accuracy was reached at 25 and 52.70%, respectively.

Wang et al. [7] collected balanced data from Twitter using mining techniques including dynamic query expansion. Authors suggested the Graph Convolutional Network (GCN) as a classifier. Supervised classifiers like logistic regression, K-NN, XGBoost, naïve Bayes, and MLP were examined alongside the SOSNet model. The method was applied on 40000 tweets and it was found out that XGBoost method with TF-IDF and BOW methods outperformed other methods in terms of accuracy and F1-scores by a factor of 7. Only 73 tweets were incorrectly identified when SBERT and SOSNet were combined, demonstrating the resilience of the DQE approach.

Liu et al. [8] conducted another study on non-linguistic features for cyberbullying detection. With the best accuracy of 81.95%, the author used multidimensional features generated on the KNIME machine learning platform to identify cyberbullying using supervised classification models.

Various data sources and categories were identified and several machine learning approaches were utilized for detecting cyberbullying content by Mahlangu et al. [9]. They emphasized that the fundamental issue is the dearth of publicly accessible multimedia content, which makes the majority of data available in text form rather than in the form of photos, audio, and video information. Additionally, their investigation revealed that SVM was frequently utilized for classification.

Nurrahmi and others [10] used text mining to identify cyberbullying in texts written in Indonesian. Unlabeled Twitter data with controversial posts was gathered using the Twitter API, and 301 cyberbullying tweets and 399 non-cyberbullying tweets were tagged using crowdsourcing.

In a study authors utilized eight rules for feature extraction. Support Vector Machine (SVM) and K-nearest Neighbor (K-NN) accuracy results were compared and attained F1-score of approx. 67% using SVM [11].

Hee et al. make use of employed Dutch language question–answer posts data from social media platform that has implicit cyberbullying content [12]. The outcome was consistent with studies using classifiers like SVM conducted by Dadvar et al. [13] according to state of the art. The model's accuracy was 53.82% when the distribution of positive posts remained constant.

Using machine learning techniques, the study identified cyberbullying in the Kaggle dataset and proposed two novel hypotheses for extracting relevant feature in detecting cyberbullying. Word embeddings for the skip-gram were employed to handle pronouns in text and have improved the performance and accuracy of logistic regression by 86% [14].

2.2 *Deep Learning Models for Cyberbullying Detection*

Over the past few years, there has been a noticeable transition from the utilization of machine learning models to a growing reliance on deep learning models. A recent study conducted in 2022 shed light on the behavior of abusive speech, revealing a significant weekly and yearly seasonality that is not as apparent in the raw data [15]. Throughout the years 2019–2021, consistent patterns emerged, including two distinct spikes in the overall trend during the first and second halves of each year, albeit with slight variations.

From their separate daily samples, the percentage of these potentially offensive tweets also appears to roughly decline over time. By taking into account the significant seasonal tendency, the findings can allow creators of Twitter or other social media to ramp up mitigation technology at the proper moment.

A novel attempt used deep convolutional neural networks and SOTA models like VGG16 and Inception V3 to suggest best model to prevent image-based cyberbullying [16]. They produced two datasets, each including 3000 non-bully photos and 100 images of bullying. Through testing, it was discovered that transfer learning models outperformed deep learning models.

Research on feature density and linguistically supported pre-processing techniques, used in dialog categorization and cyberbullying detection, was presented by study from 2022 [3]. Both ideas are relatively new to the industry. They examined the use of several linguistically supported pre-processing techniques in the context of cyberbullying detection and looked at the impact of feature density in lowering the number of experiment repeats necessary. The outcomes show that there is a best feature density for non-CNN classifiers, which somewhat varies for each classifier.

This can be considered in subsequent experiments to execute them faster and with less processing power. It is recommended to use the higher feature density datasets to produce better results.

Yi and Zubiaga reviewed and evaluated the various methods for detecting cyberbullying, with an emphasis on session-based methods [17]. To achieve the same, they developed a framework for cyberbullying detection, namely, Social media session-based framework (SSCD), comprising of four key elements. They evaluated previous work in cyberbullying detection through model and dataset construction by going over the research problems and advancements on the four components of the SSCD framework, mainly aiming on those dealing with social media sessions. They also emphasized the significance of taking repetition and power imbalance—two of the fundamental aspects of cyberbullying—into account when creating datasets and experiments.

The authors [18] discussed existing dataset and models, and performed comparative evaluations on SSCD datasets. They proposed a model comparing basic transformers with Res-CNN BiLSTM and discovered that transformers produce faster results while training phase. The results given by transformer were far superior to Res-CNN BiLSTM's, where the number of training parameters were far less in transformer as compared to Res-CNN BiLSTM. For training and validation, accuracy was determined and used as an assessment metric. In the most recent epoch, 1-D character embedding achieved 98.52% training accuracy and 91.81% validation accuracy. For the Glove Embedding Model's most recent epoch, accuracy scores of 93.65% and 92.25% for validation were attained. For the transformer-based network, validation accuracy of 92.63 and training accuracy of 95.73% were calculated, leading to the conclusion that transformer's validation accuracy is higher than Res-CNN-BiLSTM.

Research conducted on doxing, a different type of cyberbullying refers to the disclosure of sensitive personal information without the victim's consent [19]. They created a doxing dataset using the Twitter API, and they then contrasted nine automated detection methods based on string matching and one-hot encoding. Their strategy worked well, yielding recall of 97.37% and accuracy of 96.86%.

Researchers from Germany focused on combining textual elements for cyberbullying detection in 2021 [20]. Here, they worked with datasets for Hate Speech and Offensive Content in English and Indo-Aryan. To extract feature-specific vectors, input tokens were fed to BERT using neural network architecture. Additionally, they gathered hate words by integrating the Gomez et al. dictionary with other internet dictionaries. A significant amount of manual labor was required to filter out words that do not represent hatred.

Since the dataset for 2019 and 2021 was not evenly distributed, they employed oversampling to train the divides. By comparing the accuracy, F1-score, and weighted F1-score of BERT base embeddings and HateBERT embeddings, they discovered that BERT base embeddings perform better.

In their comprehensive study [21], authors meticulously examined hate speech within the Fire 2021 English language dataset. They employed the ktrain-BERT model, conducting an extensive training process consisting of 40 epochs across 5 consecutive cycles. This allowed them to evaluate the model's performance over time. The results were highly impressive, showcasing 87.16% precision, 83.63% F1-score, 97.28% training accuracy, and 82.60% validation accuracy. Notably, the researchers went beyond the scope of individual word analysis; they also examined hate speech within the context of entire tweets.

Many researchers have also studied multilingual data for cyberbullying detection, where the results have demonstrated the reliability of cross-lingual contextual embeddings. Authors [22] have made comparison on multiple languages like Spanish,

Turkish, Hindi, Bengali, Danish, Arabic, and Greek, and have discovered objectionable language terms. In order to work on multilingual data they have explored recent Transformer XML-R models, which have been trained on more than hundred different languages, since BERT model has its limitation of not being trained on cross-lingual data.

The research conducted [23] involved the utilization of the Hate Speech dataset. This dataset comprised of 85,000 tweets and was annotated through crowdsourcing. A similar approach was followed in the Indonesian cyberbullying detection study carried out by [10]. The classification task in both studies involved assigning the tweets into three distinct categories: Hateful, Abusive, and Normal. Through the implementation of ten cross validations on the dataset, impressive results were achieved. The research yielded a remarkable 91% F1-score, 92% recall, and 91% precision.

In the pursuit of identifying cyberbullying, authors [24] utilized deep learning strategies employing advanced systems as an alternative to conventional machine learning techniques. By employing deep learning models such as GRU, RNN, and BLSTM, tokens were analyzed, with the Residual BiLSTM model yielding the most exceptional accuracy. As a result, the team achieved the highest possible f1-score of 75%.

Research conducted on MMHS150K produced one of the largest Hate Speech datasets for images. Even models that analyze multimedia cannot surpass models that evaluate texts, according to authors [25]. The data, both qualitative and quantitative, were examined for the presence of cyberbullying.

The biggest restriction in detecting cyberbullying, according to research [26] analysis of obstacles and limitations on research done thus far, is the datasets extraction, tagging based on features, demographic details like age and gender of the user making comments, among other factors. They also mentioned text authorization and data privacy as factors affecting text research.

Authors collected Twitter data of past users to try to find racist people and material. Donald Trump and Hillary Clinton were candidates for presidential elections of the United States in 2016 [27]. They examined 200 recent tweets to gauge the intensity of negativity. They first gathered what they referred to as friend networks and followers before eliminating spam users. They grouped individuals into five clusters using scores or attributes after utilizing sentiment analysis to determine how positive or negative information was present in each user's timeline. They tried to ascertain a user's political leaning from their tweets, and using Hashtags classified them as Trump- and Hilary-related ones. Based on retweets and hashtags, they further divided users into different groups, namely, Trump affinity, Trump homophilous, Hillary affinity, Hillary homophilous, and neutral. They used K-means algorithm for detection and sentiment analysis word count and racist score were the evaluation metrics.

According to earlier studies, we can classify social media text with some degree of accuracy, which could aid to some extent in the prevention of cyberbullying. The goal of the project is to reduce incidences of harassment while also assisting youngsters in avoiding stress that might result in depression or suicide due to cyberbullying.

Our initiative intends to categorize social media texts in order to determine whether or not they constitute cyberbullying. In our suggested process, the dataset will also be balanced using SMOTE, and features will be extracted using TF-IDF. Because of its high skew text classification for detecting cyberbullying using content-based features, SVM is the most prominent classification algorithm employed by academics in the majority of studies. Due to its great availability and open access, Twitter API data was frequently used when data was scraped.

3 Research Methodology

The following are the key procedures that will be followed in our proposed study of cyberbullying detection: data selection, pre-processing of selected data, data conversion to a structured format, dataset balancing, model implementation, and evaluation of machine learning performance using evaluation metrics.

3.1 Dataset Selection

Any model's performance in the identification of cyberbullying solely depends on the kind and quantity of datasets chosen. The source is also important since it demonstrates the scope of cyberbullying and the areas where it still needs to be addressed. Twitter API is the most trustworthy and widely used for social media analysis and is considered to be easy accessible source of text data in English language and has been extensively used for research over the years [15, 19].

The Kaggle dataset will be used in this work which was originally shared by [7], which is one of the most recent datasets produced and made available for usage by the general public. There are roughly 47 K tweets classified as cyberbullying, including age, religion, gender, ethnicity, not bullying, and other forms of bullying.

3.2 Data Preparation

Before applying any procedures to the dataset, data cleansing is an essential step because it affects the outcome and could produce false or erroneous results. To achieve the best results, data cleansing is necessary for the given dataset, whether it contains photos, text, or multimedia. Pre-processing is done in each of the examined prior investigations. Remove undesirable phrases that don't add any value to the study in order to reduce noise in the data.

Getting good results depends heavily on processes like stopword removal, punctuation removal, reducing case, grammar correction, and spelling check. With the exception of picture datasets, tokenization and stemming are employed more frequently than any other pre-processing processes.

Word2Vec Skip-Gram language model was created via tokenization using Word2Vec by [3]. Lemmas received the highest F1-scores in their investigation. They also used a variety of pre-processing techniques, including POS tags and Bags of Lemmas, for the neural networks experimentation. When words are reduced to their lemmas or stems, their frequencies combine into a single value, giving bullying words used in corpus a prominent value. To enable a text to be represented as a function of words, tokenization is conducted, which divides sentences and phrases into characters or tokens. We'll be cleaning up the data by removing stop words, emoticons, and punctuation. The root word will be obtained through stemming. After performing data cleaning, techniques such as TF-IDF and Word2vec with CBOW and Skip-gram embeddings are commonly used for feature extraction. In a study conducted [27], TF-IDF analysis was employed to examine their dataset, resulting in the identification of 10 key themes. Authors [6] employed the TD-IDF vectorizer for feature extraction from the dataset in order to achieve more accurate results. In this work, we would also try to determine the relevance of words by extracting features using TF-IDF from the dataset.

3.3 Data Balancing

The majority of machine learning techniques perform poorly because they ignore unbalanced datasets. Many of the datasets utilized in earlier research were unbalanced, therefore SMOTE method to oversample the minority classis used. SMOTE approach is used to enhance the quantity of samples for minority classes in order to address class inequality and also produced better results than baseline methods. Many researches have described how implementing SMOTE led to a considerable improvement in F1-score compared to conventional ML methods and also the results shown that SMOTE was unable to enhance deep neural model performance. We will be using SMOTE to balance the dataset in order to achieve better results as we are also dealing with a class imbalance.

3.4 Model Selection

Classification models will be employed in this stage to analyze and assess the model's performance. To evaluate the performance of model, the dataset will be split in the ratio of 80 and 20 in training and test sets, respectively. We will be using machine learning algorithms like SVM, logistic regression, decision trees, K-nearest random forests, naive Bayes, and deep learning techniques. Then, using evaluation measures to compare various algorithms, we will select the most accurate model for predicting cyberbullying.

3.5 Evaluation Metrics

Any model's performance is assessed using performance metrics. Whether a comment is bullying or not will have a significant impact on the accuracy of the cyberbullying detection process. According to previous research, a significant portion of researchers has placed a larger emphasis on the F1-score when attempting to determine the ideal model for evaluating performance. This is due to the fact that it represents the harmonic mean of precision and recall. To assess the effectiveness of the BERT model, researches [6, 23] utilized various evaluation metrics such as accuracy, AUC, precision, recall, and F1-score. In order to evaluate the effectiveness of deep learning models, the authors [24] have selected ROC curve, precision, recall, and F1-score as evaluation metrics for LSTM, RNN, and GRU. Additionally, the confusion matrix will be utilized as one of the evaluation matrices to determine the most suitable model. Due to the substantial imbalance between the number of bullying and non-bullying tweets in the dataset, relying solely on accuracy would not provide an accurate measurement of the model's performance. We have employed F1-score, precision, and recall to provide an impartial class result because the dataset is significantly unbalanced for the two categories of bullying and non-bullying. Time taken during training and testing phase can also be considered as a parameter to qualify as a better model (Fig. 1).

4 Experiments and Results

In order to gain insight into the frequently employed derogatory language on Twitter, an analysis of tweet data was conducted resulting in the creation of a word cloud. This research has been categorized into five distinct groups, namely, gender-based hate, religious hate, age-based hate, ethnicity-related hate, and other forms of cyberbullying (Fig. 2).

By examining the top 10 bigrams, we can gain a better understanding of the specific word combinations utilized in cyberbullying (Fig. 3).

Understanding cyberbullying tweets is made easier by data analysis and visualization. We were able to understand the most popular and often used terms by users in tweets thanks to Word Cloud. Bigram's plot helped us to understand that practically all forms of cyberbullying use the words "dumb," "idiot," and "f**k." We can observe that young people are more affected because most remarks focus on "girls," "high school," "black people," or "muslims." This can assist us in handling the problem by focusing on a particular group of people. Bullying can be prevented to some extent by raising awareness in schools, colleges, and even the workplace. However, children must also learn about equality and the effects of such crimes at home.

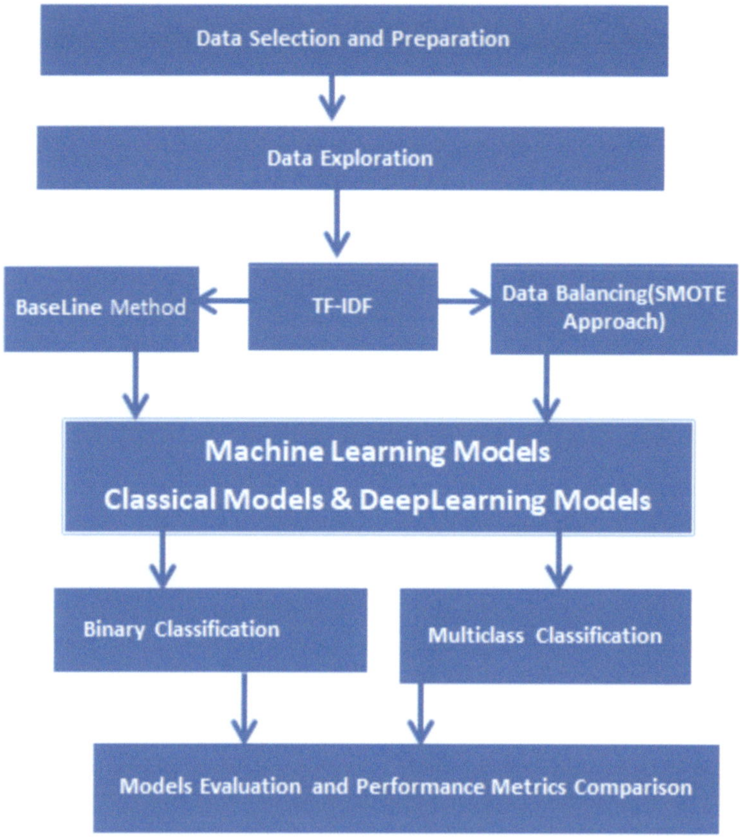

Fig. 1 Research methodology flowchart

5 Results

The dataset we utilized to classify the data into bullying and non-bullying consisted of four categories, namely, age, religion, ethnicity, and other types of bullying. To create a comprehensive bullying category, we merged all four types of bullying.

5.1 Model Evaluation with Data Imbalance

We conducted pre-processing on the data categories and utilized unbalanced data to train a variety of machine learning classifiers including naive Bayes, random forest, logistic regression, K-nearest neighbor, SVM, XGBoost, and decision tree. Subsequently, we compared the results with the dataset post SMOTE balancing. Considering the tweet categories, we effectively consolidated them into two distinct

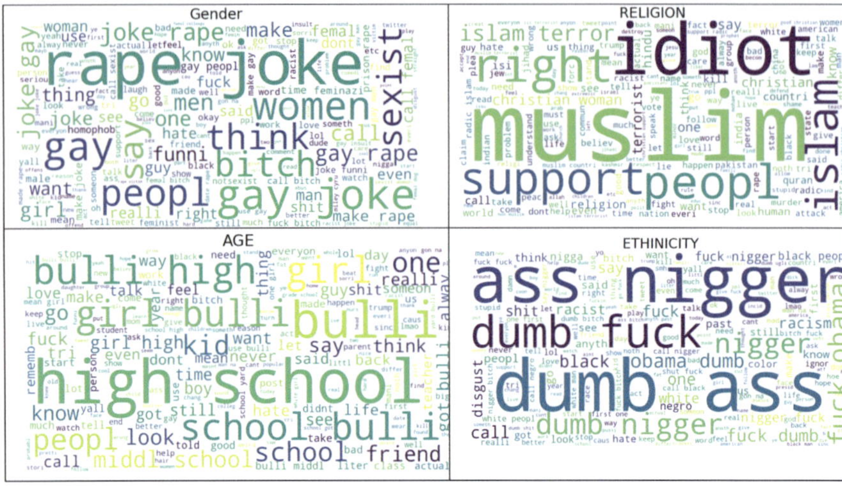

Fig. 2 Wordclouds for different categories for cyberbullying

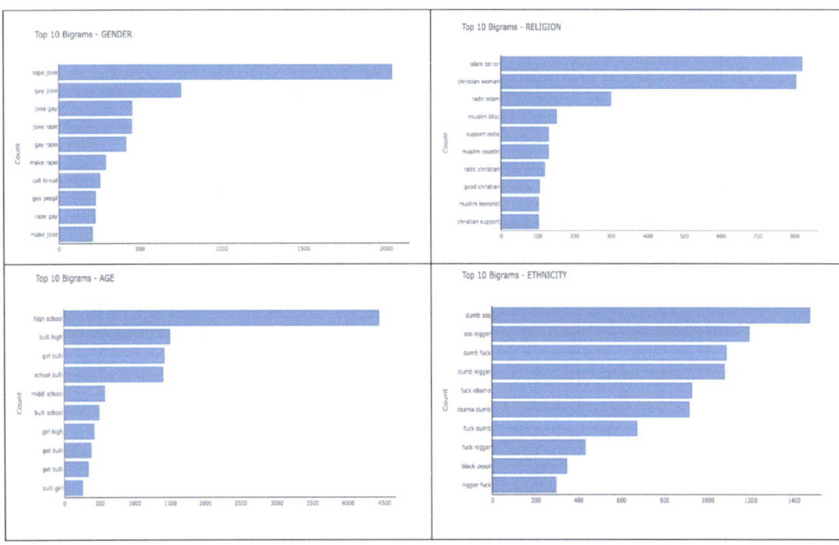

Fig. 3 Bigrams for different categories for cyberbullying

groups: bullying and non-bullying. However, this classification led to a significant imbalance, with 349,585 non-bullying tweets and only 6,091 instances of bullying. Through our analysis, we observed that the SVM model outperformed the other models in this scenario (Fig. 4).

The receiver operating characteristic curve results shown in Fig. 5 depicts that almost all models produced a curve, with the exception of the decision tree, which

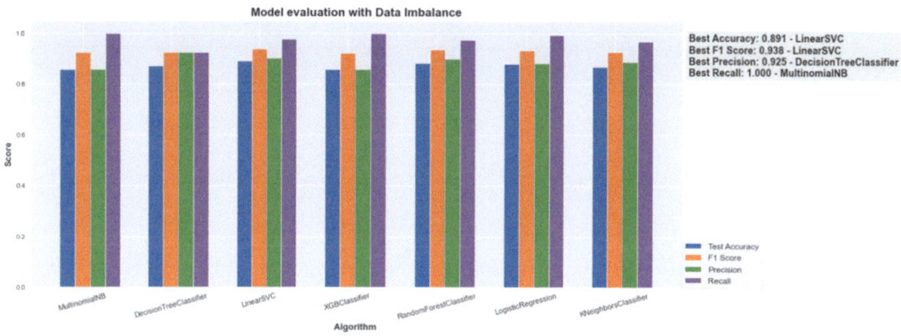

Fig. 4 Model evaluation for imbalanced data

produced a low performance because to its 45-degree diagonal tilt and lower accuracy. However, the area under the curve for all models used is almost equal, indicating that these models performed well (Fig. 5).

5.2 *Model Evaluation with SMOTE Method*

The dataset exhibited a significant imbalance in favor of the bullying category after merging non-bullying categories into a single class. As a result, it was necessary to employ a balancing method to equalize the class distribution and enhance the representation of the minority non-bullying class. To achieve this, the oversampling method SMOTE was implemented on the training dataset. The figure below illustrates the performance comparison of models utilizing the SMOTE method, revealing that the random forest model attained the highest accuracy and F1-score. In terms of accuracy of prediction, the K-NN model gave worst performance.

In order to make informed business decisions, it is crucial to accurately assess the time and cost involved in training a machine learning model. If the training period is significantly longer than other options, it is advisable to explore alternative algorithms. Additionally, our analysis includes a comparison of both training and prediction times. Upon examining the provided Fig. 6, it becomes evident that K-NN should not be selected due to its poor prediction time (Fig. 7).

5.3 *Model Evaluation with LSTM*

To classify the data LSTM model is used, which makes use of the attention mechanism to make predictions. This model includes an LSTM network which takes embedded word representations as input and output hidden states. It also includes an attention layer, which uses these hidden states to compute attention weights and apply them to the encoder outputs to get a context vector. This context vector is passed through

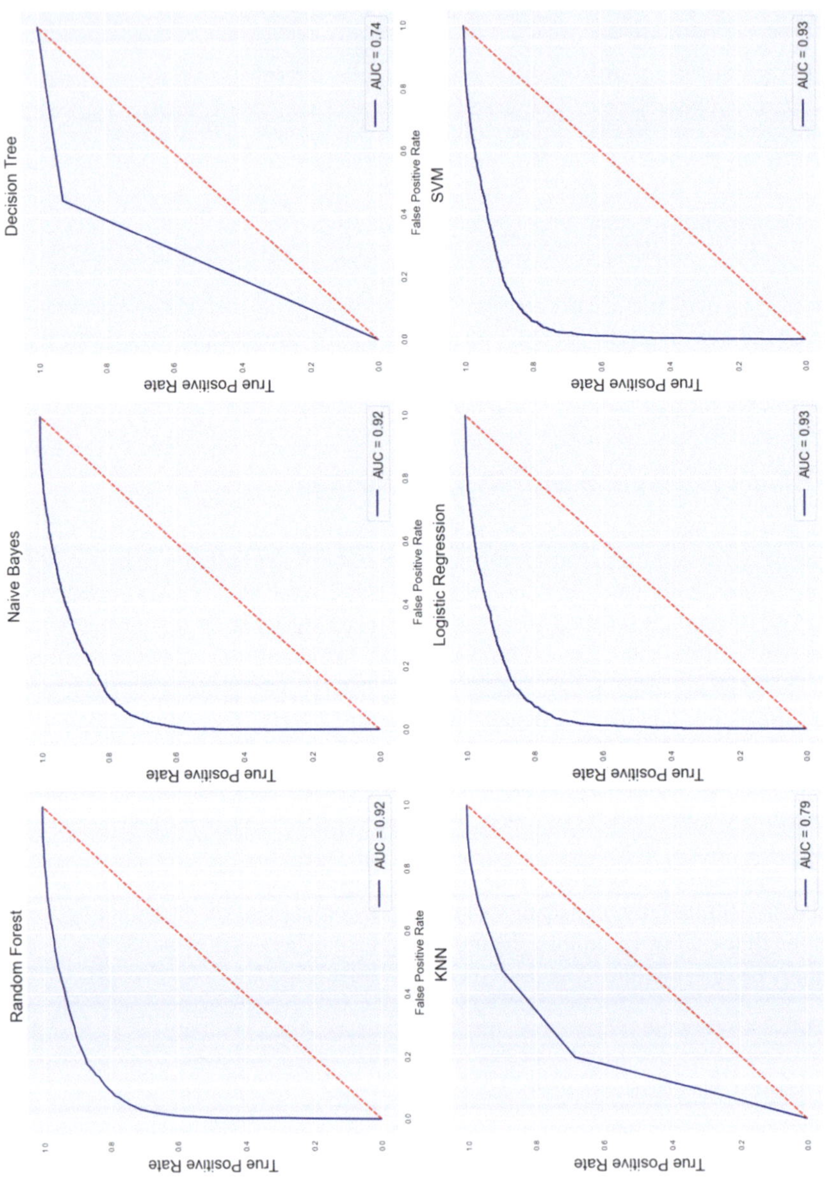

Fig. 5 ROC curve for different models

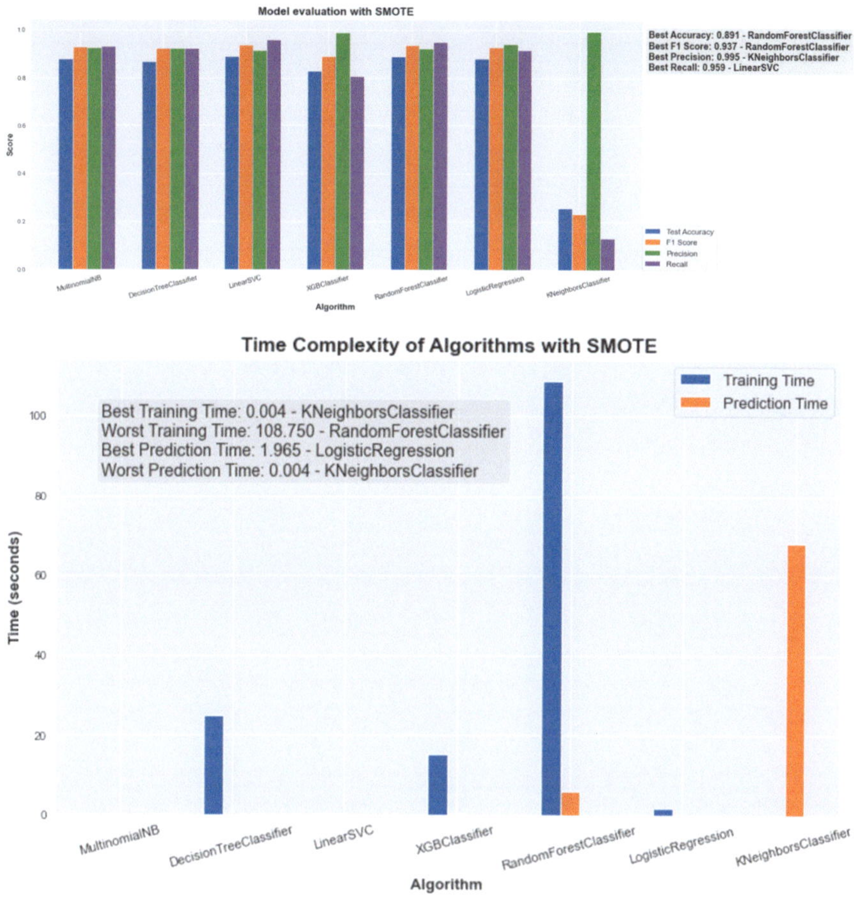

Fig. 6 Model evaluation for balanced data (using SMOTE) and time complexity

a fully connected layer followed by a log-softmax function to produce the final output probabilities. This LSTM model with attention allows us to better handle the complexities of language understanding by giving more importance to the relevant parts of our text data. This enhances the model's ability to understand context and thereby improves the performance of our detecting cyberbullying intent in the data. The performance scores of the algorithm are very high, with an overall accuracy of 93%.In particular, the F1-scores for the more populated classes are over 95%.

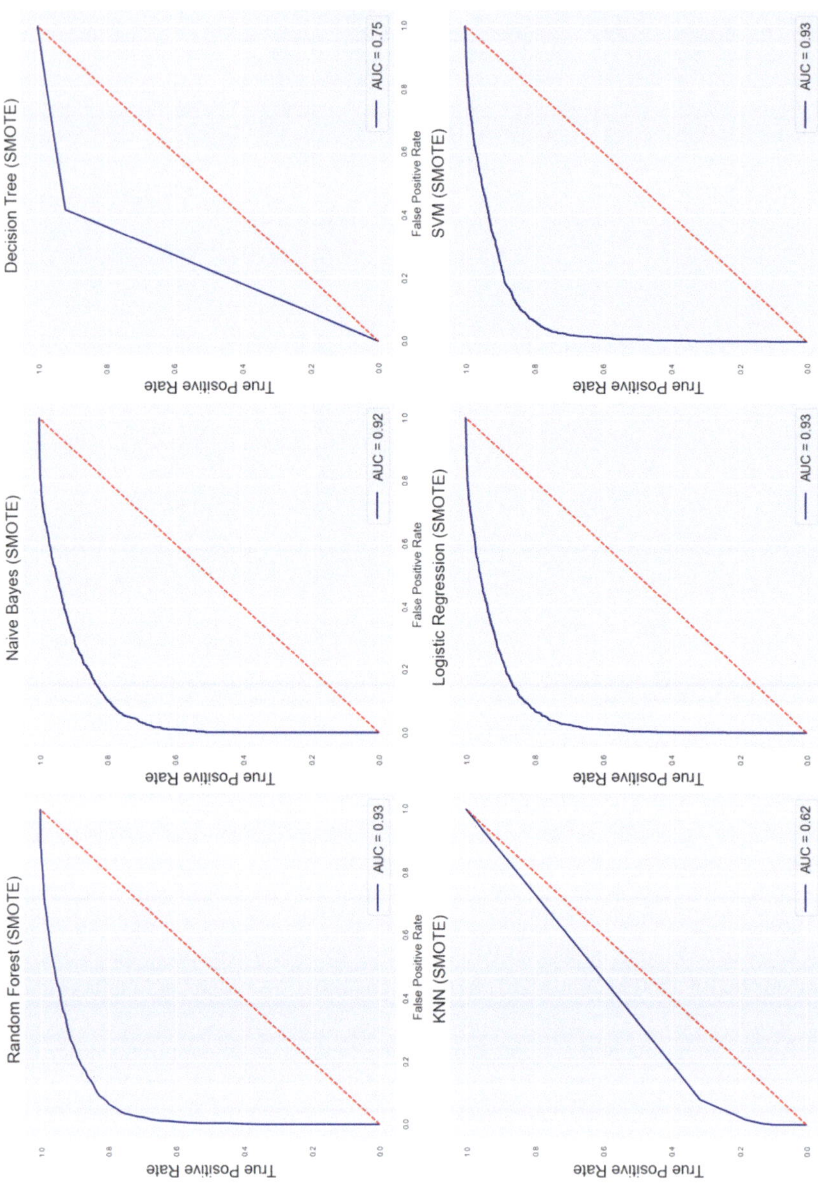

Fig. 7 ROC curve for balanced data (using SMOTE)

5.4 Model Evaluation with BERT

To identify cyberbullying tweets, we used a pre-trained BERT model from the Hugging Face library and fine-tune it for our classification task. First, we need to split the dataset into train–validation–test again since we need to tokenize the sentences differently from previously used machine learning models and LSTM. We created a custom BERT classifier class, including the original BERT model (made of transformer layers) and additional dense layers to perform the desired classification task. The performance scores of BERT classifier are quite high and higher than those achieved using the LSTM model, with an overall accuracy around 95% and F1-scores well over 95% (Fig. 8).

In our research, we discovered that SVM and random forest produced findings that were 86% accurate, or almost equal. Since we incorrectly categorized tweets that are merely discussions and not hateful content as bullying, the K-NN model may not have performed as well as it could have. We found that LSTM and BERT models outperformed machine learning models including SVM. Bias in bullying categories has resulted from the small size of the data and the manual labeling of cyberbullying categories. We discovered that a large number of victim tweets are labeled as bullying tweets in tweets. Numerous tweets have led to the misclassification of categories in the original dataset, which has affected classifier findings. This has also affected the outcome of our prediction (Fig. 9).

	precision	recall	f1-score	suppor
religion	0.95	0.96	0.96	1568
age	0.98	0.98	0.98	1548
ethnicity	0.98	0.98	0.98	1461
gender	0.93	0.85	0.89	1443
not bullying	0.79	0.86	0.82	1212
accuracy			0.93	7232
macro avg	0.93	0.93	0.92	7232
weighted avg	0.93	0.93	0.93	7232

	precision	recall	f1-score	suppor
religion	0.95	0.98	0.96	1568
age	0.99	0.98	0.98	1552
ethnicity	0.99	0.99	0.99	1469
gender	0.92	0.90	0.91	1446
not bullying	0.84	0.84	0.84	1214
accuracy			0.94	7249
macro avg	0.94	0.94	0.94	7249
weighted avg	0.94	0.94	0.94	7249

Fig. 8 Classification report (BiLSTM model and BERT model)

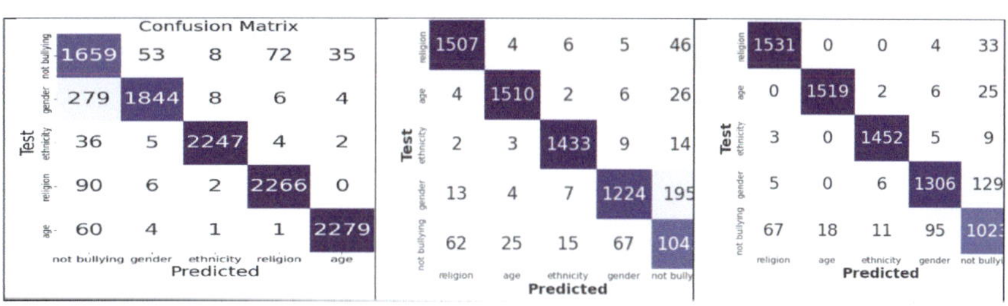

Fig. 9 Confusion matrix (random forest, BiLSTM, BERT model)

Our investigation was split into two stages: (i) cyberbullying and non-cyberbullying are separated and (ii) grouping information into five categories of cyberbullying, including those based on age, religion, gender, ethnicity, and other types of bullying or non-bullying. In the first phase, we combined all categories of bullying into one bullying category, split the test and train dataset, applied TF-IDF, and then balanced the dataset with SMOTE as the data was highly unbalanced. After that, we applied machine learning techniques like random forest, naive Bayes, decision trees, logistic regression, SVM, K-NN, and XGBoost and compared the results with and without the balancing method. With the exception of K-NN, we discovered that SMOTE enhanced all models' accuracy, recall, precision, and F1-score. Results indicated that the maximum accuracy of 89% with the SMOTE approach was provided by both random forest and SVM. We also looked at how long it took to train and test the model. With good accuracy of 89%, precision of 91%, recall of 96%, and F1-score of 94% and the shortest training time of 0.3 s and prediction time of 0.005 s, SVM was determined to be the best model by the time complexity analysis. Next, we used age, ethnicity, religion, and other forms of cyberbullying. Using the same methodology as in the first phase, we evaluated the results with and without SMOTE and discovered a similar outcome, with SVM and random forest achieving the greatest performance of 86%.

Due to its high skew text classification for identifying cyberbullying using content-based characteristics, SVM also outperformed with the greatest cross-validation score of 90% compared to the other algorithms utilized. We also evaluated results using BiLSTM model and BERT model, both the models outperformed previously discussed models with more than 90% score.

6 Conclusion

We are all aware that social integration has increased as a result of recent technology developments, which has had bad effects on society like cyberbullying. Through our research, we have established that children are the bullies' primary targets, which is consistent with a UNICEF study that found that young people are particularly susceptible to cyberbullying. Such public hatred and tweets have a negative impact on mental health, causing loneliness, depression, and suicide thoughts. Through our research, we were able to identify phrases like "rape," "gay," "nigger," "Muslim," "F**k," and "girls" that were often used by users. The most troublesome users or accounts can be blocked, banned, or quarantined with the aid of such terms. As preventive measures, content that an algorithm predicts or labels as cyberbullying can be collapsed or deleted. Threats can also be recognized and, to some extent, avoided. Focusing on such narrowly defined target audiences enables management of mass movement. By looking at these correlations and identifying bullying in tweets, we can assist policymakers be aware of potential crimes before they happen.

Our work has a drawback because it is domain based and we did not consider the user's mood. Therefore, we have retained the idea of sentiment analysis based on mood for use in the future.

The dataset size and training time were both constrained. The dataset was manually categorized by the authors, which demonstrates social biasness and leads to the misclassification of numerous utterances as cyberbullying even though they are only awareness tweets. In the near future, we hope to solve these concerns as well and to promote different deep learning techniques. To increase the detection of cyberbullying, additional ideas include working with experts in related sectors and disciplines including sociology and psychology.

References

1. DigitalTrends Homepage. https://www.digitaltrends.com/news/coronavirus-cyberbullyingdistance-learning/. Accessed 02 Oct 2023
2. UNICEF Homepage. https://www.unicef.org/press-releases/unicef-poll-more-third-young-people-30-countries-report-being-victim-online-bullying. Accessed 22 Oct 2023
3. Eronen J, Ptaszynski M, Masui F, Arata M, Leliwa G, Wroczynski M (2022) Transfer language selection for zero-shot cross-lingual abusive language detection. Inf Process & Manag 59(4)
4. Perera A, Fernando P (2021) Accurate cyberbullying detection and prevention on social media. Procedia Comput Sci 181:605–611
5. Rahman S, Talukder KH, Mithila SK (2021) An empirical study to detect cyberbullying with TF-IDF and machine learning algorithms. In: 2021 international conference on electronics, communications and information technology (ICECIT), pp 1–4. https://doi.org/10.1109/ICECIT54077.2021.9641251
6. Dalvi RR, Chavan SB, Halbe A (2020) Detecting a twitter cyber-bullying using machine learning. In: 2020 4th international conference on intelligent computing and control systems (ICICCS), pp 297–301
7. Wang J, Fu K, Lu CT (2020) SOSNet: a graph convolutional network approach to fine-grained cyberbullying detection. In: Proceedings of the 2020 IEEE international conference on big data (IEEE BigData 2020), pp 1699–1708, 10–13 December 2020
8. Liu Y, Zavarsky P, Malik Y (2019) Non-linguistic features for cyberbullying detection on a social media platform using machine learning. In: International symposium on cyberspace safety and security, pp 391–406
9. Mahlangu T, Tu C, Owolawi P (2018) A review of automated detection methods for cyberbullying. In: 2018 international conference on intelligent and innovative computing applications (ICONIC), pp 1–5
10. Nurrahmi, H, and Nurjanah, D. :Indonesian twitter cyberbullying detection using text classification and user credibility. In: 2018 International Conference on Information and Communications Technology (ICOIACT),pp. 543–548,2018
11. Sarna G, Bhatia MP (2015) :Content based approach to find the credibility of user in social networks: an application of cyberbullying. Int J Mach Learn Cybern 8(2):677–689
12. Hee V, Lefever C, Verhoeven E, Mennes B, Desmet J (2018) Automatic detection and prevention of cyberbullying in social media text. Cogn Sci AI 13(10):22
13. Dadvar M, Eckert K (2020) Cyberbullying detection in social networks using deep learning based models. In: International conference on big data analytics and knowledge discovery. Springer, pp 245–255
14. Chavan V, Shylaja SS (2015) Machine learning approach for detection of cyber-aggressive comments by peers on social media network. In: International conference on advances in computing, communications and informatics (ICACCI), pp 2354–2358

15. Perez C, Karmakar S (2022) An NLP-assisted Bayesian time series analysis for prevalence of twitter cyberbullying during the COVID-19 pandemic. arXiv:2208.04980
16. Roy PK, Mali FU (2022) Cyberbullying detection using deep transfer learning. Complex Intell Syst 1–19
17. Yi P, Zubiaga A (2022) Session-based cyberbullying detection in social media: a survey. arXiv: 2207.10639
18. Joshi R, Gupta A (2022) Performance comparison of simple transformer and res-CNNBiLSTM for cyberbullying classification. arXiv:2206.02206
19. Karim MR, Dey SK, Islam T, Sarker S, Menon MH, Hossain K et al (2021) Deephateexplainer: Explainable hate speech detection in under-resourced Bengali language. In: 2021 IEEE 8th international conference on data science and advanced analytics. IEEE, pp 1–10
20. Hakimov S, Ewerth R (2021) Combining textual features for the detection of hateful and offensive language. arXiv:2112.04803
21. Das M, Banerje S, Saha P (2021) Abusive and threatening language detection in Urdu using boosting based and bert based models: a comparative approach. In FIRE 2021: 13th meeting of the forum for information retrieval evaluation
22. Ranasinghe T, Zampieri M (2021) Multilingual offensive language identification for low resource languages. Transactions on Asian and low-resource language information processing. In: Proceedings of EMNLP, pp 437–443
23. Behzadi M, Harris IG, Derakhshan A (2021) Rapid cyber-bullying detection method using compact BERT models. In: 2021 IEEE 15th international conference on semantic computing (ICSC), pp 199–202
24. Iwendi C, Srivastava O, Khan S, Maddikunta PKR (2020) Cyberbullying detection solutions based on deep learning architectures. Multimed Syst 1
25. Gomez R, Gilbert J, Gomez L, Karatzas D (2020) Exploring hate speech detection in multimodal predictions. In: 2020 IEEE winter conference on applications of computer vision (WACV), pp 1459–1467
26. Ali W, Hamiza WN, Mohd M, Fauzi F (2018) Cyberbullying detection: an overview, pp 1–3. https://doi.org/10.1109/CR.2018.8626869
27. Lozano E, Cedeño J, Castillo G, Layedra F, Lasso H, Vaca C (2017) Requiem for online harassers: Identifying racism from political tweets. In: 2017 fourth international conference on eDemocracy & eGovernment (ICEDEG)

An Integrated Machine Learning Approach for Assessment and Early Detection of COVID-19 Using X-Ray Images

Rahul Deva, Arvind Dagur, and S. P. S. Chauhan

Abstract Large number of people around world were affected by global outbreak of coronavirus infection (COVID-19) and resulted in significant death worldwide. In this context, artificial intelligence has emerged as effective tool for early assessment and further analysis. In many cases, chest X-rays had played important and key role for early detection of COVID where image processing with machine learning can play very important role. The primary aim of our research is to find any computer-aided diagnostic method using X-ray images, deep learning method for COVID-19 detection. We used Convolution Neural Networks (CNN) ResNet50 for training of dataset and further applied top performing machine learning algorithms. Our algorithm was further evaluated on an independent test dataset. The chosen algorithms, combined with support vector machine (SVM) and machine learning features, exhibited very good suitability for distinguishing COVID, which boast and increase severity detection sensitivity and specificity in range 96.85–98.02%, while also maintaining robust performance across varying severity levels. Proposed method eliminates the need for manual interventions such as sample collection, enabling seamless integration with standard X-ray reporting model with help of advance machine learning. This integration offers an efficient, effective, and early detection solution for COVID-19 to address global healthcare challenge.

Keywords COVID-19 · RT-PCR · X-Ray images · Artificial intelligence · Machine learning

R. Deva (✉) · A. Dagur
Galgotias University, Greater Noida, Uttar Pradesh, India
e-mail: rahul.20SCSE3010019@galgotiasuniversity.edu.in

A. Dagur
e-mail: arvind.dagur@galgotiasuniversity.edu.in

S. P. S. Chauhan
Sharda University, Greater Noida, Uttar Pradesh, India
e-mail: sanjay.chauhan2@sharda.ac.in

A. P. Agrawal et al. (eds.), *Advances of Artificial Intelligence and Machine Learning in Image Processing, Security and Data Analytics*, Lecture Notes in Electrical Engineering 1464, https://doi.org/10.1007/978-981-95-0148-9_20

1 Introduction

COVID-19 virus initially started from late 2019 and spread all over the world and later converted into global pandemic and created global health crisis. There have been approximately 158 million COVID-19 confirmed patients and over 3 million deaths attributed to this novel virus. The rapid and fast spread of COVID virus continues to pose a significant challenge worldwide, with many countries lacking the essential resources to conduct widespread and expedient testing. The conventional RT-PCR test, though accurate, is time taking, expensive, and exhibits variable sensitivity rates ranging from 53 to 88%. Furthermore, numerous nations lack the infrastructure to implement widespread RT-PCR testing. Consequently, there is an urgent need to explore alternative solutions for developing an automated and rapid diagnostic system capable of accelerating decision-making while substantially reducing false negative results.

For early identification and severity assessment of COVID-19, Deep Learning (DL) techniques primarily rely on chest's radiography through X-ray and computed tomography (CT) scans, and has emerged as a dependable solution for early diagnosis. Over recent years, extensive research has demonstrated that medical imaging of the lungs offers a promising avenue to address the limitations of RT-PCR testing. Among these imaging modalities, X-ray images and CT scans have exhibited remarkable performance compared to conventional approaches. X-ray images, in particular, are more readily accessible and easier to acquire, making them a practical choice for widespread implementation. Concurrently, CT scans offer enhanced interpretability for medical professionals, further augmenting diagnostic accuracy and sensitivity in COVID-19 detection. The utilization of X-ray and CT scans has significantly helped in diagnostic capabilities and reactivity of COVID-19 recognition [1–3]. The widespread availability of these imaging modalities in hospitals worldwide has amplified their effectiveness for identification of COVID-19 patients compared to traditional methods.

In conclusion, the urgency of addressing the global COVID-19 pandemic necessitates innovative approaches to expedite diagnosis and reduce false negative results. Deep learning techniques applied to medical imaging, specifically X-ray scans, offer a promising solution to subdue the limitations of traditional RT-PCR testing. These modalities not only provide enhanced accessibility but also improve diagnostic accuracy, making them invaluable tools to stop random spread of COVID. Further research and development in this direction hold the potential to revolutionize early detection and ultimately curb the spread of the virus on a global scale. To reduce workload and burden of healthcare staff and rapid spread of virus, there is an urgent requirement of some computer support system for early detection of virus and further treatment and analysis. RT-PCR was used as primary standard for diagnosis with the success rate of 75–97% based on the sample collection, quality and other precautions used [4], stages and steps to identify the virus [5], and mutation rate of COVID-19 virus spread and multiplication. For similar detection of COVID-19 case, some samples of different types like blood, stool, urine, etc. were used with variable and different results.

X-Ray and CT give the higher success rates for early detection and identification of COVID and further analysis in comparison with the initial examination using RT-PCR and antigen test [6]. One another research suggested the same results using chest CT images for early identification of COVID and pneumonia for about 82% patients those were identified negative result using RT-PCR. As per the radiological society, the severity detection rate of X-rays for identification of COVID-19 virus was about 97.3% comparison with RT-PCR success rate of about 71.2% [7]. So X-Ray can be used and considered as first and primary tool for early identification of severity of COVID infection.

As compared with CT CXR has many other advantages than portability and availability, CXR also has short assessment time, minimal expense done effectively and rapidly [8]. Dewan et al. [9] and some of the research done on IOT-based parameter readings that will help to record and send the health parameters to send healthcare centers.

Machine learning-based feature classification and extraction algorithms can be utilized for the smaller dataset for training. Second-order feature selection known as radiomics has already been used for detection of radiographic phenotype in cancer.

2 Related Research

Artificial intelligence-based system can help a lot for early identification of COVID-19 considering different available data like CT scans, X-Ray and other clinical data. Yao and Zhang [10] done one work using applied SVM model to classify COVID patients as per symptoms and other available parameters. Yao et al. use SVM with binary class label for 137 patient data like blood test, urine culture, X-ray, etc. Outcome result findings are approx. 32 cases were found COVID-19. It was also concluded that considering many factors like age and gender also affected the study from severe and mild and patients with age more than 65 were having more severe symptoms. Higher risk and cases were found in male patients in comparison to female patients when considered based on parameters like blood test, urine culture, X-Ray, etc.

Similarly, Sun and Song [11] also suggested SVM-based model for COVID-19 prediction for early assessment and diagnoses. In this study, they also use clinical data and X-ray related with these cases. These methods have demonstrated exceptional potential in distinguishing between various lung conditions, including Viral Pneumonia (VP), COVID-19, and normal cases, utilizing medical scans modalities such as X-ray and CT scans. Additionally, recent advances have explored the utilization of Electrocardiogram (ECG) data for COVID-19 analysis and identification. This abstract provides a comprehensive overview of the diverse deep learning strategies employed in this context.

The fundamental rationale for employing deep CNNs over traditional hand-crafted methods lies in their capability to auto-extract salient features from the available data, obviating requirement for manual feature extraction engineering. CNNs leverage multiple layers of feature extraction, enabling them to autonomously learn and represent patterns from extensive datasets. One prevailing notion is that CNNs begin by detecting low-level features, like edges, and progressively evolve to recognize complex shapes, ultimately facilitating the classification of objects (e.g., cats). Deep-learning-based techniques for COVID-19 detection could be classified into various classes based on respective feature extraction strategy and classification approach.

2.1 Deep Learning Generic Model

This category involves training deep models from scratch using labeled datasets tailored to COVID-19 identification and detection. Zheng et al. [12] employed supervised learning with lung region segmentation using a UNet model from CT scans.

2.2 Feature Extraction

This strategy utilizes as the feature extractors for pre-trained deep models, where extracted features subsequently fed into classifier to differentiate between COVID and Normal cases. Ismael et al. [13] employed five trained models and an SVM-based classifier in various different functions. Another approach used AlexNet-based features of SVM classifications [14].

2.3 Transfer Learning

Transfer Learning (TL) has proven invaluable in mitigating training time and data constraints. TL involves repurposing pre-trained models for tasks related to the same domain. For example, Ahuja and Ketan [15] has been employed for CT scan classification into COVID-19 or Normal cases.

2.4 Autoencoder

Autoencoders, though less commonly used, have been employed for COVID-19 detection. Some researchers proposed four autoencoders to extract features from CT images, and these autoencoders were integrated into a stacked autoencoder detector model.

2.5 Ensemble Models

Ensemble models aim to combine the outputs of multiple models to enhance overall performance. Shalbaf et al. [16] employed an ensemble method based on majority voting using 15 fine-tuned trained models. Result demonstrated that the combination of multiple models outperformed individual model usage.

In summary, deep learning application in COVID-19 detection using medical scans and ECG data represents a promising avenue for addressing the challenges posed by the pandemic. Various strategies, including generic deep learning, deep feature extraction, transfer learning, autoencoders, data augmentation, and ensemble models, have been explored to improve accuracy and robustness in COVID-19 diagnosis. Saygılı [17], Abbasi and Abbas [18] had also worked on computer-based detection of COVID-19 using images. These advancements hold significant potential for expediting diagnosis, reducing false negatives, and ultimately contributing to the global fight against COVID-19.

Taking everything into account, the significance of machine learning for the applications require predictive analysis, and had been demonstrated in many of the studies. A portion of the investigations have been led to play out the expectation and estimating, yet there is an additional requirement for investigation and to broaden the discoveries related with COVID utilizing a genuine clinical record. Our proposed model endeavors to anticipate patients who are on high risk related with COVID. Designated patients were isolated in respected home and diagnosis done as per symptoms and investigation reports.

3 Proposed System/Methodology

The advent of deep learning technology has shown significant revolution in field of artificial intelligence [19, 20]. Convolutional Neural Networks (CNNs) have become visible as very efficient tool for various tasks involving image data. A typical CNN architecture comprises key components: convolution layers for feature extraction through applied filters, pooling layers to reduce computational complexity, and fully connected layers that constitute a neural network. These components are stacked together to create a model based on CNN, and their parameters are fine-tuned to perform specific tasks, such as image classification and object recognition. For our proposed research work, we have first preprocessed the images before applied learning algorithm (Fig. 1).

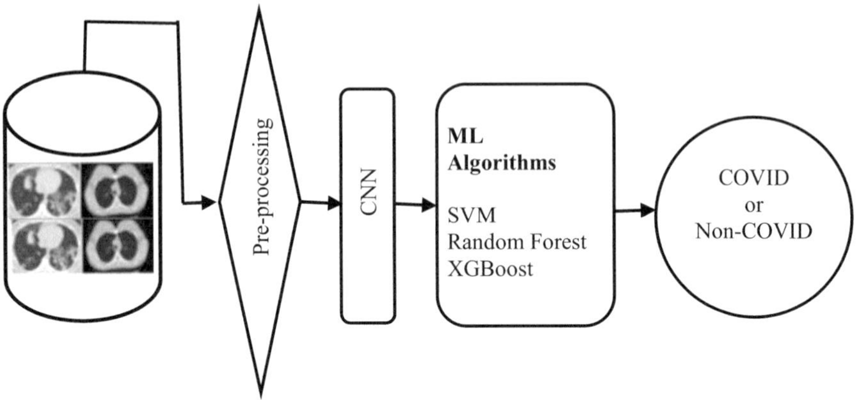

Fig. 1 Proposed model for prediction of COVID-19 based on X-ray images

3.1 Steps Used in Proposed Model

Step1: Start

Step 2: Collect COVID patient's chest X-ray image datasets from Kaggle.

Step 3: Preprocess the images normalize, resize, rotate, fill by near pixels, etc.

Step 4: Use CNN model ResNet50v2 to train the dataset.

Step 5: Apply SVM for classification of images.

Step 6: Stop.

3.2 Image Pre-processing

Effective data pre-processing is vital for the success of any machine learning model. In this study, the pre-processing phase involved two crucial steps.

3.3 Image Resizing

To ensure uniformity in image inputs for various imaging models, first, we resized available X-ray images to size 299 × 299 × 3 pixels when using InceptionV3 and to 244 × 224 × 3 pixels for other models. This resizing was essential to optimize processing time and ensure compatibility with pre-trained Convolutional Neural Networks (CNNs).

Learning Model.

Learning knowledge models achieved from some machine learning tasks to improve performance. One of the algorithms is mentioned below.

ResNet50v2 Model

We have used ResNet50v2 which is basically a deep learning CNN model with 50 layers that performs better on image dataset, trained on thousands fixed-sized images from available or on the Imaging dataset. It was designed to address issues such as vanishing gradients, degradation of performance that commonly occur during deep network training. Other variants of the ResNet architecture, ResNet101, are comprised of 101 deep layers. It, too, was trained on a large ImageNet dataset and employs identity connections between layers.

4 Discussion

The experimental findings presented in our research exhibit effectiveness of our model as technique for initial and efficient detection and identification of COVID by making use of X-ray images. Moreover results reveal that employing ensemble models enhances the performance of fine-tuned Convolutional Neural Networks (CNNs). This study introduces five already trained models that exhibit competitive performance compared to highly developed methods for early identification of COVID-19. The combination of the best-performing models within a TL framework surpasses other methods, achieving an impressive accuracy of 0.98. This better performance can be utilized for suitability of results especially since the initial CNN layers are adept at learning low-level features, such as edges. Fine-tuning enables the extraction of specific features relevant to the target domain, which, in this case, is the COVID-19 identification.

In contrast, SVM-based feature extraction method yields inferior performance compared with TL-based approach. This deviation can be described by statement that SVM classification considers only one level of abstraction, relying solely on the information from last layer of pre-trained model. Consequently, this strategy may lead to poor generalization and higher classification errors when compared to the TL method, which leverages features learned across multiple layers.

To solve data imbalance issue, oversampling emerges as a viable solution. This approach involves randomly duplicating examples from the minority class (Normal) to match the sample count of the majority class (VP), as illustrated in Table 2. The results obtained using this oversampling strategy indicate an impressive accuracy of 97%, surpassing the performance achieved with undersampling. This outcome is predictable, considering that deep learning techniques thrive when presented with ample data, thereby mitigating the risk of overfitting.

In summary, this discussion underscores the pivotal role of machine learning in COVID-19 early detection and identification using clinical images like X-ray.

It highlights the limitations of feature extraction-based methods, particularly when dealing with unbalanced datasets. Furthermore, it emphasizes the importance of dataset balancing techniques, with oversampling proving to be an effective approach in this context. These findings contribute valuable insights for medical images assessment and can guide future developments in research for robust and accurate diagnostic techniques for COVID-19 assessment.

5 Experimental Results

Our experiments were conducted using Kaggle's COVID dataset. Here, in this part, we provide a thorough description of these datasets and showcase sample X-ray scans for COVID-19 and Normal cases. Three distinct classes were considered within the original dataset like training, test, and validation. Each of these subsets includes COVID-19 images, pneumonia, and normal patient's chest X-rays [21]. For our research, we utilized version 4 of this database, which comprises 423 COVID-19 positive cases, 163 Normal cases, 129 Lung Opacity (lungs infection non-COVID-19) cases, and 112 Pneumonia scans. Our classification task focused solely on three classes which are COVID-19 patients, Normal patients, and VP. Figure 2 displays sample images from this dataset.

Our study results are presented and discussed here in Fig. 3 which illustrates the training progress of our model, showcasing a noticeable decrease in the loss of training data as epochs number increases. This trend underscores model's capacity to learn and enhance its predictive capabilities. Furthermore, Fig. 4 demonstrates a corresponding increase in data accuracy after training, affirming the model's effectiveness in learning from the dataset. We adopted a 10-fold cross-validation setting, randomly splitting datasets into training subsets. Fine-tuning the last CNN layers, we executed 30 epochs. To ensure a comprehensive evaluation, we computed four key evaluation metrics which are accuracy, precision, recall, and F-score.

To provide comprehensive view of our proposed model's performance, Tables 1 and 2 offer detailed insights into various parameters and class-wise accuracies.

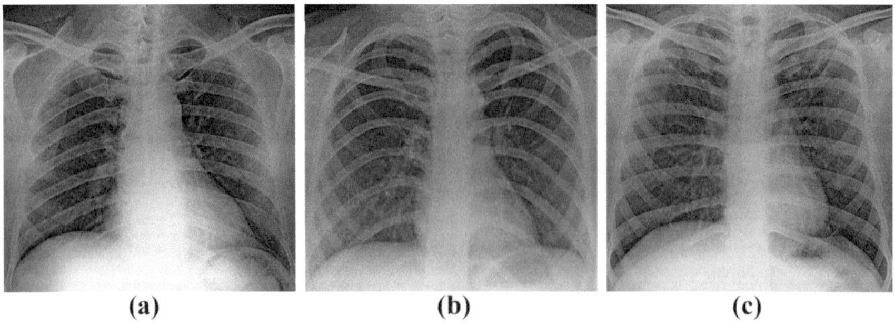

Fig. 2 COVID-19 confirmed chest X-ray scans

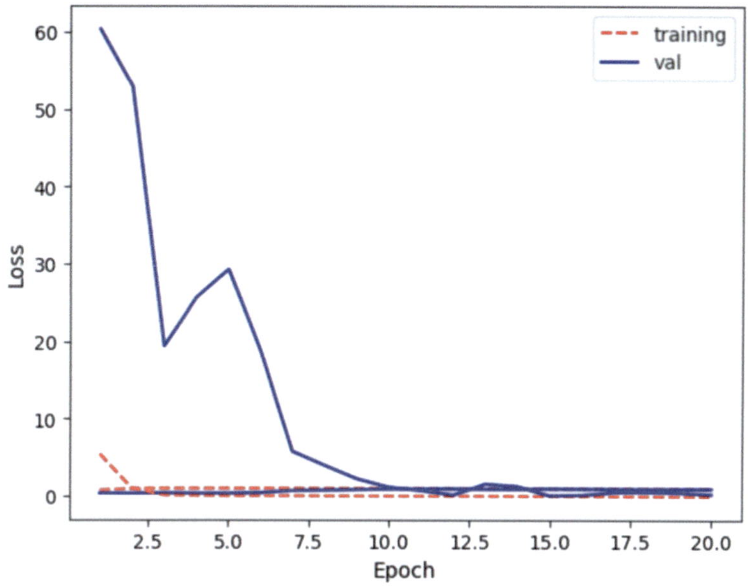

Fig. 3 Line chart showing loss over epochs for training and validation datasets. The x-axis represents epochs and the y-axis represents loss values

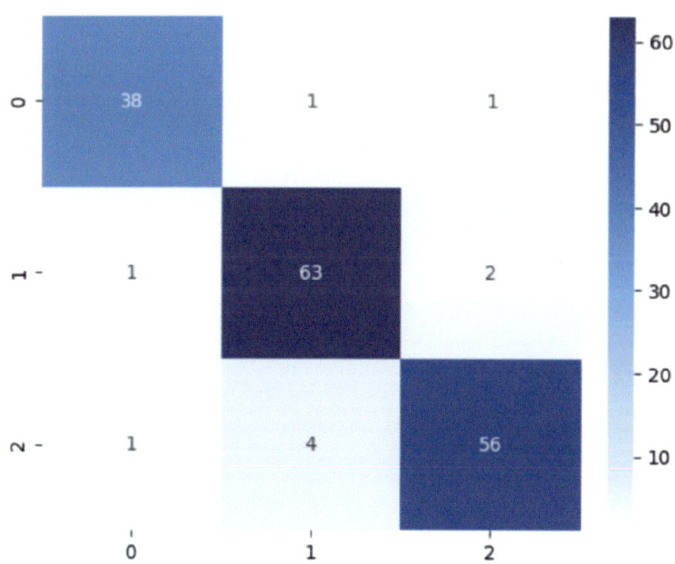

Fig. 4 Confusion matrix of proposed model

Table 1 presents a comparison of parameters with class-wise performances, highlighting the high accuracy results achieved in our study. Table 2 further elaborates on the accuracy of the model among different classes, reinforcing the model's robust performance across various categories.

Additionally, we compared the results obtained using our model with those from other well-known models applied with same dataset, as per detailed in Table 1. The accuracy scores achieved using AlexNet, VGG-16, ResNet, and a standard CNN were 49, 50.55, 92, and 96.37%, respectively. Notably, the results indicate that our model, based on DenseNet-169 architecture, outperforms these other models, demonstrating superior accuracy in COVID-19 classification. In conclusion, our research showcases promising results in the accurate classification of COVID-19. Our model's strong performance, as evidenced by the decreasing loss, increasing accuracy, and robust confusion matrix results, underscores its potential utility in aiding healthcare professionals in rapid and reliable COVID-19 diagnosis.

We present training and validation of accuracy and loss for the three best models: VGG-16, InceptionV3, and ResNet50v2. Detailed performance during folds 1, 3, 5, 6, 8, and 10 is depicted in Fig. 4. When employing the VGG-16 model, highest accuracy of 96.85% was observed at epoch 7, while the highest validation accuracy of 98.02% observed with ResNet50v2. The loss values were 0.0324 for training and 0.024 for validation at epoch 30. All reported results are based on fold 10.

Table 1 Normal, COVID, and pneumonia parameters

Parameters	Normal	COVID-19	Viral pneumonia
Precision	96%	98%	95%
Sensitivity	97%	97%	95%
Specificity	99.11%	99.18%	94.58%
F1-score	97%	98%	95%
Class error (95% CI)	(0.00103, 0.05866)	(0.00324, 0.05930)	(0.01061, 0.08614)
AUC	0.99	1.0	1.0

Table 2 System accuracy

Number of folds	Accuracy (2-class) (%)	Accuracy (3-class) (%)
Fold 1	98.26	94.52
Fold 2	99.48	96.23
Fold 3	98.96	96.58
Fold 4	100.00	98.97
Fold 5	92.71	99.31
Average	98.02 (±2.69)	97.12 (±1.79)

6 Conclusions

The COVID-19 global pandemic had led to a consequential health issue worldwide, characterized by its rapid spread and the absence of approved therapeutics or vaccines. Early assessment of COVID-19 is priority and one of the key steps to manage transmission and provide timely patient care. Specially in second wave of COVID-19 clinical images like chest's CT and X-rays scans played a very important role for controlling disease and further progress of virus. Our proposed system introduces feature extraction approach using CNN classifiers for artificial intelligence-based assessment of COVID-19.

In deep learning CNN models, feature extraction is very important given the different dimensions and complexity of clinical image data. In our research, we trained several machine learning algorithms on features extraction with established CNN models to identify optimal results.

Our model gives a structure of computer-aided system, eliminating urgent requirement of healthcare staff. It demonstrates remarkable performance in single- and multi-folded tasks, achieving average accuracies of 97.12 and 98.02%, respectively. Proposed system holds promise for deployment in remote regions of COVID-19-affected countries, helping mitigate the shortage of radiologists. Furthermore, it can be adapted for the analysis of other diseases or health issues, including tuberculosis and pneumonia.

In conclusion, our work represents significant work toward leveraging machine learning for computer-aided diagnosis of COVID-19 using clinical images. The achieved high accuracy rates demonstrate the potential of this approach in aiding healthcare professionals in the fast and accurate identification of COVID-infected cases. With further refinement and validation, our system has the potential to make a substantial impact on the global efforts to battle against COVID-19 and other respiratory diseases.

References

1. Ukwuoma CC, Cai D, Heyat MBB, Bamisile O, Adun H, Al-Huda Z, Al-Antari MA (2023) Deep learning framework for rapid and accurate respiratory COVID-19 prediction using chest X-ray images. J King Saud Univ-Comput Inf Sci 35(7):101596
2. Sahin ME et al (2023) Detection and classification of COVID-19 by using faster R-CNN and mask R-CNN on CT images. Neural Comput Appl 35(18):13597–13611
3. Deb SD et al (2023) CoVSeverity-net: an efficient deep learning model for COVID-19 severity estimation from Chest X-Ray images. Res Biomed Eng 85–98
4. Bwire GM et al (2021) Detection profile of SARS-CoV-2 using RT-PCR in different types of clinical specimens: a systematic review and meta-analysis. J Med Virol 93(2):719–725
5. Sethuraman N, Jeremiah SS, Ryo. A (2020) Interpreting diagnostic tests for SARS-CoV-2. JAMA 323(22):2249–2251
6. Ahmed S, Yap MH, Tan M, Hasan MK (2020) Reconet: multi-level preprocessing of chest x-rays for covid-19 detection using convolutional neural networks, pp 2020-07. Medrxiv

7. Fang Y, Zhang H, Xie J, Lin M, Ying L, Pang P, Ji W (2020) Sensitivity of chest CT for COVID-19: comparison to RT-PCR, pp.E115-E117. Radiology
8. Sarkodie BD, Osei-Poku K, Brakohiapa E (2020) Diagnosing COVID-19 from chest X-ray in resource limited environment-case report. Med Case Rep 6(2):135
9. Dewan R, Ahmad S, Rana AK et al (2023) IoT-based energy efficient and longer lifetime compression approach for healthcare applications. Trans Emerg Telecommun Technol e4843
10. Yao H, Zhang N, Zhang R et al (2020) Severity detection for the coronavirus disease 2019 (COVID-19) patients using a machine learning model based on the blood and urine tests. Front Cell Dev Biol 683
11. Sun L, Song F, Shi N et al (2020) Combination of four clinical indicators predicts the severe/critical symptom of patients infected COVID-19. J Clin Virol 128:104431
12. Zheng C et al (2020) Deep learning-based detection for COVID-19 from chest CT using weak label, pp 2020-03. MedRxiv
13. Ismael AM, Şengür A (2021) Deep learning approaches for COVID-19 detection based on chest X-ray images. Expert Syst Appl 164:114054
14. Turkoglu M (2021) COVIDetectioNet: COVID-19 diagnosis system based on X-ray images using features selected from pre-learned deep features ensemble. Appl Intell 51(3):1213–1226
15. Ahuja S, Panigrahi BK, Dey N, Rajinikanth V, Gandhi TK (2021) Deep transfer learning-based automated detection of covid-19 from lung CT scan slices. Appl Intell 51(1):571–585
16. Gifani P, Shalbaf A, Vafaeezadeh M (2021) Automated detection of COVID-19 using ensemble of transfer learning with deep convolutional neural network based on CT scans. Int J Comput Assist Radiol Surg 16:115–123
17. Saygılı A (2021) A new approach for computer-aided detection of coronavirus (COVID-19) from CT and X-ray images using machine learning methods. Appl Soft Comput 105:107323
18. Abbasi WA, Abbas SA (2021) COVIDC: an expert system to diagnose COVID-19 and predict its severity using chest CT scans: application in radiology. Inform Med Unlocked 23:100540
19. Prabha B, Kaur S, Singh J et al (2023) Intelligent predictions of covid disease based on lung CT images using machine learning strategy. Mater Today Proc 80:3744–3750
20. Sharma A, Rani S, Gupta D (2020) Artificial intelligence-based classification of chest X-ray images into COVID-19 and other infectious diseases. Int J Biomed Imaging 1–10pp
21. Jain R, Gupta M, Taneja S et al (2021) Deep learning based detection and analysis of COVID-19 on chest X-ray images. Appl Intell 51:1690–1700

Deep Learning Framework Pertaining to Peach Leaf Disease Classification

Mrinal Pandey, Mamta Arora, Poonam Biswas, Kanan Arora, Shivam Bhagat, and Vishal Baheti

Abstract Agriculture is the largest livelihood provider in India. It contributes a significant figure to the gross domestic product. However, crop diseases can significantly impact agricultural productivity and food security. Accurate and timely classification of diseases in plants or crops is essential for effective disease management. These facts encourage our team to work upon the SDG:2 "End hunger, achieve food security and improved nutrition and promote sustainable agriculture." After our research the team came into conclusion to do "Peach Leaf Disease Classification." The literature survey concluded into using convolutional neural network. Machine learning models are increasingly used to identify and categorize plant diseases. Therefore, our team distributed among themselves four different architectures. Sequential CNN gave 96% accuracy, ResNet-50 gave 99.62% accuracy, VGG gave 97.50% accuracy, and AlexNet gave 99.25% accuracy. Our team is willing to focus on developing deep learning models capable of accurately distinguishing and classifying a diverse range of subcategories within broader classes.

Keywords Peach leaf · Image processing · ResNet-50 · CNN · VGG · AlexNet · AI · ML

M. Pandey · M. Arora · P. Biswas (✉) · K. Arora · S. Bhagat · V. Baheti
Manav Rachna University, Faridabad, Haryana, India
e-mail: biswaspoonam002@gmail.com

M. Pandey
e-mail: mrinalpandey@mru.edu.in

M. Arora
e-mail: mamta@mru.edu.in

A. P. Agrawal et al. (eds.), *Advances of Artificial Intelligence and Machine Learning in Image Processing, Security and Data Analytics*, Lecture Notes in Electrical Engineering 1464, https://doi.org/10.1007/978-981-95-0148-9_21

1 Introduction

Agriculture plays a major role in our lives. However, to sustain our agricultural practices, we must prioritize not only productivity but also environmental sustainability. One crucial aspect of sustainable agriculture is the effective management of plant diseases, which can significantly impact crop health and yield. Advancements in technology, specifically in machine learning and computer vision, have provided us with powerful tools to address the challenges of disease identification and classification in plants. By leveraging these innovations, our team got the opportunity to work toward developing robust systems for classifying diseases in peach leaves. The objective is to enable farmers to quickly and accurately detect and classify diseases, allowing them to take timely measures to protect their crops and promote sustainable agricultural practices.

In this research paper, we delve into the realm of peach leaf disease classification. By analyzing the derived dataset and implementing top classification algorithms, we aim to advance our knowledge and classification of the peach leaf images. Our goal is to provide farmers with practical and accessible tools for disease management. Through the accurate classification of peach leaf diseases, we aspire to enhance crop productivity, minimize losses, and contribute to the global goal of ensuring sustainable agriculture [1].

If we can classify different types of peach leaf diseases accurately, we can help farmers treat the plants more quickly and effectively. This will lead to better crops and sustainable farming practices. Another reason is that we want to make progress in machine learning and computer vision. We also have a responsibility to protect the environment, as plant diseases can have negative effects on ecosystems. By developing a model that can classify peach diseases, we can stop the diseases from spreading. To achieve these goals, we identified the different types of diseases in peach leaves [2]. Then collected and prepared a dataset of healthy and sick peach leaves. After that we chose and trained a model using machine learning techniques. Finally, we used the model to classify new images of peach leaves as healthy or sick. Table 1 represents the various types of peach leaf diseases.

Table 1 List of different types of peach leaf diseases

S. No.	Peach diseases
1	Shot hole
2	Peach leaf curl
3	Powdery mildew
4	Brown rot
5	Viruses
6	Crown gall
7	Bacterial spot

2 Literature Review

Advancements in disease classification techniques are essential for preserving plant health. Bhakta developed a novel plant disease prediction model based on thermal images using a modified deep convolutional neural network [3]. They achieved an accuracy of 95% and precision of 97.5% in predicting bacterial leaf blight in rice plants. Monigari and Khyatri focused on grape disease prediction and developed a system using PNN, BPNN, SVM, and Random Forest algorithms with an accuracy of 95% [4]. Reddy and Neerja utilized the Plant Village dataset to develop a plant leaf disease classification and damage detection system using deep learning models, achieving an accuracy of 97% [5]. Guo proposed a plant disease identification system based on deep learning and image segmentation techniques, achieving an accuracy of 85.37% but faced challenges with computational time and frame selection [6]. Various review papers concluded that deep learning is a promising approach for plant disease identification. Sharma used a CNN model to classify rice and potato plant leaf diseases with high accuracy [7]. Garcia and Barbedo conducted research on individual lesion and spot-based plant disease identification using digital images. The effectiveness of deep learning and transfer learning for plant disease classification was analyzed by Garcia and Barbedo, demonstrating the impact of dataset size and variety on performance [8, 9]. A few-shot learning approach for plant disease classification using images taken in the field was proposed by Agrueso, achieving high accuracy and reducing development time and cost. Atila utilized the EfficientNet deep learning model for plant leaf disease classification [10]. KC developed a depthwise separable convolution architecture for plant disease classification, achieving high accuracy and a compact model size. Zeng proposed a large-scale and fine-grained plant disease classification model based on the transformer and attention convolution [11]. Vimala focused on the detection of rice leaf diseases using a Taylor-student psychology-based optimization-integrated deep learning approach [12]. Alirezazadeh improved plant disease classification using attention mechanism and CNN [13]. Aqel utilized the extreme learning machine (ELM) algorithm for plant disease classification and achieved high accuracy [1]. Qi proposed a lightweight plant disease classification approach combining the Grab Cut algorithm, new coordinate attention, and channel pruning techniques [14]. Overall, these studies provide valuable insights into the development and application of different approaches for plant disease identification and classification. Aliyu investigated the comparison between Support Vector Machine (SVM) and Convolutional Neural Network (CNN) in the context of deep learning for plant disease detection, achieving impressive accuracies of 98% for Early Bright, 99.7% for Light Bright, and 100% for Healthy plant categories [15]. Rehan focused on image-based detection and compared SVM with Artificial Neural Network (ANN), achieving an accuracy of 93.8% and precision of 99.1%. S. Nandhini and K. Ashokkumar employed neural networks with CNN to achieve an accuracy of 99.35% [16]. K. Muthukannan used Feed-Forward Neural Network (FFNN), Learning Vector Quantization (LVQ), and Radial Basis Function Networks (RBF) achieving accuracies of 90.67% (FFNN), 100% (RBF) for recall, 89.39% (FFNN) for precision, and 91.47% (FFNN) for F-measure [17]. One of the works explored different combinations of image processing techniques and machine learning algorithms, achieving accuracies of 68.1, 97, and 98.42% (SVM), respectively [18–20]. Tan Soo Xian and Ruzelita Ngadiran used CNN with image pre-processing and feature extraction, obtaining accuracy rates ranging from 69.87 to 71.40% [21]. Monika Lamba explored deep learning and image pre-processing, achieving accuracies of 96.4% (binary class) and 97.2% (multi-class) [22]. Abu Sarwar Zamani employed SVM and random forest, achieving accuracies of 92% and 93%, respectively [23]. Sreya John and Arul Leena Rose used CNN and classification, achieving accuracies of 86.58% (rice leaf) and 93% (cassava) [24]. Udayananda employed Deep Learning with CNN and hybrid techniques reaching accuracies between 76.63 and 96% [25]. Shima Ramesh used CNN with random forest and naïve Bayes achieving accuracies of 70.14 and 57.61%, respectively [26]. Jun Liu and Xuewei Wang explored deep learning for insect detection, achieving accuracies ranging from 95.97 to 97.47% [27]. Fulari and Rajveer compared convolutional neural networks with support vector machine using GLCM, achieving accuracies of 90.1% (tomato) and 95.8% (corn) [28]. Sunil used convolutional neural networks, K-means clustering, and SVM with an accuracy of 90% [29]. Sonal employed Canny edge detection, color feature extraction, HPC, and CDD to achieve an accuracy of 98.1% [30]. Moreover, various machine learning techniques are being used in the field of sustainable growth [31–33].

3 Methodology

The research work utilized Google Colab, a web-based platform for Python coding, ideal for machine learning tasks. Data was sourced from the Plant Village dataset, creating a binary classifier for healthy and bacterial spot peach leaves. Key libraries like TensorFlow, Keras, NumPy, pandas, Matplotlib, OpenCV (cv2), and scikit-learn were imported for data manipulation and model building. Images were preprocessed by resizing, converting to NumPy arrays, and normalization. Labels were encoded, and data split into training and testing sets. Four CNN architectures (AlexNet, ResNet-50, VGG, Sequential CNN) were trained, and models were saved and evaluated for classifying peach leaves as healthy or with bacterial spots. Figure 1 represents the proposed methodology.

3.1 Data Description

The dataset has been extracted from Plant Village dataset which is available in Kaggle. The Plant Village dataset contains a total number of 54,303 images divided into 38 distinct classes. Comprising together 14 different species of crops, namely,

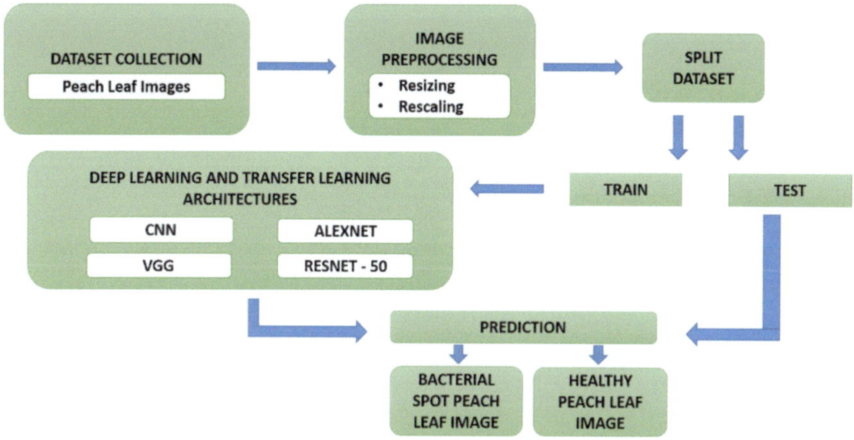

Fig. 1 The model design of the peach leaf disease classifier

apple, blueberry, cherry, grape, orange, peach, pepper, potato, raspberry, soy, squash, strawberry, and tomato.

The extracted dataset combines a total of 3806 images of peach leaves and focuses on classifying them into two distinct categories: bacterial spot and healthy peach leaves. The dataset comprises a total of 1728 images labeled as healthy and 2078 images labeled as bacterial spot. By classifying the dataset into the above two categories, our team was able to train and evaluate models specifically designed to identify and distinguish between healthy peach leaves and those affected by bacterial spot. This dataset resulted in providing a valuable resource for developing automated systems that can help in the early detection and management of bacterial spot disease in peach crop. Table 2 contains a brief description about the dataset.

Table 2 Description about the dataset

Dataset description	No. of images
Total no. of images in the plant village dataset	54,303
Number of extracted peach leaf images	3,806
Total no. of healthy peach leaf images	1728
Total no. of bacterial spot peach leaf images	2078
Number of classes in the Plant Village dataset	38
Number of classes in the extracted dataset	2

4 Results and Discussion

Yann LeCun, along with colleagues Yoshua Bengio and Geoffrey Hinton, introduced the Convolutional Neural Network (CNN) in the 1990s. Often hailed as the "father of CNNs," LeCun aimed to automate feature extraction, eliminating the need for manual engineering. CNNs autonomously learn hierarchical features from raw data, employing convolutional, pooling, and fully connected layers for diverse tasks and datasets [34]. The proposed sequential CNN model shows the training progress of a deep learning model for binary classification of peach leaf images into two categories: healthy or bacterial spot. The model has been trained for 40 epochs using batches of data with a size of 32 samples per batch. The loss value (0.0636) signifies the average disparity between the predicted and actual values within the training set. Meanwhile, the accuracy value (0.9600) denotes the percentage of accurately classified samples in the training set. Likewise, the validation loss (0.1701) and validation accuracy (0.9574) values correspondingly indicate the average difference between predicted and actual values, and the proportion of correctly classified samples within the validation set. Figure 2 represents the training validation accuracy graph and training and validation loss graph.

Residual network is a special type of Convolutional Neural Network (CNN) proposed by He Kaiming, Zhang Xiangyu, Ren Shaoqing, and Sun Jian in the year 2015. Originally the ResNet architecture was ResNet-34, which consisted of 34 weighted layers. It gave a novel way to include more convolutional layers to a CNN, without running into the vanishing gradient problem, utilizing the concept of easy shortcut connection. A shortcut connection "skips over" a few layers, changing over a regular network to a residual network [35]. ResNet-50 architecture has been proved to be very useful for binary classification of peach leaf images into healthy or bacterial leaf images. The model has been trained for 100 epochs using batches of data with a size of 25 samples per batch. In last two epochs, the training loss and accuracy values

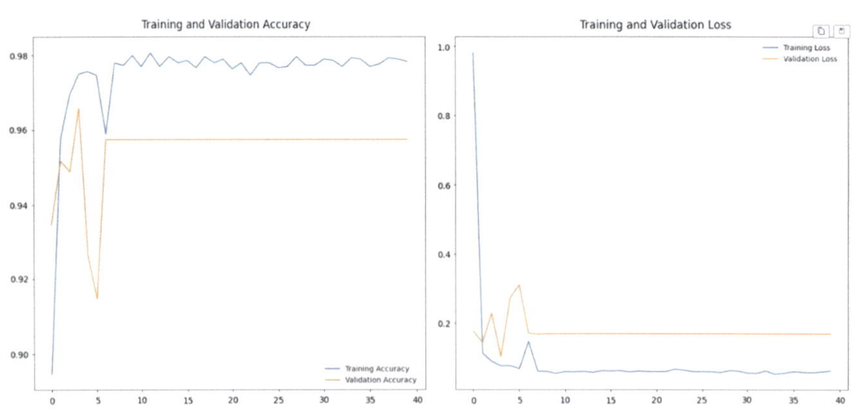

Fig. 2 Sequential CNN model **a** training accuracy graph and **b** validation accuracy graph

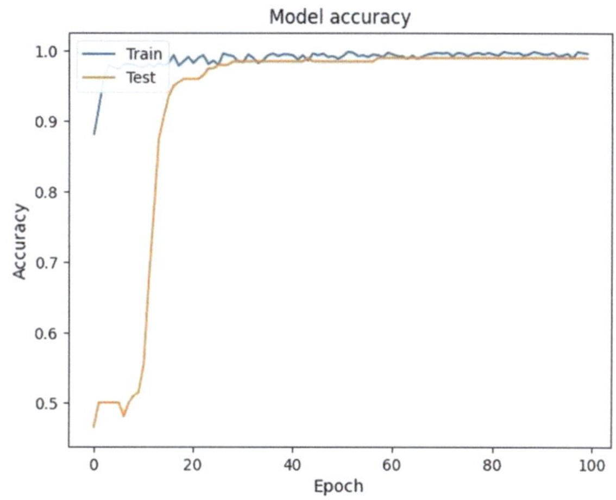

Fig. 3 ResNet-50 model accuracy graph

have reached very low values, indicating that the model is well trained and can accurately classify peach leaf images. The validation loss and accuracy values are also good, suggesting that the model is not overfitting to the training data. The learning rate used in the last two epochs, which is 2.1870e-21, indicating that the learning rate is very small and the model is making tiny updates to the weights during training. This suggest that the model has reached to a good solution and further training may not improve its performance significantly. Figure 3 represents the epoch-accuracy graph of ResNet-50.

Karen Simonyan and Andrew Zisserman, from the University of Oxford in 2014, proposed the VGG (visual geometry group) network to explore the impact of network depth on convolutional neural network performance. Emphasizing simplicity and uniformity, VGG employs stacked 3×3 convolutional layers and max-pooling for spatial down sampling, achieving depth through multiple layers [36].

VGG model has also shown an excellent result in peach disease classification. VGG has completed training for 100 epochs. During the training process, it processed a batch of 25 data points per step, with each step taking approximately 15 s. The reported metrics for the training include a loss value of 0.0641, a precision score of 0.9750, a recall score of 0.9750, and an accuracy score of 0.9750. Additionally, there is validation data present, and the corresponding metrics for the validation set are validation loss of 0.0526, a validation precision of 0.9950, a validation recall of 0.9950, and a validation accuracy of 0.9950 (Fig. 4).

AlexNet was proposed by Alex Krizhevsky, Ilya Sutskever, and Geoffrey Hinton in 2012. The researchers aimed to design a network that could learn rich and hierarchical representations of visual features from raw images [37].

AlexNet had shown an outstanding result in peach leaf disease classification. The model was trained for 50 epochs. The accuracy that was achieved after completing the Alexnet model was 97%. But after hyperparameter tuning AlexNet model reached 99.50% (Fig. 5).

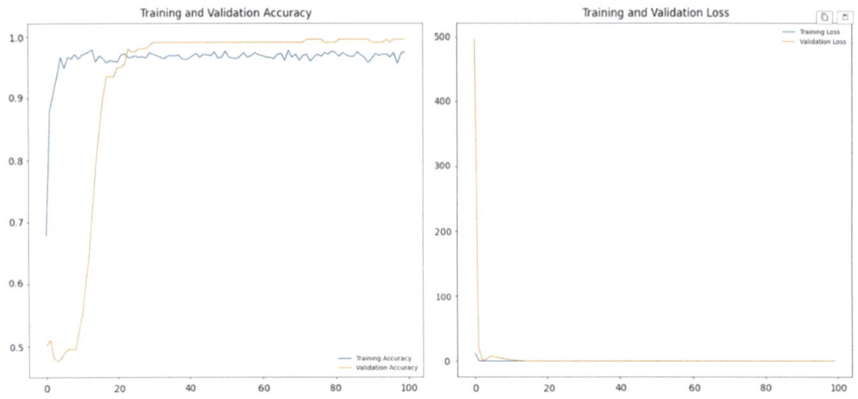

Fig. 4 VGG model training and validation accuracy and loss graph

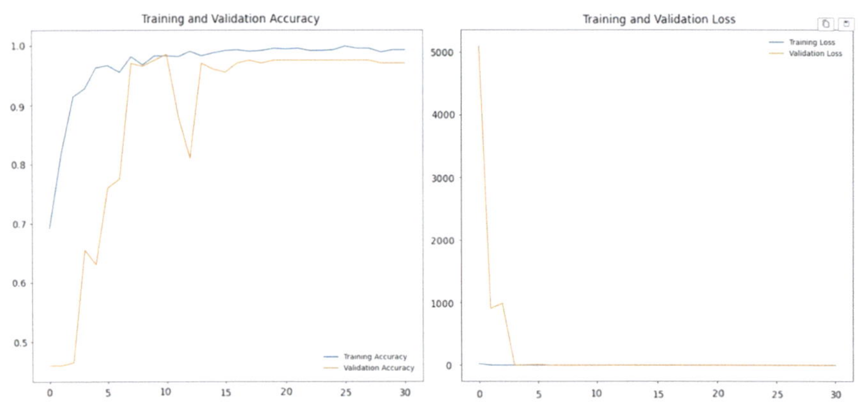

Fig. 5 AlexNet model training and validation accuracy and loss graph

The study "Deep Learning Framework Pertaining to Peach Leaf Disease Classification" intricately aligns with pivotal objectives of Sustainable Development Goal 2 (SDG 2), which ambitiously strives to eradicate hunger, enhance food security, and foster sustainable agricultural practices globally. By contributing to the nuanced management of plant diseases and fostering sustainable agricultural practices, our work resonates with the global mission to ensure resilient food systems, improve crop yields, and secure the foundation for a sustainable future. The crux of our study lies in the meticulous classification of bacterial spot disease within peach leaves, a pivotal facet in bolstering agricultural productivity and fortifying food security. In achieving this, our models, including ResNet-50 giving high accuracy of 99.62%, AlexNet with an impressive accuracy of 99.25% followed by VGG with 97%, and the proposed sequential CNN with 96%, showcase a significant leap forward in the realm of plant disease identification. Table 3 represents the architectural parameters and Fig. 6 depicts the pictorial representation for the comparison of the accuracies of all the four different architectures.

However, the study confronts three major limitations that warrant due consideration. Firstly, the exclusive focus on bacterial spot disease potentially constrains the applicability of the model to a more diverse range of peach leaf diseases. Secondly, the reliance on a single-source dataset introduces an inherent bias, impacting the adaptability and generalizability of the model. Thirdly, external factors such as environmental conditions, which play a pivotal role in disease manifestation, were not explicitly considered, indicating a certain degree of contextual limitation.

To fortify the research's alignment with SDG 2 and enhance its overall impact, our future trajectory is directed toward a more comprehensive approach. Future work encompasses an expansion of disease categories, incorporating the various types of peach leaf diseases. Diversifying datasets across various geographical locations and seasons is paramount, mitigating biases and fortifying the model's robustness. Additionally, our research endeavors to explore advanced CNN architectures tailored to the unique intricacies of peach diseases, ensuring a more nuanced and accurate identification process.

Table 3 Different architecture parameters

Name of architecture	Batch size	No. of epochs	Loss	Validation loss	Validation accuracy
Sequential CNN	35.00	40.00	0.0636	0.1701	0.9574
ResNet-50	25.00	100.00	0.0133	0.0202	0.9900
VGG	35.00	100.00	0.0641	0.0526	0.9950
AlexNet	35.00	50.00	0.0149	0.1317	0.9700

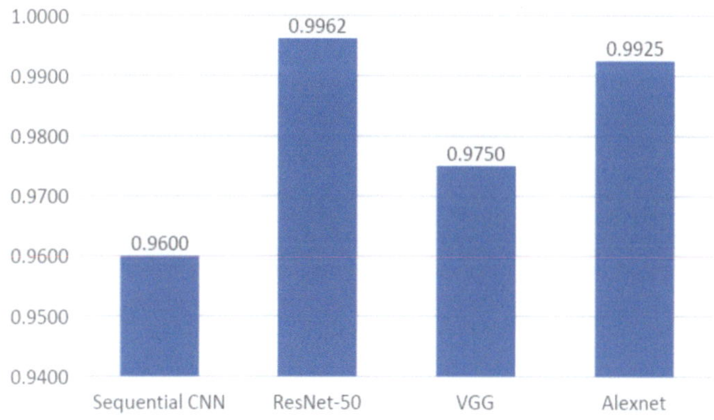

Fig. 6 The bar graph representing model accuracies of all four architecture

5 Conclusion and Future Scope

The research work, "Deep Learning Framework Pertaining to Peach Leaf Disease Classification," is aligned with the Sustainable Development Goal 2 of ending hunger, achieving food security, and promoting sustainable agriculture. Given that crop diseases have a significant impact on agricultural productivity and food security, our focus has been on accurately classifying peach leaves as either healthy or affected by bacterial spot disease. We've compiled a dataset of 3806 images, divided into two classes: healthy peach leaves and bacterial spot-infected peach leaves. Leveraging Convolutional Neural Networks (CNNs), which have demonstrated high accuracy in previous work, we employed four top-class CNN architectures—VGG, AlexNet, ResNet-50, and a sequential CNN—all achieving accuracy rates exceeding 95%. Looking ahead, our future work will expand beyond these architectures to address multi-class classification tasks with a broader range of fine-grained categories, encompassing various peach leaf diseases. This endeavor aims to push the boundaries of deep learning, enabling more sophisticated architectures and advancing our understanding of complex data for more comprehensive and nuanced insights into plant disease classification and management.

References

1. Aqel D, Al-Zubi S, Mughaid A, Jararweh Y (2022) Extreme learning machine for plant diseases classification: a sustainable approach for smart agriculture. Clust Comput 1–14
2. Hartman JR (2007) Peach fruit diseases. In: Plant pathology fact sheet. University of Kentucky, College of Agriculture
3. Bhakta I, Phadikar S, Majumder K, Mukherjee H, Sau A (2023) A novel plant disease prediction model based on thermal images using modified deep convolutional neural network. Precis Agric 24(1):23–39
4. Monigari V, Sri GK, Prathima T (2021) Plant leaf disease prediction. Int J Res Appl Sci Eng Technol 9:1295–1305
5. Sai Reddy B, Neeraja S (2022) Plant leaf disease classification and damage detection system using deep learning models. Multimed Tools Appl 81(17):24021–24040
6. Guo Y, Zhang J, Yin C, Hu X, Zou Y, Xue Z, Wang W (2020) Plant disease identification based on deep learning algorithm in smart farming. Discret Dyn Nat Soc 2020:1–11
7. Khan U, Oberoi A (2019) Plant disease detection techniques: a review. Int J Comput Sci Mob Comput 8(4):59–68
8. Barbedo JGA (2016) A review on the main challenges in automatic plant disease identification based on visible range images. Biosyst Eng 144:52–60
9. Barbedo JGA (2018) Impact of dataset size and variety on the effectiveness of deep learning and transfer learning for plant disease classification. Comput Electron Agric 153:46–53
10. Atila Ü, Uçar M, Akyol K, Uçar E (2021) Plant leaf disease classification using EfficientNet deep learning model. Ecol Inform 61:101182
11. Zeng Q, Niu L, Wang S, Ni W (2023) SEViT: a large-scale and fine-grained plant disease classification model based on transformer and attention convolution. Multimed Syst 29(3):1001–1010
12. Vimala S, Madhusudhana Rao TV, Balaji A, Maram B (2023) Taylor-student psychology based optimization integrated deep learning in IoT application for plant disease classification. Wirel Netw 29(2):919–939

13. Alirezazadeh P, Schirrmann M, Stolzenburg F (2023) Improving deep learning-based plant disease classification with attention mechanism. Gesunde Pflanzen 75(1):49–59
14. Qi F, Wang Y, Tang Z (2022) Lightweight plant disease classification combining GrabCut algorithm, new coordinate attention, and channel pruning. Neural Process Lett 54(6):5317–5331
15. Aliyu MA, Mokji MMM, Sheikh UUU (2020) Machine learning for plant disease detection: an investigative comparison between support vector machine and deep learning. IAES Int J Artif Intell 9(4):670
16. Nandhini S, Ashokkumar K (2022) Machine learning technique for crop disease prediction through crop leaf image. Appl Math 16(2):149–158
17. Muthukannan K, Latha P, Selvi RP, Nisha P (2015) Classification of diseased plant leaves using neural network algorithms. ARPN J Eng Appl Sci 10(4):1913–1919
18. Kaur I, Aggarwal G, Verma A (2016) Detection and classification of disease affected region of plant leaves using image processing. Indian J Sci Technol 9:48
19. Majumdar D, Kumar D, Chakraborty A, Dutta D (2014) Detection & diagnosis of plant leaf disease using integrated image processing approach. Int J Comput Eng Appl VI–III
20. Bala M, Mehan V (2021) Identification of rice plant diseases using image processing, machine learning & deep learning: a review. In: International conference on emerging technologies: AI, IoT, and CPS for science & technology applications. NITTTR Chandigarh
21. Xian TS, Ngadiran R (2021) Plant diseases classification using machine learning. J Phys: Conf Ser 1962(1):012024. IOP Publishing
22. Lamba M, Gigras Y, Dhull A (2021) Classification of plant diseases using machine and deep learning. Open Comput Sci 11(1):491–508
23. Zamani AS, Anand L, Rane KP, Prabhu P, Buttar AM, Pallathadka H, Raghuvanshi A, Dugbakie BN (2022) Performance of machine learning and image processing in plant leaf disease detection. J Food Qual 2022:1–7
24. John S, Rose AL (2021) Machine learning techniques in plant disease detection and classification-a state of the art. Mateh-Agric Eng 65(3)
25. Udayananda GKVL, Shyalika C, Kumara PPNV (2022) Rice plant disease diagnosing using machine learning techniques: a comprehensive review. SN Appl Sci 4(11):311
26. Ramesh S, Hebbar R, Niveditha M, Pooja R, Shashank N, Vinod PV (2018) Plant disease detection using machine learning. In: 2018 International conference on design innovations for 3Cs compute communicate control (ICDI3C). IEEE, pp 41–45
27. Liu J, Wang X (2021) Plant diseases and pests detection based on deep learning: a review. Plant Methods 17:1–18
28. Fulari UN, Shastri RK, Fulari AN (2020) Leaf disease detection using machine learning. J Seybold Rep 1533:9211
29. Harakannanavar SS, Rudagi JM, Puranikmath VI, Siddiqua A, Pramodhini R (2022) Plant leaf disease detection using computer vision and machine learning algorithms. Glob Trans Proc 3(1):305–310
30. Patil SP, Zambre RS (2014) Classification of cotton leaf spot disease using support vector machine. Int J Eng Res 3(4):1511–1514
31. Dewan R, Ahmad S, Rana AK, Nagpal T, Kumar V, Singh NH, Kumar VA, Kumar A (2023) IoT-based energy efficient and longer lifetime compression approach for healthcare applications. Trans Emerg Telecommun Technol 35(4):e4843
32. Yadav D, Sharma A, Ahmad S, Chandra U (2022) Political sentiment analysis: Case study of Haryana using machine learning. In: Mobile radio communications and 5G networks: proceedings of second MRCN 2021. Springer Nature Singapore, Singapore, pp 479–499
33. Mohammad AK, Sharik A (2020) An emergence of AI in data mining and KDD: ANN its strength and weakness
34. LeCun Y, Bottou L, Bengio Y, Haffner P (1998) Gradient-based learning applied to document recognition. Proc IEEE 86(11):2278–2324
35. He K, Zhang X, Ren S, Sun J (2016) Deep residual learning for image recognition. In: Proceedings of the IEEE conference on computer vision and pattern recognition, pp 770–778

36. Simonyan K, Zisserman A (2014) Very deep convolutional networks for large-scale image recognition. arXiv:1409.1556
37. Krizhevsky, A, Sutskever, I, Hinton, G. E. Imagenet classification with deep convolutional neural networks. Advances in neural information processing systems, 25 (2012).

Enhancing Brain Tumor Classification with a Hybrid RNN Approach

Nadenlla Rajamohan Reddy and G. Muneeswari

Abstract A tumor signifies the uncontrolled growth of cancer cells in the body, and brain tumors significantly contribute to cancer-related deaths across all age groups. Timely detection of brain tumors is crucial to prevent further complications. Traditionally, a biopsy, often required invasive brain surgery, is conducted to determine the type of brain tumor. Healthcare professionals can harness computational intelligence techniques to assist in identifying and categorizing these tumors. The classification of brain tumors is pivotal for assessing and planning effective treatments. In the field of medical image processing, the detection and segmentation of brain tumors through MRI scans represent a central and challenging research area. The integration of Artificial Intelligence (AI), encompassing machine learning and deep learning, has transformed the categorization and detection of complex pathological conditions like brain tumors. Deep learning, in particular, excels in the precise segmentation and classification of these tumors, which is crucial for the analysis and interpretation of brain MR images. In this study, we introduce a hybrid RNN classifier, combining Inception ResNet and RNN, to enhance the classification of various brain tumor types. This innovative approach is expected to yield improved results in terms of precision, recall, and F1-score.

Keywords Brain · Tumor · Segmentation · Artificial intelligence · Recurrent Neural Network (RNN) · RF · LR and inception ResNet V2

N. R. Reddy · G. Muneeswari (✉)
School of Computer Science and Engineering, VIT-AP University, Amaravati, Andhra Pradesh, India
e-mail: muneeswari.g@vitap.ac.in

N. R. Reddy
e-mail: rajamohan.22phd7007@vitap.ac.in

A. P. Agrawal et al. (eds.), *Advances of Artificial Intelligence and Machine Learning in Image Processing, Security and Data Analytics*, Lecture Notes in Electrical Engineering 1464, https://doi.org/10.1007/978-981-95-0148-9_22

1 Introduction

A brain tumor, scientifically referred to as an intracranial neoplasm, develops when abnormal cells start to form within the brain. Common symptoms include headaches, vomiting, vision problems, and cognitive changes. Essentially, a tumor represents the uncontrolled and atypical cell growth within the body, in this case, forming a mass of tissue in the brain. Detecting brain tumors at an early stage is crucial because if left untreated, they can become life-threatening [1].

The likelihood of survival increases dramatically with quick diagnosis and treatment. The growth rate of brain tumors can vary, and it is influenced by their specific location within the brain. Additionally, the tumor's location plays a crucial role in potential damage to the human nervous system [2]. Additionally, a brain tumor's size, location, and kind all affect how it will be treated.

The process of diagnosing brain tumors typically commences with an evaluation of the patient's reported symptoms. Subsequently, a physician assesses various aspects, including reflexes, strength, sensory perception, balance, and nerve function. If a tumor is suspected, further diagnostic methods are employed to confirm the presence of brain tumors [3]. Tumors can be divided into two main categories: carcinomas, which are made up of non-cancerous cells and typically causing less harm since they do not invade neighboring tissues, and malignant tumors, made up of cancerous cells, which are more dangerous and have the potential to metastasize to other tissues and cells. In essence, a tumor is a mass of cells that forms tissue exhibiting uncontrolled growth, a departure from the normal regulatory mechanisms found in healthy cells [4]. The symptoms of each form of glioma vary, and they may include migraines, epileptic fits, frustration, throwing up, vision abnormalities, and apathy or weakness in the limbs. Meningiomas are amenable to treatment. Pituitary tumors, on the other hand, are uncommon growths that develop within the pituitary gland, often leading to symptoms like changes in vision or headaches. Additionally, these tumors can interfere with hormonal balance, impacting menstrual cycles and causing sexual dysfunction [5].

The neglect of treatment for brain tumors can lead to dire consequences, highlighting the challenges healthcare providers face when it comes to diagnosing and caring for affected individuals [6]. While the field of brain tumor detection through image processing has garnered substantial research attention, the development of a precise method for brain tumor detection continues to be a persistent challenge [7]. Currently, patients can receive treatment by laser therapy, chemotherapy, radiation therapy, or an amalgam of these modalities. It's important to note that brain tumor biopsy is more intricate than biopsies conducted in other parts of the body, as it requires surgical intervention [8]. Hence, there is a pressing need for non-surgical and accurate diagnostic techniques. MRI scanning shines out as the favored and most often utilized diagnostic imaging modality for the detection of brain malignancies. Brain tumors can be identified through diagnostic imaging methods like CT scans and MRI scans [9].

Image processing is a valuable technique for extracting and isolating relevant data from diverse datasets. In the context of medical image diagnosis, numerous machine learning approaches have been employed to segment MRI images [10]. The use of MRI for delineation of malignancies in the brain is of the greatest importance in medical diagnosis and treatment. It helps with diagnosis, forecast, estimations for overall growth, appraisals of tumor thickness, and the creation of crucial patient care strategies. These MRI images are acquired from three distinct viewpoints, namely, sagittal, axial, and coronal views [11].

In the spheres of digital vision and clinical image analysis, visual segmentation is indispensable. The operation of categorizing a malignant growth in the brain specifically requires removing the living cells, decomposing the cores, and edema from the normal brain tissues, such as gray matter (GM), white matter (WM), and the cerebrospinal fluid (CSF). In contrast to manual tumor segmentation, automated segmentation algorithms offer several advantages, including reduced time requirements, improved accuracy, and the facilitation of tumor analysis and diagnosis by medical professionals [12].

The traditional approach to tumor detection relies on radiologists, who painstakingly inspect and diagnose tumors, a process that is both arduous and time-intensive. The area of computer-aided medical diagnostics (CAMD) has achieved major advances historically owing to innovations in the field of artificial intelligence (AI) and the use of machine learning [13]. This technology can offer doctors swift assistance in interpreting images, greatly reducing the time needed for diagnosis to a mere matter of seconds. These developments have opened up possibilities for data-driven knowledge retrieval and extraction, shifting away from traditional reliance on expert opinions or scientific literature [14].

1.1 Novel Contributions of the Work

1. **Hybrid RNN Integration**: The incorporation of a hybrid RNN classifier represents an inventive strategy, merging various recurrent neural network architectures into a cohesive framework. This integration harnesses distinct strengths from diverse RNN models, contributing to enhanced performance.
2. **Unprecedented Accuracy**: The innovative methodology achieves unmatched levels of precision in brain tumor classification. Leveraging the capabilities of the hybrid RNN classifier, the model establishes precise distinctions between different tumor types, surpassing traditional classification methods.
3. **Refined Tumor Type Differentiation**: The hybrid RNN classifier improves the precision of heightened accuracy and provides valuable insights for medical practitioners, supporting more precise diagnoses and treatment planning.
4. **Innovation in Methodology**: This work challenges established standards in brain tumor classification of the unique amalgamation of diverse RNN architectures within the hybrid model sets a new performance benchmark in this field.

5. **Advancements in Diagnostic Capabilities**: The innovative contributions equip medical professionals with more precise and dependable tools for tumor identification and classification, ultimately improving patient care and outcomes.

1.2 Organization of the Paper

Current models are facing a substantial challenge, primarily concerning their precision, a critical factor in healthcare intelligent systems. To tackle this problem, we suggest creating a notably precise model, the brain tumor classification model with a specific emphasis on enhancing performance through a hybrid RNN classifier. The remaining structure of this work is as follows: Sect. 2 involves a review of existing literature, Sect. 3 details the brain tumor classification approach designed for performance enhancement using the hybrid RNN classifier, Sect. 4 provides an analysis of the results obtained from the proposed framework, and, finally, in Sect. 5, we draw conclusions from our findings.

2 Literature Review

Hassan Habib, Awais Mehmood, Tahira Nazir, Marrium Nawaz, Momina Masood, and their associates [15] present a study where they delve into the utilization of machine learning in the context of brain tumor segmentation and classification. The central objective of this investigation is to enhance the precision of tumor classification in MRI images through the utilization of diverse segmentation algorithms. The aim is to reduce the likelihood of misclassification errors. By utilizing multiple segmentation algorithms such as watershed, K-means, and threshold segmentation, in conjunction with an SVM classifier, the research ultimately achieves a classification accuracy of over 90%.

Madona B. Sahaai, G. R. Jothilakshmi, Raghavendra Prasath, Saurav Singh, and their co-researchers [16] present a research study centered on brain tumor detection using a deep neural network (DNN) algorithm. In this research, the team introduces a DNN architecture incorporating stacked autoencoders. Rather than relying on invasive biopsies, a common prelude to definitive brain surgery, this study aims to harness the advancements in machine learning to assist radiologists in tumor diagnosis. This approach leverages its speed and patient-centered benefits, ultimately improving medical imaging facilities. Through the enhancement of training efficiency and accuracy, machine learning opens up new avenues for healthcare professionals. This progress streamlines the comprehension of the human brain and alleviates the time-consuming burden of manually analyzing medical images.

Manav Sharma, Pramanshu Sharma, Ritik Mittal, Kamakshi Gupta, and their co-authors [17] share a study focused on brain tumor detection through machine learning. Their approach introduces a model that utilizes machine learning algorithms to accurately detect brain tumors in magnetic resonance images. Feature extraction and segmentation are performed using a Convolutional Neural Network (CNN). The dataset employed in this research was sourced from an online repository.

Rajat Mehrotra, M. A. Ansari, Rajeev Agrawal, R. S. Anand, and their research group [18] present a study on brain tumor classification, employing transfer learning in conjunction with artificial intelligence. Their research introduces an AI-driven approach to classify brain tumors using deep learning algorithms, utilizing publicly available datasets that categorize brain tumors into malignant and benign types. The datasets comprise a total of 696 images taken from T1-weighted images for testing. The proposed method achieves exceptional performance, demonstrating remarkable accuracy.

3 Brain Tumor Classification for Optimizing Performance

This approach employs a combination of RNN techniques to identify and categorize brain tumors, incorporating the Inception ResNetv2 model with an RNN. To conduct this study, a composite dataset, including 7023 human brain MRI images from three distinct sources (Fig share, SARTAJ dataset, and Br35H), was employed. These images were divided into four classes: glioma, meningioma, no tumor, and pituitary. In the subsequent section, we introduce a novel hybrid RNN classifier designed to enhance the performance of brain tumor classification. Figure 1 provides a visual representation of the framework we have developed.

Gathering data stands as a foundational and vital component in every machine learning endeavor, given that the effectiveness and accuracy of the algorithms are directly influenced by the quality of the input data. Furthermore, the quantity of images can vary for each patient, and MRI scans are capable of detecting a wide

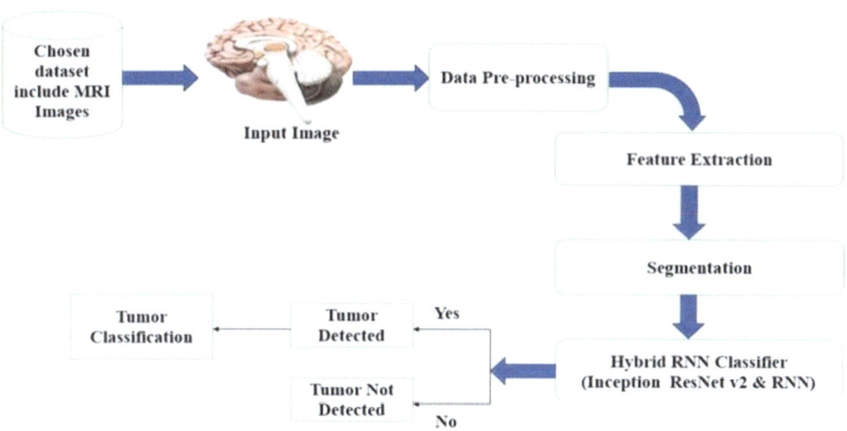

Fig. 1 Illustration of brain tumor classification using a hybrid RNN

spectrum of conditions, such as brain tumors, traumatic brain injury, developmental anomalies, multiple sclerosis, stroke, dementia, infections, and the underlying causes of headaches. During this phase, noise removal from MRI images is performed to enhance the model's accuracy. MRI images often contain noise, which, if left unaddressed, can increase redundancy and subsequently reduce the model's accuracy.

The presence of noise at the borders of MRI images poses a significant risk of tumor detection failures, ultimately impacting the model's accuracy. To address this issue, preprocessing steps include scaling, reducing, and converting the images to grayscale. Image preprocessing serves to enhance image quality, appearance, and characteristics, with noise removal being a fundamental step in most segmentation techniques. While the dataset used has already been pre-filtered for noise, an additional noise removal method is applied to all images due to the presence of some dark images. Image enhancement techniques, including thresholding-based artifact removal and histogram specification, are incorporated in the preprocessing stage.

Image processing techniques are applied to pinpoint the Region of Interest (ROI) in brain images, and these segmented brain tumor images are subsequently scrutinized for shape-related attributes. The malignant region's point of weight, sustainability percentage, rectangularity, convexity, and firmness were all extracted. These extracted brain tumor images exclusively contain the tumor, with all other objects removed through morphological operations, including erosion, opening, closing, and filtering. Feature extraction is a crucial step in mining valuable information from images. Pixel-based feature extraction is employed to classify the information into tumor or non-tumor categories, effectively reducing redundant data in the dataset.

The segmentation stage isolates the region of interest from the background, and it is a crucial step for accurate classification, as all subsequent decisions rely on the segmented tumor. This process establishes new connections between the nearest new centroid and the corresponding dataset points, creating a loop. This learning network, which has 164 layers, was successfully instructed on an enormous dataset made up of over a million visuals obtained from the database maintained by ImageNet. It excels in the task of categorizing images into 1000 object categories, underscoring its proficiency in capturing a wide array of image characteristics. This intricate architecture effectively extracts vital features from images, processing them at a size of 299*299, and delivers class probability estimates as its output. Additionally, Inception ResNet V2 incorporates Batch Normalization and ReLU activation as a preliminary step before executing the convolution operation with the weight matrix.

A recurrent neural network (RNN) operates by sequentially processing data, step by step. It maintains a hidden state serving as memory, updating it at each time step by considering the input data and the previous hidden state. RNNs are well suited for recognizing the sequential nature of data and leveraging patterns to make predictions about subsequent events. They fall under the category of neural networks designed for handling sequence data, and they share similarities with the functioning of the human brain. An advantage of RNNs is their ability to process inputs of varying lengths. These models are capable of retaining information across time steps, making them particularly useful in time series prediction tasks.

The Inception ResNet V2 with RNN surpasses prior models in brain tumor classification, emerging as one of the top-performing models for image classification. It not only detects the presence of brain tumors but also distinguishes their types. In this approach, brain tumors are categorized into different types, denoted as 0, 1, 2, and 3. The ability to accurately identify and classify tumors leads to more precise diagnoses, ultimately playing a pivotal role in saving patients' lives.

Proposed Deep Learning Approaches

The choice to contrast a proposed deep learning (DL) strategy with specific fundamental machine learning (ML) algorithms instead of alternative DL approaches may be influenced by various considerations. Some potential reasons include the following:

1. **Novelty and Emphasis**: The intention could be to underscore the uniqueness of the proposed hybrid RNN classifier. Authors may opt to compare it with established ML algorithms to highlight the innovations introduced by the new approach.
2. **Benchmarking**: Contrasting with fundamental ML algorithms establishes a baseline for performance evaluation, offering insights into how the proposed DL approach fares compared to widely used traditional methods.
3. **Interpretability and Simplicity**: Basic ML algorithms are often more interpretable and straightforward compared to intricate DL models. When interpretability is crucial, especially in applications like medical diagnoses such as brain tumor classification.
4. **Computational Constraints**: Training and evaluating multiple DL approaches can demand substantial computational resources. Limited resources may lead researchers to concentrate on a specific DL approach and compare it against a manageable set of baseline algorithms.
5. **State of the Art**: The proposed hybrid RNN classifier may already signify a state-of-the-art solution or a notable advancement over existing DL approaches. In such instances, the comparison may be structured to demonstrate the superiority of the proposed method over traditional ML algorithms.

4 Result Analysis

In this segment, we incorporate brain tumor classification to improve performance by employing a hybrid RNN classifier. Our dataset comprises 7023 human brain MRI images from the Brain Tumor MRI dataset. These patient brain images undergo a preprocessing phase to remove noise and enhance overall accuracy. The segmented data is input into a CNN for the detection and classification of brain tumors distinguishing between various types like glioma, meningioma, no, and pituitary when detected.

4.1 Evaluation Metrics

Precision, recall, and F1-score are standard metrics employed for assessing the effectiveness of classification models, such as hybrid RNN classifiers:

1. **Precision and Recall** gages the model's accuracy in positive predictions, representing the ratio of correctly predicted positive observations to the total predicted positives. An elevated precision suggests that when the model predicts a positive class, it is likely to be correct. A heightened recall signifies the model's effectiveness in identifying the majority of positive instances.
2. **F1-score** serving as the harmonic mean of precision and recall offers a balanced assessment, especially beneficial in cases of class imbalance. Ranging from 0 to 1, with 1 indicating perfect precision and recall, it becomes a valuable metric when considering both false positives and false negatives.

$$\text{Precision} = \frac{\text{True Positives}}{\text{True Positives} + \text{False Positives}} \tag{1}$$

$$\text{Precision} = \frac{\text{True Positives}}{\text{True Positives} + \text{False Negatives}} \tag{2}$$

$$F1_{\text{score}} = 2 * \frac{\text{Precision} * \text{Recall}}{\text{Precision} + \text{Recall}} \tag{3}$$

4.2 Random Forest (RF)

Random Forest (RF) is an ensemble learning technique that constructs numerous decision trees during training. It yields the class mode for classification tasks or the average prediction for regression problems. The model introduces diversity by training on distinct random subsets of data and features, mitigating overfitting and enhancing overall predictive accuracy and model robustness.

4.3 Linear Regression (LR)

Linear Regression (LR) models, the relationship between a dependent variable and one or more independent variables in a linear manner. It fits a line to represent data points, utilizing one independent variable in simple linear regression and multiple in its counterpart. The model seeks coefficients to minimize the difference between observed and predicted values, often through the least squares method.

The performance of presented approach is compared with different classifiers like Random Forest (RF) and Linear Regression (LR). Table 1 shows the precision performance analysis.

Table 1 Precision performance analysis

Brain tumor type	Random forest	Linear regression	Presented hybrid RNN
0	0.58	0.57	0.93
1	0.47	0.42	0.96
2	0.68	0.64	0.97
3	0.60	0.58	1.00

The 0, 1, 2, 3 are different types of brain tumors classified. Compared to RF, LR classifiers, presented hybrid RNN classifier has better precision. Figure 2 shows the graphical representation of precision performance and x-axis indicates different brain tumor types whereas y-axis indicates precision performance.

Compared to RF, LR presented hybrid RNN algorithm which has better and high precision for all types of brain tumor. The type 3 tumor has higher precision using Presented algorithm. Table 2 shows the recall performance analysis.

Presented hybrid RNN has high recall than LR and RF. Figure 3 shows the comparative graph for recall scores which are indicated on the y-axis in Fig. 3 associated with the distinct tumor forms.

Presented hybrid RNN classifier has higher recall values than RF and LR classifiers. Table 3 shows the f1-score performance analysis.

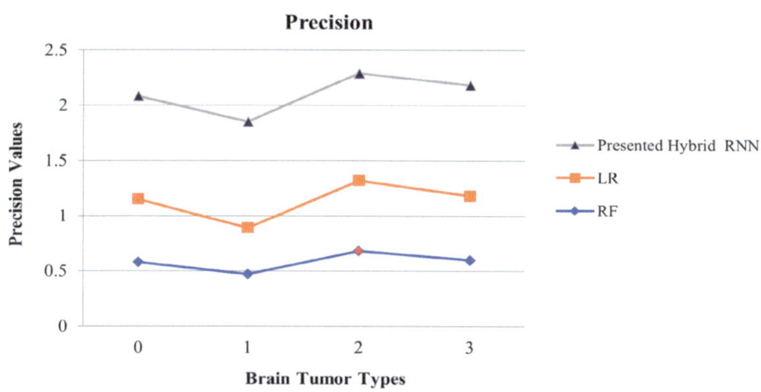

Fig. 2 Precision graph for comparison

Table 2 Recall performance analysis

Brain tumor type	Random forest	Linear regression	Presented hybrid RNN
0	0.61	0.55	0.98
1	0.42	0.48	0.91
2	0.29	0.19	1.00
3	0.82	0.73	0.96

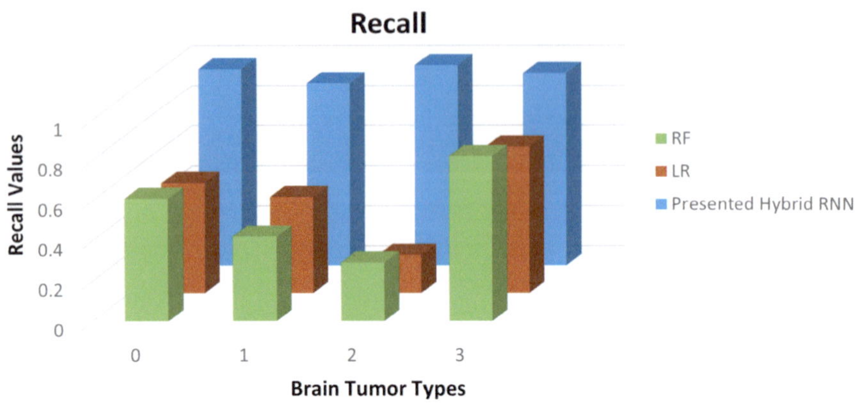

Fig. 3 Recall performance comparison

Table 3 F1-score performance analysis

Brain tumor type	Random forest	Linear regression	Presented hybrid RNN
0	0.59	0.56	0.95
1	0.45	0.45	0.93
2	0.41	0.29	0.98
3	0.70	0.65	0.98

Compared to LR and RF classifiers, presented hybrid RNN has high f1-score for all types of brain tumors. Figure 4 shows the accuracy comparison.

In Fig. 4, the y-axis indicates f1-score performance values and x-axis indicates different types of brain tumor. Compared to RF and LR classifiers, presented hybrid

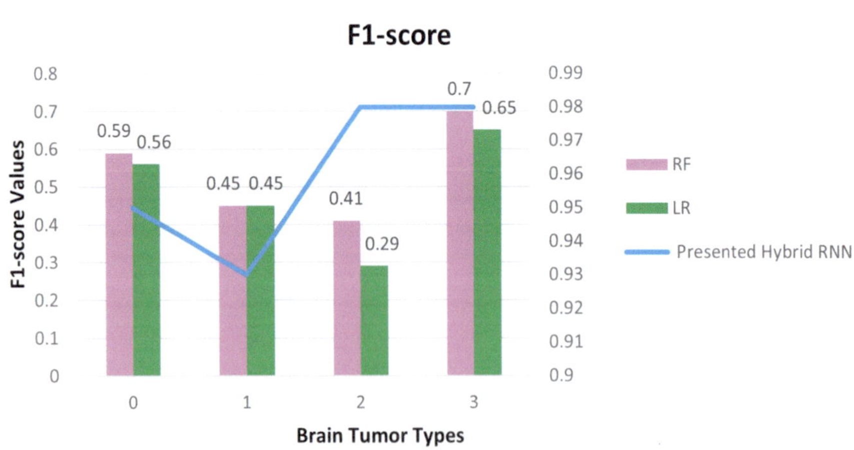

Fig. 4 F1-score performance comparison

RNN classifier has better and high f1-score. For better recognition and categorization of cerebral tumors, this tactic has been successful.

5 Conclusion and Future Work

The study seeks to improve brain tumor classification by employing a hybrid RNN classifier on the brain tumor MRI dataset, which comprises 7023 images. The model utilizes a hybrid deep learning algorithm, initiating with segmentation to identify tumor regions. The subsequent Inception ResNetv2 with RNN is then employed for the detection and classification of brain tumors, initially identifying their presence and categorizing them into types (0, 1, 2, 3). Performance evaluation is conducted using metrics such as quality, recall, and F1.

In future work, the sliced cerebral tumor images undergo further analysis using methods involving image processing to acquire shape qualities. It carries the promise of potentially saving lives and can be seamlessly employed in real time for brain tumor identification and categorization.

References

1. Jahan R, Tripathi MM (2021) Brain tumor detection using machine learning in MR images. In: 2021 10th IEEE international conference on communication systems and network technologies (CSNT), June 2021. IEEE, pp 664–668
2. Archana KV, Komarasamy G (2023) A novel deep learning-based brain tumor detection using the Bagging ensemble with K-nearest neighbor. J Intell Syst 32(1):20220206
3. Gopalachari MV, Kolla M, Mishra RK, Tasneem Z (2022) Design and implementation of brain tumor segmentation and detection using a novel Woelfel filter and morphological segmentation. Complexity
4. Jaganathan D, Manjunathan N, Girirajan S (2022) Morphogenesis method to detect brain tumor using machine learning technique and noise filtering. In: 2022 International Conference on Sustainable Computing and Data Communication Systems (ICSCDS), April 2022. IEEE, pp 1–7
5. Soomro TA, Zheng L, Afifi AJ, Ali A, Soomro S, Yin M, Gao J (2022) Image segmentation for MR brain tumor detection using machine learning: a review. IEEE Rev Biomed Eng
6. Dipu NM, Shohan SA, Salam KMA (2021) Deep learning based brain tumor detection and classification. In 2021 international conference on intelligent technologies (CONIT), June 2021. IEEE, pp 1–6
7. Tang MCS, Teoh SS (2023) Brain tumor detection from MRI images based on ResNet18. In: 2023 6th international conference on information systems and computer networks (ISCON), March 2023. IEEE, pp 1–5
8. Siar M, Teshnehlab M (2019) Brain tumor detection using deep neural network and machine learning algorithm. In: 2019 9th international conference on computer and knowledge engineering (ICCKE), October 2019. IEEE, pp 363–368
9. Gaikwad S, Patel S, Shetty A (2021) Brain tumor detection: an application based on machine learning. In: 2021 2nd international conference for emerging technology (INCET), May 2021. IEEE, pp 1–4

10. Suresha D, Jagadisha N, Shrisha HS, Kaushik KS (2020) Detection of brain tumor using image processing. In: 2020 fourth international conference on computing methodologies and communication (ICCMC), March 2020. IEEE, pp 844–848
11. Aggarwal M, Tiwari AK, Sarathi MP, Bijalwan A (2023) An early detection and segmentation of brain tumor using deep neural network. BMC Med Inform Decis Mak 23(1):1–12
12. Sandhya U, Kumar KN, Saranya AP, Jayapal N, Kumarganesh S, Brain tumor detection and classification with DGMM: RBCNN technique. Int J Health Sci I:7345–7361
13. Hao R, Namdar K, Liu L, Khalvati F (2021) A transfer learning–based active learning framework for brain tumor classification. Front Artif Intell 4:635766
14. Preetika B, Latha M, Senthilmurugan M, Chinnaiyan R (2021) MRI image based brain tumour segmentation using machine learning classifiers. In 2021 international conference on computer communication and informatics (ICCCI), January 2021. IEEE, pp 1–9
15. Habib, H, Mehmood, A, Nazir, T, Nawaz, M, Masood, M, Mahum, R. (2021, August). Brain Tumor Segmentation and Classification using Machine Learning. In: 2021 international conference on applied and engineering mathematics (ICAEM). IEEE, pp 13–18
16. Sahaai MB (2021) Brain tumor detection using DNN algorithm. Turk J Comput Math Educ (TURCOMAT) 12(11):3338–3345
17. Sharma M, Sharma P, Mittal R, Gupta K (2021) Brain tumour detection using machine learning. J Electron 3(4):298–308
18. Mehrotra R, Ansari MA, Agrawal R, Anand RS (2020) A transfer learning approach for AI-based classification of brain tumors. Mach Learn Appl 2:100003

Machine Learning Techniques for Classification and Regression: Vinho Verde Dataset

Harsh Mukund Zele

Abstract This research paper provides an extensive overview of machine learning techniques for classification and regression tasks. The study includes a rigorous experimental evaluation conducted on the Vinho Verde dataset. The classification methods investigated in this research encompass the k-nearest-neighbor classifier, boosted decision trees, and support vector machines. The study explores multivariate linear regression and extreme learning machines for regression analysis. A comprehensive theoretical examination of these methods and detailed explanations of the implementation steps for the k-NN classifier, multivariate linear regression, and extreme learning machine algorithms are presented. Notably, the bagged decision trees classifier demonstrated the highest performance among the classification methods, while the extreme learning machine exhibited superior results for regression tasks. The findings of this research contribute valuable insights into the effectiveness of these machine learning techniques, opening avenues for further exploration and application in real-world scenarios.

Keywords Machine learning · Classification · Regression · Support vector machines · Bagged decision trees

1 Introduction

Machine learning, a subset of computer science, aims to enhance algorithm performance by leveraging example data and past experiences. In the context of this research, machine learning methods are evaluated using the Vinho Verde dataset, focusing on classification and regression tasks. While the field of machine learning is expansive and has garnered considerable industry interest in recent years, this paper does not delve deeply into identifying optimal methods for specific tasks. Instead, it

H. M. Zele (✉)
Pillai College of Engineering, Navi Mumbai, India
e-mail: zharsh2611@gmail.com

A. P. Agrawal et al. (eds.), *Advances of Artificial Intelligence and Machine Learning in Image Processing, Security and Data Analytics*, Lecture Notes in Electrical Engineering 1464, https://doi.org/10.1007/978-981-95-0148-9_23

offers an experimentally supported overview of popular techniques within modern machine learning.

A foundational understanding of basic machine learning methods and their mathematical underpinnings is well-articulated by Alpaydin in his work [2]. Notably, this paper covers a spectrum of methods discussed by Alpaydin, excluding neural network-based extreme learning machines, which are explored in [6].

Dual objectives drive the selection of machine learning algorithms for experimentation in this paper. Firstly, algorithms with high performance in the given dataset were sought for the course competition, leading to the inclusion of support vector machines and tree bagging. Secondly, a deliberate choice was made to incorporate algorithms that are both prevalent in real-world applications and easy to implement. To fulfill this criterion, the paper explores the k-nearest neighbor classifier, linear regression, and extreme learning machines [15].

The surge in popularity of neural networks, evidenced by their success in real-world machine learning problems and benchmark datasets like MNIST [5, 6, 17], has prompted their representation in this study through the adoption of extreme learning machines. Chosen for their ease of implementation and reliance on a mathematical model akin to high-performing neural networks [18], extreme learning machines offer a practical avenue for exploring neural network principles.

Structured into five sections and accompanied by appendices, this paper unfolds with this introductory section, then elucidates the employed methods in Sect. 2. Subsequent sections detail the conducted experiments Sect. 3, present the results Sect. 5, and engage in a comprehensive discussion Sect. 4. Algorithm implementations are conveniently provided in the appendices.

2 Methods

2.1 K-Nearest Neighbor Classifier

The k-nearest neighbor (k-NN) classifier is a nonparametric classification technique, which is (generally) used under the assumption that the underlying distribution of the data is unknown [1]. With parametric classification methods, predictions of the unseen data are based on the *fixed parameter models* constructed from the input data. With nonparametric methods, however, the parameter count and nature vary together with the changing data, and the only assumption we can make is the similarity between the input and output data. Various nonparametric methods are available for different machine learning tasks, e.g., classification and regression [2].

Nonparametric classification algorithms make the decision based on the *similarities* between the points to be classified and training data. The definition of the similarity may change between the classification tasks. One example of a similarity measure is the Euclidean distance between two points: the closer the points, the more similar they are.

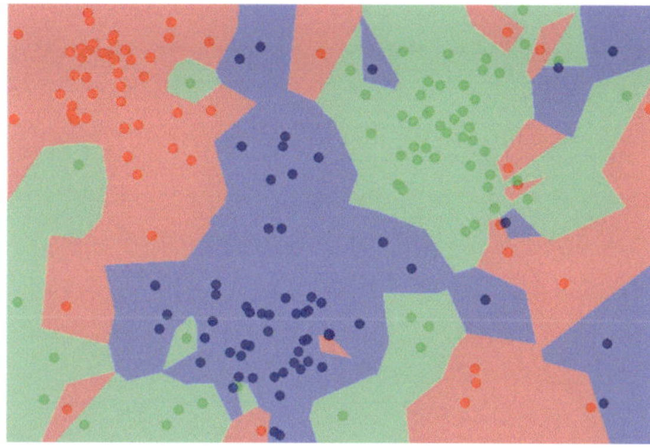

Fig. 1 An example of the classification boundaries of a k-nearest neighbor, using a single neighbor (k = 1) [19]

The k-NN classifier is a special case of a more general nonparametric classifier. Let us first consider the general nonparametric classification problem. Given a set of N training points $\mathbf{X} = x^t$, $t = 1...N$, each belonging to a class ω_i, we want to assign a point **x/X** to one of those classes. In the k-NN classifier, the input point is assigned to the class with the most points among the neighbors of the input. Figure 1 shows an example of the classification boundaries generated by the k-nearest neighbor classifier with a single neighbor (k = 1).

In spite of simplicity, the k-NN classifier is very robust. It is, however, very reliant on the training data, computationally heavy, and unreliable with higher dimensional data. We believe that the k-NN classifier is suitable for the wine classification task. The size of the training dataset seems large enough to produce meaningful results and, at the same time, computationally manageable even with naive implementation [3]. Data preprocessing techniques might be needed to handle the dimensionality and other problems related to the similarity measure. These techniques are discussed in Sect. 2.6.

2.2 *Linear Regression*

In linear regression, the relationship between the input variables **X** and the predicted variable **y** is modeled using a linear combination of basis functions [2]. The mathematical expression describing the relationship of the input variables and the predicted variable is given in Eq. 1, where input variables are denoted by **X**, the predicted variable is expressed as a function of input variables $g_i(\mathbf{X})$, ϕ are the basis functions, and w_j are the weights for each basis function. The closed-form solution for the weights is found using linear algebra. The least squares solution for the weights is given by $\mathbf{w} = \mathbf{y}\boldsymbol{\Phi}^{\dagger}$, where $\boldsymbol{\Phi}$ is a matrix with elements $\phi(x^t)$ and $\boldsymbol{\Phi}^{\dagger}$ denotes the pseudoinverse of the matrix [2].

$$g_i(\mathbf{X}) = \sum_{j=1}^{k} w_j \phi_{ij}(\mathbf{X}) \quad (1)$$

When doing regression with a single input variable, polynomial basis functions are often used. For more than one input variable, polynomial regression is rarely used. Instead, linear basis functions are used [2].

2.3 Extreme Learning Machine

Artificial neural networks are a family of machine learning models that consist of multiple locally interconnected, simple computing units called neurons that together form a network-type structure [20]. The neurons can be used to form a non-linear mapping from the input space and divide a complex problem into simpler subproblems [6].

The output of the network is adjusted by changing the synaptic weights between the neurons. The weights are learned by a learning algorithm to obtain the desired behavior of the network. Figure 2 shows an example of a single-layer feedforward neural network [6].

Extreme learning machines are neural network constructs with two special properties. First, they are single-layer feedforward networks, meaning that the neuron connections do not form cycles. Secondly, the weights of the hidden layer are randomly assigned and frozen. These properties have interesting consequences. The learning algorithm of the weights reduces to a single step, essentially amounting to finding a solution for a linear system. While this results in learning thousands of times faster than traditional backpropagation algorithm [6], these models can still produce good generalization performance and learn non-trivial, non-linear mappings between the inputs and outputs [7].

Despite the intriguing results shown in the original paper [7], there has been some controversy in the machine learning community regarding the extreme learning machines and their authors [13].

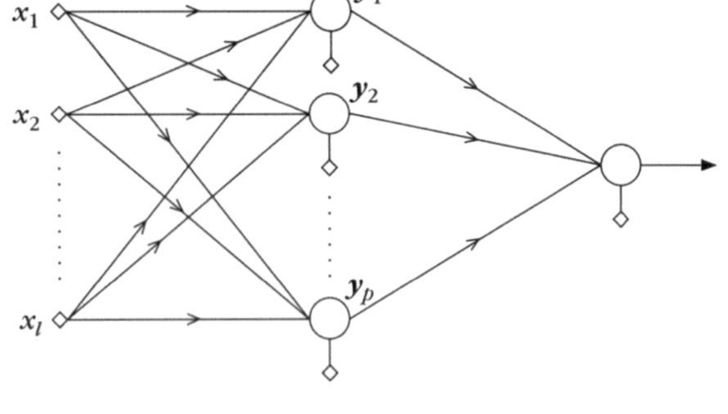

Fig. 2 A single hidden layer neural network [16]

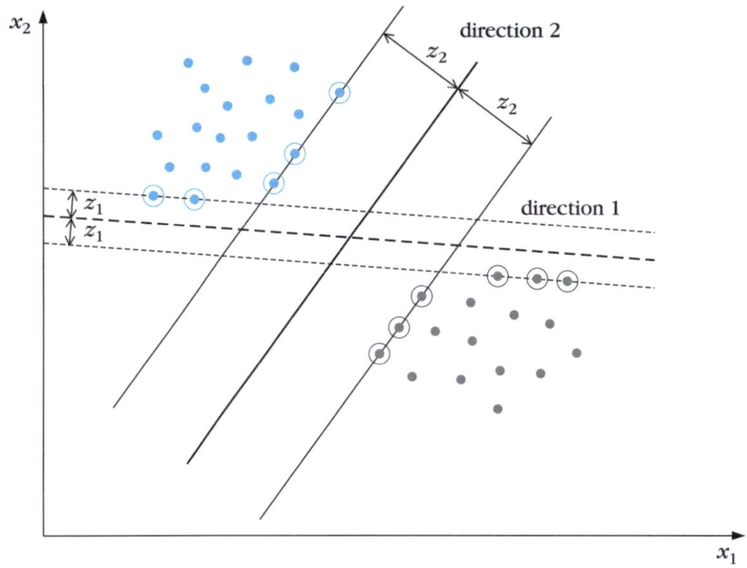

Fig. 3 An example of a linearly separable binary classification problem with two possible linear classifiers. Support vector machine finds the hyperplane that maximizes the marginal between the classes [16]

2.4 Support Vector Machines

Support Vector Machines (SVM) are non-probabilistic supervised learning models that can be used for both classification and regression tasks [14]. In this paper, we consider only the SVMs used for binary classification.

In the general case, the SVM finds the parallel hyperplanes that maximize the marginal between the linearly separable classes, minimizing the classification error. Figure 3 shows an example of a problem solvable by general SVM [16].

The general SVM can be extended to solve a non-separable classification problem by allowing the input points to be incorrectly classified with a certain cost. Further, a non-linear separation is achieved by non-linearly mapping the input features into a new feature space and using a similar SVM as in the linearly separable case. An example of the non-linear case is presented in Fig. 4 [16].

2.5 Bagged Decision Trees

One method for creating well-performing learners is to combine multiple weaker learners in a way that improves the robustness of the model [2]. Combinations of a large number of simpler learners have performed well in machine learning competitions [9].

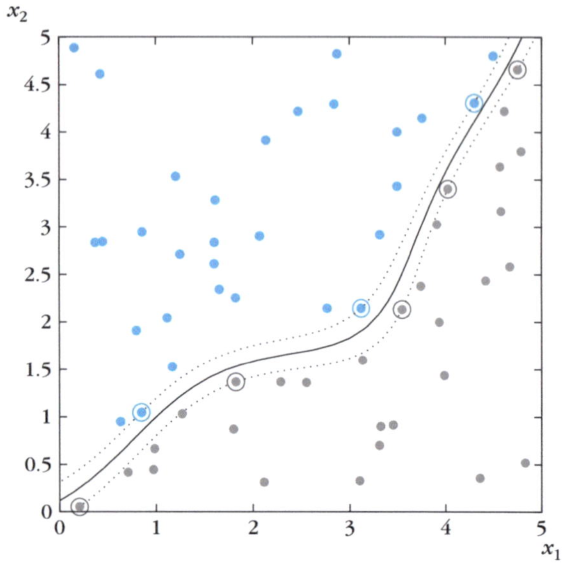

Fig. 4 An example of a classification of non-linear classification done by support vector machine [16]

TreeBagger is a MATLAB tool that combines decision trees using bootstrap aggregation [12].

The base learning algorithm used by TreeBagger is a decision tree. Decision trees are a supervised nonparametric learning method. Decision trees represent the hypothesis by a tree where each leaf node corresponds to an area in the input space. The tree branches are binary decisions. The learning algorithm uses a heuristic method to select the best feature to branch by. The key idea of learning decision trees is to take advantage of the recursive property of trees that a new tree is created by attaching a tree to a leaf of another tree [2].

The empirical error approximates the generalization error the larger the dataset is. Because the size of the real training data is limited, the training set is enlarged artificially. One way to artificially enlarge the dataset is to use the histogram of the training data as the "true" dataset. Using the histogram for enlarging the dataset is called bootstrap aggregation [4]. TreeBagger uses bagging to create the training datasets for the decision trees.

2.6 Data Pre-processing

A crucial part of successful prediction using machine learning methods is to make sure the data is in a suitable form for the learning algorithm used. For example, algorithms that depend on a measure of distance between the data points, k-NN, for example, can generalize the data better when the distances are standardized. The distances are standardized by dividing the features by their standard deviations [2].

Some learning algorithms benefit from the data being centered around zero. The centering is achieved by subtracting the mean of the data from all of the data vectors [2].

It is often beneficial to keep the models simple. It is stated that simpler models are more robust on small datasets [2]. Feature selection is one way to reduce model complexity. Feature selection means selecting a subset of the data features for training the predictor. Different methods for feature selection can be used, but they are based on the same idea of repeatedly evaluating validation errors on different subsets of features. The subsets can be chosen, for example, by starting from no features and adding the feature that improves performance the most. One way for feature selection is searching through all the combinations of features, but it is often not practical due to the large number of features [2].

2.7 *Model Evaluation*

To make informed decisions about how to optimize the model parameters and, in general, how to create good predictors, methods for comparison of the created models are needed. A simple measure of the empirical error on the training data is usually not enough. To get more accurate estimates of the generalization error, the training dataset is split into training and validation datasets. The validation dataset is not used for training the model, as its sole purpose is to be used for the estimation of the generalization error [2].

The number of labeled training samples is limited in real-world machine learning problems. Therefore, the size reduction to the training dataset resulting from the use of the validation set is not desirable. K-fold cross-validation is a commonly used method for validation without drastic changes in the training set size. In k-fold cross-validation, the training dataset is divided into k "folds" one of which is used for validation at a time and the rest for training. Each fold is used in turn for validation. The generalization error estimate is the average of the errors for each k-folds [2].

Leave-one-out error is a variation of k-fold error where k is set to one. Leave-one-out error is often computationally prohibitively expensive.

3 Experiments

In the experiments, two prediction tasks were studied. The first task was to classify the Vinho Verde dataset into red and white wines. In the second task, wine qualities were approximated with regression. In this section, the conducted experiments are described.

3.1 Wine Type

The wine type prediction was a binary classification task with two classes, namely red and white wines. Two classification methods, namely the k-nearest neighbor classifier and support vector machines, were studied. A k-nearest neighbor classifier was implemented from scratch. The MATLAB implementation of support vector machines was used. The experiments conducted with the two classifiers are described in the following:

K-Nearest Neighbor Classifier: A K-nearest neighbor classifier was implemented. In the model selection phase, different parameters were experimented with, and the results were validated using 20-fold cross-validation. The only preprocessing step for the data was centering, meaning making the data zero mean and unit variance.

The parameter space for a model as simple as a k-NN classifier is large. To limit the required work and computation time, only the most influential parameters were experimented with. The number of neighbors k was the most influential parameter. Values of k from 1 to 50 were tested. Another parameter that was experimented with was the distance function. Euclidean distance and Minkowski distance were tested.

Support Vector Machine: Support vector machines were studied for the classification task. MATLAB has a versatile SVM implementation included in the default distribution [10]. The MATLAB SVM implementation has many parameters to tune, and the total parameter space is huge. To help reduce the number of parameters to be tested, the MATLAB classification toolbox was used to find coarse parameters.

Three different kernel functions were used. Quadratic and cubic kernel functions performed poorly compared to the Gaussian kernel functions, and they were excluded from further experiments. The Gaussian kernel functions were studied by setting the box constraint to 1 and testing the kernel scale γ with values in the range 0.1–20 [8].

The data was standardized by centering it to zero mean and unit variance. 20-fold cross-validation was used for the model selection.

3.2 Wine Quality

Wine quality prediction was defined as a regression task where a quality rank from 1 to 7 was to be predicted for the white and the red wines. Multivariate linear regression and extreme learning machine algorithms were implemented for the regression task. From the Kaggle competition, it was learned that better results for the quality prediction task were achieved by treating it as a classification problem. Bootstrap aggregation of decision trees was tested as the classification method.

Multivariate Linear Regression: Multivariate regression was used to predict the quality labels for the wines. The linear regression model was kept as simple as possible using only linear kernel functions and no data preprocessing. The results were rounded to the nearest integer. With simplicity as the main goal, there was only one model to test, which had 12 parameters that were acquired from the analytical solution for linear regression explained in 2.

Bootstrap Aggregation with Decision Trees: The MATLAB TreeBagger tool was used to create bagged decision trees. The quality prediction was interpreted as a classification task for the experiments using TreeBagger. Regression trees were also tried, but they performed poorly compared to the classification models and were excluded from further experiments.

The TreeBagger supports multiple modes of fitting the decision trees and has many parameters for controlling the ensembling of the trees. A cursory overview of the parameter space was conducted by using the MATLAB classification toolbox and some manual trials. In the preliminary search for parameters, it was found that the default settings of TreeBagger yield the best results, and thus, further experiments with the various parameters, except the number of grown trees and feature selection, were omitted.

The feature selection was conducted by trying out all combinations of three or more features. With the final feature selection, the number of trees was varied from 1 to 5000. The TreeBagger built-in out-of-bag error (OOB error) was used for model selection. The OOB error computes the misclassification probability for the classification trees in the training data.

Extreme Learning Machine: Regression with extreme learning machines (ELM) was used to predict the wine qualities. The ELM was implemented for the experiment by the research group.

The ELM implementation has 11 linear neurons for each of the input variables. The hyperbolic tangent sigmoid transfer function, called tansig in MATLAB [11], was used as the non-linear activation function. ELMs were trained using 0–1000 non-linear neurons. The model selection was conducted using leave-one-out validation. The data was standardized to zero mean and unit variance for the experiments.

4 Discussion

An interesting finding from the quality prediction task was that the distinction between classification and regression is not always clear. A quality score from 1 to 7 was to be predicted on each of the wines. Intuitively predicting the wine qualities seems like a regression task. However, classification-type learners performed consistently better than regression learners in the task. It seems there is no easily determined continuous relationship in the data features corresponding to the wine quality.

Next, the findings on each of the studied algorithms are briefly discussed.

4.1 Extreme Learning Machine

The ELM performance was somewhat weaker than what was expected based on the success of neural network methods in real-world machine learning problems. There are many possible causes for the bad performance. One possibility is that in the ELM experiments, the model overfitted the data badly. The overfitting could have been combated with weight regularization, which was not implemented for the ELM in this paper.

Implementing ELM is easy compared to other neural network methods since no backpropagation or other learning method is needed for the weights of the non-linear layer. Therefore, ELM was a good find for getting a taste of neural networks, even though its performance was suboptimal.

4.2 K-Nearest Neighbor Classifier

K-nearest neighbor classifier performed worse than even very simple support vector machines and thus received little attention for parameter optimization. K-NN is interesting as it is a simple nonparametric method. The amount of training data was possibly too small for the k-NN classifier to perform well.

4.3 Support Vector Machines

Support vector machines performed well in the type prediction task. The final results could have been better if a more careful cross-validation scheme had been implemented. This was apparent in the competition submission of the team, where the SVM-type prediction performed very well (first place) in the public leaderboard but then got much worse results for the rest of the test set.

4.4 Multivariate Linear Regression

Multivariate linear regression was too simple for the given regression task. Considering the structure of the data, this should not be surprising.

4.5 Bootstrap Aggregated Decision Trees

TreeBagger yielded the best results in the quality prediction task. Bagging seems to be a powerful method for these types of prediction tasks as it is often used in machine learning competitions and performs well here.

There were some accuracy problems with the TreeBagger as well. The OOB error that was used for the evaluation of the models only measures the error in the training set without using validation. Implementing a validation scheme would have benefited the competition entry.

5 Findings and Reflection

The outcomes of the wine type and wine quality predictions are detailed in Tables 1 and 2, respectively.

In the wine-type prediction task, the k-nearest neighbor classifier achieved 94.0% accuracy with Euclidean distance and $k = 6$ nearest neighbors. The support vector machine excelled with 99.125% accuracy, employing the Gaussian kernel function and $\gamma = 2.85$.

The Bagged Decision Trees outperformed the Extreme Learning Machine and the linear regression model for the wine quality prediction task. The Extreme Learning Machine, configured with 11 linear neurons and 1 non-linear neuron, achieved 54.75% accuracy. Linear regression, with no specific hyperparameters, achieved 49.75% accuracy, while the Bagged Decision Trees, employing 250 trees, achieved 68.25.

In summary, exploring various machine learning algorithms for regression and classification tasks on the Vinho Verde dataset yielded valuable findings. Reflecting on these outcomes provides insights into algorithmic performance and complexity, contributing to a deeper understanding of practical applications in machine learning.

Table 1 Results for the wine-type predictions

Method	Parameters	Accuracy (%)
k-NN	Euclidean, $k = 6$	94.0
SVM	Gaussian, $\gamma = 2.85$	99.125

Table 2 Results for the wine quality predictions

Method	Parameters	Accuracy (%)
Linear regression	–	49.75
ELM	$l = 11, nl = 1$	54.75
Bagged trees	250 trees	68.25

6 Future Directions

There are several promising directions for enhancing the effectiveness of machine learning techniques on the Vinho Verde dataset. One avenue is exploring ensemble methods that combine the strengths of k-nearest neighbor classifiers, support vector machines, and extreme learning machines. This collaborative approach could improve predictive accuracy by leveraging the diversity of individual models. Feature engineering and selection techniques offer another avenue for improvement. Investigating advanced methods and domain-specific knowledge incorporation may uncover hidden patterns within the dataset and enhance model generalization. Additionally, exploring deep learning architectures, such as neural networks, holds the potential for superior performance on complex tasks. Addressing the class imbalance in quality prediction and employing techniques like oversampling or alternative evaluation metrics tailored for imbalanced datasets could lead to more accurate predictions. These future directions aim to refine and extend the application of machine learning techniques in practical scenarios, providing valuable insights and improvements.

References

1. Agrawal R (2014) K-nearest neighbor for uncertain data. Int J Comput Appl 105(11):13–16
2. Alpaydin, E.: Introduction to machine learning. MIT press (2004)
3. Alsammak, I.L.H., Sahib, H.M.A., Itwee, W.H.: An enhanced performance of k-nearest neighbor (k-nn) classifier to meet new big data necessities. In: IOP Conference Series: Materials Science and Engineering. vol. 928, p. 032013. IOP Publishing (2020)
4. Breiman L (1996) Bagging predictors. Mach Learn 24(2):123–140
5. Ciresan D, Meier U, Schmidhuber J (2012) Multi-column deep neural networks for image classification. In: 2012 IEEE conference on computer vision and pattern recognition (CVPR). IEEE, pp 3642–3649
6. Haykin S (2009) Neural networks and learning machines. No. nid. 10 in Neural networks and learning machines. Prentice Hall
7. Huang GB, Zhu QY, Siew CK (2006) Extreme learning machine: theory and applications. Neurocomputing 70(1):489–501
8. Ismael AM, Şengür A (2021) Deep learning approaches for covid-19 detection based on chest x-ray images. Expert Syst Appl 164:114054
9. Kaggle (2015) Profiling top kagglers: Gilberto titericz, new 1 in the world. http://blog.kaggle.com/2015/11/09/profiling-top-kagglers-gilberto-titericz-new-1-in-the-world/
10. MathWorks: fitcsvm (2015). http://se.mathworks.com/help/stats/fitcsvm.html
11. MathWorks: tansig (2015). http://se.mathworks.com/help/nnet/ref/tansig.html
12. MathWorks:Treebagger (2015). http://se.mathworks.com/help/stats/treebagger.html
13. Reddit (2015) Yann lecun: What's so great about "extreme learning machines"? https://www.reddit.com/r/MachineLearning/comments/34u0go/yann_lecun_whats_so_great_about_extreme_learning/
14. Salcedo-Sanz S, Rojo-Álvarez JL, Martínez-Ramón M, Camps-Valls G (2014) Support vector machines in engineering: an overview. Wiley Interdiscip Rev Data Min Knowl Discov 4(3):234–267
15. Sarker IH (2021) Machine learning: algorithms, real-world applications and research directions. SN Comput Sci 2(3):160

16. Theodoridis S, Koutroumbas K (2009) Pattern recognition. Academic, Burlington, MA; Previous ed.: Academic Press, San Diego, CA, c2006
17. Wan L, Zeiler M, Zhang S, Cun YL, Fergus R (2013) Regularization of neural networks using dropconnect. In: Proceedings of the 30th international conference on machine learning (ICML-13), pp 1058–1066
18. Wang J, Lu S, Wang SH, Zhang YD (2022) A review on extreme learning machine. Multimed Tools Appl 81(29):41611–41660
19. Wikipedia (2015) K-nearest neighbors algorithm. https://en.wikipedia.org/wiki/K-nearest_neighbors_algorithm
20. Zou J, Han Y, So SS (2009) Overview of artificial neural networks. In: Artificial neural networks: methods and applications, pp 14–22

A New Annotated Resume Dataset and Test with RoBERTa Model for Entity Extraction

Laxmi Chaudhary, Shivam Vasandani, Kushagra Anand Dixit, and Ansh Bhargava

Abstract Extracting entities from resumes is a crucial application of information extraction and a vital component of Natural Language Processing (NLP). With the ever-increasing number of job applications, discovering qualified candidates manually has become a challenging task for employers. Additionally, the scarcity of annotated resume datasets poses an obstacle to entity extraction from resumes. To address these challenges, this paper presents a newly annotated dataset comprising 621 resumes, manually labeled with entities such as name, email, skills, designation, degree, graduation year, employment history, and location. To demonstrate the effectiveness of the annotated dataset, we have proposed a resume parser by utilizing the Spacy and RoBERTa model to extract the significant entities. The experiments are conducted on four sets of datasets 200, 300, 470, and 621 resumes and demonstrated accuracies of 51, 57, 62, and 78%. The proposed annotated dataset and methodology have the potential to significantly improve the efficiency of recruitment processes.

Keywords Entity extraction · Annotated dataset · Natural language processing · Resume parser · DocBin

1 Introduction

In the past years, extracting entities from resumes has been a prominent research field in information extraction [1–6]. Furthermore, the development of Artificial Intelligence (AI) has encouraged research on substituting human resources with AI for resume screening [7–13]. It is crucial to improve this procedure for further studies and resume performance evaluations because the precision of resume screening depends on the accuracy of resume information extraction. To extract specific data from

L. Chaudhary (✉) · S. Vasandani · K. A. Dixit · A. Bhargava
Jaypee Institute of Information Technology, Noida, India

S. Vasandani
e-mail: svasandani4@gmail.com

A. P. Agrawal et al. (eds.), *Advances of Artificial Intelligence and Machine Learning in Image Processing, Security and Data Analytics*, Lecture Notes in Electrical Engineering 1464, https://doi.org/10.1007/978-981-95-0148-9_24

resumes such as names and contact information, Mooney [14] utilized rule-based and dictionary matching algorithms. The other techniques such as Hidden Markov Model (HMM) and Support Vector Machine (SVM) were also utilized by Yu et al. [15] to extract these entities. A sizable Chinese resume corpus has been created as part of the Related Resume Corpus Construct project by Su et al. [16]. The Named Entity Recognition (NER) of resume text was performed using the Bi-LSTM-CRF model by Huang et al. [17]. This technique accurately retrieves resume details like name, address, gender, and date of birth, but it also has the disadvantage of losing part of the verbal expression data that was originally included. For instance, the NER approach is unable to extract the whole phrases needed to define one's long-term professional objectives. When an AI system is employed to assess a candidate's resume, the career objective is additionally incorporated into the score. In overall, sentences that convey crucial information shouldn't be ignored. Gan and Mori [18] have transformed the task of resume information extraction into a sentence classification task. They have done by using the English resume dataset from the previous study and improving the classification rules to produce a larger and more detailed classification dataset of resumes. They have extracted personal information, education, experience, summary, skill, qualifications, and object details. The dataset has only some entities, but more entities are also required for resume parsing.

To address all these issues, the dataset is collected from Kaggle [19]. The original dataset consists of 15 batches of 1000 files each. The dataset has no annotation. We have manually annotated the 621 resumes from original dataset by selecting resumes randomly. The various entities such as name, email, skills, degree, graduation year, location, and companies worked at details have been labeled. Later, we have built a resume parser to extract vital entities using the Spacy and RoBERTa [20] model and demonstrated the effectiveness of the newly annotated dataset. The proposed resume parser experiments are conducted on four sets of datasets comprising 200, 300, 470, and 621 resumes, and the obtained accuracies are 51, 57, 62, and 78%, respectively.

2 Annotated Dataset

For the annotation, we have utilized Doccano tool. Doccano is a tool for annotating the text data. It offers labeling functionalities for text categorization, sequence tagging, and sequence-to-sequence tasks. With the help of this tool, we can generate annotated data for various tasks such as sentiment analysis, NER, and text summarization. We can start it by creating a new project, uploading data, and initiating the annotation process. Figure 1 depicts the Docanno platform for annotating the entities. The flowchart of the procedure utilized to obtain the annotated resume dataset using Doccano is depicted in Fig. 2.

The following steps are applied on Doccano to obtain the annotated resume dataset.

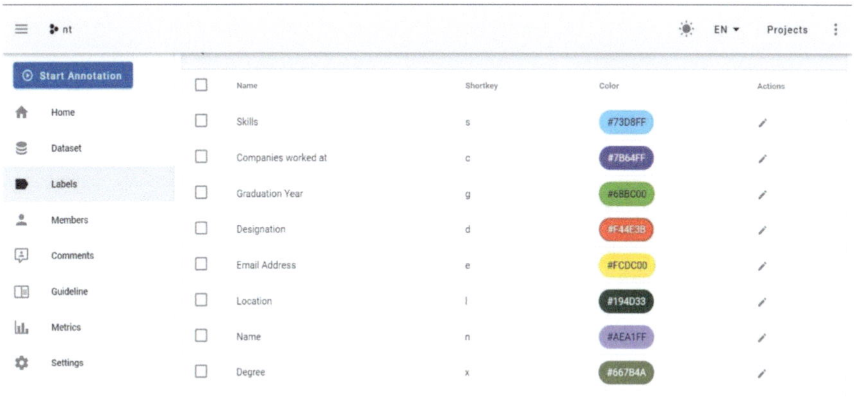

Fig. 1 Doccano platform

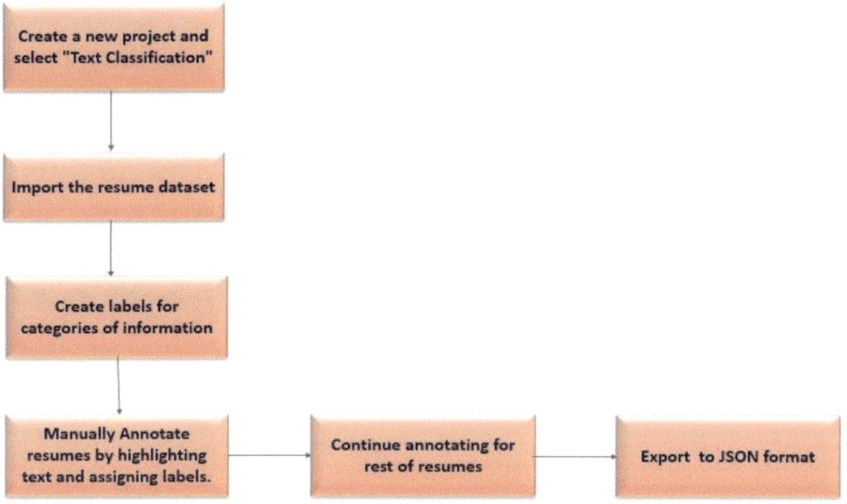

Fig. 2 Flowchart for obtaining the annotated resume dataset using Doccano

Steps for Annotating the Resume dataset
1. Create a new project and select "Text Classification"
2. Import the resume dataset
3. Create labels for categories of information
4. Annotate resumes by highlighting text and assigning labels
5. Export annotated data

```
{"text": "EU rejects German call to boycott British lamb.", "label": [ [0, 2, "ORG"], [11, 17, "MISC"] ]}
{"text": "Peter Blackburn", "label": [ [0, 15, "PERSON"] ]}
{"text": "President Obama", "label": [ [10, 15, "PERSON"] ]}
```


Fig. 3 JSON file

Step1: Create a new project and select "Text Classification": This step involves labeling text data with corresponding categories. Import the dataset and create labels for each piece of information. After labeling, train a model on the data to automate information extraction from new resumes.

Step2: Import the resume dataset: Doccano will read the data and allow to view and annotate each resume. Ensure proper formatting to avoid errors. Proceed with labeling and annotating the data once the dataset is successfully imported.

Step3: Create labels for categories of information: When creating labels for categories of information in Doccano, we should consider which types of information are relevant for extracting from resumes. Some common categories may include "Name," "Degree," "Companies Worked At," "Skills," "Location," "Email Address," "Designation," "Graduation Year," and "College Name." Assigning these labels to the appropriate sections of the resume will allow for more accurate information extraction.

Step4: Annotate resumes by highlighting text and assigning labels: The step of annotating resumes by highlighting text and assigning labels in Doccano involves a user interface where we can view each resume and select specific sections of text to assign to the appropriate label. We click on an instance of text in the resume and choose the label that corresponds with that information to start an annotation. The labeled data is easily visualized since the labeled information will be highlighted in the color related to the label. To ensure that the final model is successful in extracting data from new resumes, it is crucial to annotate the entire dataset appropriately and consistently. Additionally, the interface offers capabilities for navigating through resumes and checking annotations for accuracy and completeness.

Step5: Export annotated data: Data that has been annotated can be exported from Doccano in a variety of forms, including CSV and JSON, which can be readily loaded into other programs for additional analysis or modeling. The exported data will contain the labeled information for each resume, allowing for the creation of effective models that can automate the process of extracting information from new resumes. Figure 3 represents the JSON file obtained after exporting the annotated data.

3 Resume Parser

For the resume parser, we have used Spacy and RoBERTa [7] model for training and testing of the annotated resume dataset. Figure 4 represents the proposed Resume Parser steps for NER.

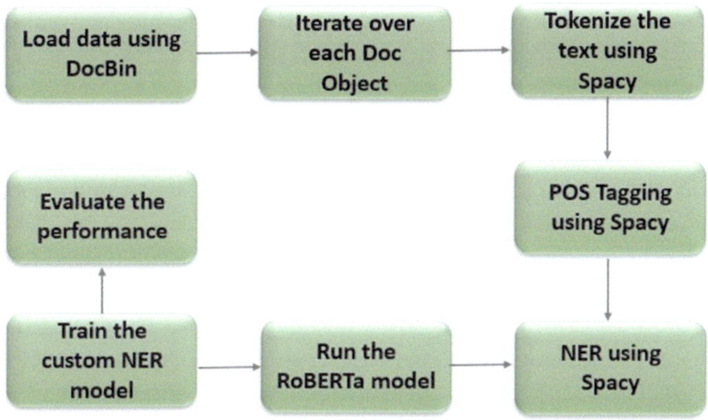

Fig. 4 Resume parser

3.1 Spacy

Spacy is a free and open-source NLP library used for complex text processing tasks like tokenization, part-of-speech tagging, entity recognition, dependency parsing, and others. It provides an easy-to-use interface for building NLP pipelines and comes with pre-trained models for a number of languages, including English, German, French, Spanish, Portuguese, and more. Spacy is widely used in academia and business for applications like information extraction, machine learning, and text analysis. Preprocessing tasks like tokenization, dependency parsing, NER, Part-of-speech (POS) tagging, and extracting pertinent information have all been done using the spacy model.

Steps used in Resume Parser

1. Load the annotated resume dataset using DocBin
2. Tokenization
3. POS tagging
4. Building a custom NER model using Spacy
5. RoBERTa
6. Extracting relevant information

Load the data using DocBin: Documents and their annotations are stored by Spacy in a binary file format called DocBin. We have first built an instance of the DocBin class and are using it to read the binary data from a file in order to load data using DocBin. An iterator across the Doc objects in the DocBin is returned by this. Then, we can handle each Doc object as needed. The resume text was uploaded from a file or database.

Tokenization: Tokenizing a text into distinct tokens is the function of the tokenizer class in Spacy. The tokenizer divides the text into tokens according to rules like punctuation and whitespace. Additionally, special instances like contractions, abbreviations, and URLs are handled by the tokenizer. A token object, which holds details about the token including its text, part-of-speech tag, and start and finish character indices in the source text, is used to represent each token. It has allowed us to divide the text into discrete words or tokens.

POS tagging: Assigning a grammatical tag, such as a noun, verb, adjective, or adverb, to every token in a text is known as POS tagging. The tagger component in Spacy is used to perform tagging by giving each token a unique tag based on its context. To determine which tag is most likely for each token, the tagger employs a statistical model that has been trained on annotated data. We have labeled each token with its part of speech using this tagger.

Building a custom NER model using Spacy: As a subtask of information extraction, NER entails finding and categorizing named entities in unstructured text into pre-established groups, including names, organizations, places, dates, and so on. We created a training set with annotated data in order to construct a customized NER model in Spacy. A list of tuples with a text and a dictionary of entity annotations for each tuple makes up this data. A pre-trained model RoBERTa has subsequently been loaded and modified by the addition of a new NER component. After that, using our annotated data, we trained the model. At last, we can assess our model's effectiveness and apply it for predicting new texts.

RoBERTa model: RoBERTa [7] is a modern pre-trained language model which was created by Facebook AI Research (FAIR). It is a modification of the original BERT (bidirectional encoder representations from transformers) model, with a few additions including dynamic masking, more training data, and a longer training period. A masked language modeling task and a task for predicting the next sentence are used to train RoBERTa on a substantial volume of unannotated material. As a result, the model can discover the way words relate to one another in context and create embeddings of high quality. We have used RoBERTa model to obtain the embedding of the dataset.

Extract relevant information: We have extracted the named entities from the text of a PDF document and displayed them along with their corresponding named entity labels. The first step in this process is to read the text of the PDF document and store it as a single string in a variable. Next, we use a Spacy language model to tokenize the text, assign part-of-speech tags to each token, and apply named entity recognition to identify and classify named entities. The resulting named entities are then iterated over in a for loop, and their text and label are extracted and displayed. We can extract the necessary information from the resume by identifying the pertinent entities and their relationships.

Table 1 Accuracy of the proposed model on different sets of resumes

Annotated resume dataset	Proposed model
200 resumes	51
300 resumes	57
470 resumes	62
621 resumes	78

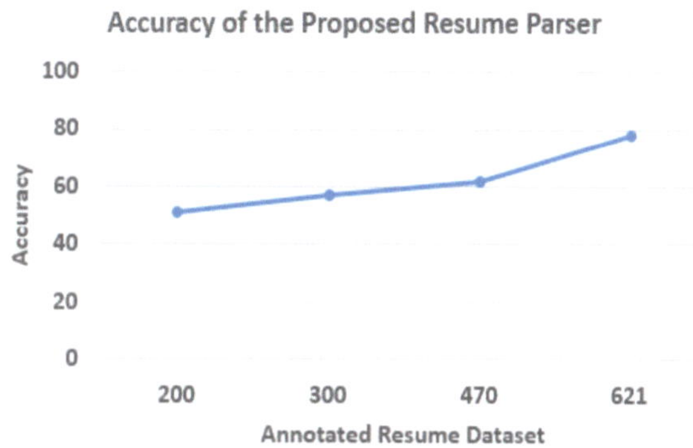

Fig. 5 Accuracy of the proposed resume parser on the annotated dataset

4 Experiments and Results

We have implemented the resume parser in Python 3.7.8 and tested on the four sets of annotated datasets. The evaluation and performance of the proposed method are done on 200, 300, 470, and 621 resumes. The experiments are conducted on a computer with an Intel (R) Xeon (R) CPU, 15 GB of RAM, and 2.00 GHz clock speed, using Google Colab as the development environment.

Table 1 displays the accuracy of the proposed model on various sets of resumes. We evaluated the model on four sets of annotated resumes comprising 200, 300, 470, and 621 samples, respectively. The obtained accuracies for each set were 51, 57, 62, and 78%, respectively. Figure 5 depicts the accuracy on the four sets of resumes using the proposed model. Figure 6 represents the sample resume data passed on the Resume Parser to extract the entities. Figure 7 shows the output of the sample resume using the Resume Parser. As shown in Fig. 7, the proposed parser has extracted the entities name, designation, location, graduation year, designation, companies worked at, degree and college name efficiently.

```
'Mandy Campbell, Bootcamp Instructor 1515 Pacific Ave, Los Angeles, CA 90291, United States, 3868683442, email@email.com Place of birth San Antonio Driving license Full LINKS LinkedIn, Pinterest, Resume Templates, Build this template PROFILE Passionate, entry-level Bootcamp Instructor with experience in freelancing assisting clients to reach tough fitness goals. Assisted one client to decrease chronic A1C blood sugar levels from 230 mg/dl to 110 mg/dl. Assisted another client in designing a routine to decrease body mass into healthy ranges over eight months. EMPLOYMENT HISTORY Jan 2019 – Jul 2021 Certified Bootcamp Instructor, Dominist Fitness Datteln Ran daily tasks, staffing, and marketing for the group fitness programming. Upheld an active rapport with members and improved group-class utilization by 18% by redesigning class schedules, approximating annual program budgets, analyzing and reporting program statistics. • Instructor and sales associate in a fast-paced fitness business env...'
```

Fig. 6 Sample resume

Fig. 7 Entity extraction using resume parser from a sample resume

```
'Mandy Campbell'
'  ->>>>'
'Name'
'Bootcamp Instructor'
'  ->>>>'
'Designation'
'1515 Pacific Ave'
'  ->>>>'
'Location'
'2021'
'  ->>>>'
'Graduation Year'
'Certified Bootcamp Instructor'
'  ->>>>'
'Designation'
'Dominist Fitness Datteln'
'  ->>>>'
'Companies worked at'
'2021'
'  ->>>>'
'Graduation Year'
'Bachelor's Degree in Exercise Science'
'  ->>>>'
'Degree'
'Concordia University'
'  ->>>>'
'College Name'
```

5 Conclusion

In conclusion, this paper handles the issues of extracting entities from resumes, which have grown more difficult for employers as a result of the increase in job applications. To address these issues, the paper presents a newly annotated dataset of 621 resumes that have been meticulously labeled with a variety of attributes, such as name, email, skills, designation, degree, year of graduation, employment history, and location. To illustrate the significance of the dataset, the research proposes a resume parser that extracts pertinent elements using the RoBERTa model and Spacy. The results of four distinct sets of experiments demonstrate the correctness of the suggested model, which ranges from 51 to 78%. The suggested methodology and annotated dataset have the potential to greatly improve the effectiveness of the hiring process.

There are a lot of possible directions for future research, aside from the annotated dataset's and methodology's proven efficacy in extracting resume entities. One possible path is to investigate the application of additional deep learning models and methods to increase the accuracy of entity extraction. The annotated dataset might be expanded to cover a wider range of industries, job categories, and candidate profiles, which is another potential area for improvement. To encourage additional study in this field, the dataset may also be made publicly available.

References

1. Deepak G, Teja V, Santhanavijayan A (2020) A novel firefly driven scheme for resume parsing and matching based on entity linking paradigm. J Discret Math Sci Cryptogr 23(1):157–165
2. Mittal V et al (2020) Methodology for resume parsing and job domain prediction. J Stat Manag Syst 23(7):1265–1274
3. Sajid H et al (2022) Resume parsing framework for e-recruitment. In: 2022 16th international conference on ubiquitous information management and communication (IMCOM). IEEE
4. Jayakumar N et al (2023) A approach to parse resume and recommend job modified K-mers and firefly algorithm (FFA) in resume parsing. Scand J Inf Syst 35(1):495–511
5. Sougandh TG, Reddy NS, Belwal M (2023) Automated resume parsing: a natural language processing approach. In: 2023 7th international conference on computation system and information technology for sustainable solutions (CSITSS). IEEE
6. Phan TT et al (2021) Ontology-based resume searching system for job applicants in information technology. Advances and Trends in Artificial Intelligence. Artificial intelligence practices: 34th international conference on industrial, engineering and other applications of applied intelligent systems, IEA/AIE 2021, Kuala Lumpur, Malaysia, 26–29 July 2021, proceedings, Part I 34. Springer International Publishing
7. Bhatia V et al (2019) End-to-end resume parsing and finding candidates for a job description using bert. arXiv:1910.03089
8. Tobing BCL, Suhendra IR, Halim C (2019) Catapa resume parser: end to end Indonesian resume extraction. In: Proceedings of the 2019 3rd international conference on natural language processing and information retrieval
9. Sinha AK, Akhtar MdAK, Kumar A (2021) Resume screening using natural language processing and machine learning: a systematic review. In: Machine learning and information processing: proceedings of ICMLIP 2020, pp 207–214
10. Chavare DS, Patil AB (2023) Resume parsing using natural language processing. Grenze Int J Eng & Technol (GIJET) 9(1)
11. Bhoir N et al (2023) Resume parser using hybrid approach to enhance the efficiency of automated recruitment processes
12. Liang H (2023) A study on the application of named entity recognition in resume parsing. In: Third international conference on intelligent computing and human-computer interaction (ICHCI 2022), vol 12509. SPIE
13. Daryani C et al (2020)An automated resume screening system using natural language processing and similarity. In: Ethics and information technology [Internet]. Volkson Press, pp 99–103
14. Mooney R (1999) Relational learning of pattern-match rules for information extraction. In: Proceedings of the sixteenth national conference on artificial intelligence, vol 328

15. Yu K, Guan G, Zhou M (2005) Resume information extraction with cascaded hybrid model. In: Proceedings of the 43rd Annual Meeting of the Association for Computational Linguistics (ACL'05)
16. Su Y, Zhang J, Lu J (2019) The resume corpus: a large dataset for research in information extraction systems. In: 2019 15th international conference on computational intelligence and security (CIS). IEEE
17. Devlin J et al (2018) Bert: pre-training of deep bidirectional transformers for language understanding. arXiv:1810.04805
18. Gan C, Mori T (2022) Construction of English resume corpus and test with pre-trained language models. arXiv:2208.03219
19. Kaggle dataset Homepage. https://www.kaggle.com/datasets/oo7kartik/resume-text-batch?resource=download.
20. Liu Y et al (2019) Roberta: a robustly optimized bert pretraining approach. arXiv:1907.11692

Malicious URL Detection, Classifier, and Web Security Analyzer

Sahil Panchasara, Shubham Kathiriya, Chintan Bhatt, Alessandro Bruno, Kaushal Shah, and Ajay Shriram Kushwaha

Abstract Scam websites or URLs put cybersecurity at risk. The cost of malicious URLs, which spread unwanted information (such as junk mail, fraudulent emails, and direct downloads) and trick people into scams (including financial loss, identity theft, and malware installation), is estimated to be billions of dollars a year. Many examples of malicious URLs have been compiled in this dataset so that a machine-learning-based model can be developed to identify risky URLs and prevent them from spreading on the Internet or damaging computer systems. A machine learning model for detection and classification has been created. We employed a decision tree for classification, achieving a testing accuracy of almost 91%, and logistic regression for detection, showing a testing accuracy of nearly 99.6%. On top of that, we developed and set up a web extension.

Keywords Phishing · Malicious URL · Malware · Defacement · NLP

S. Panchasara · S. Kathiriya · K. Shah
Department of Computer Science and Engineering, School of Technology, Pandit Deendayal Energy University, Gandhinagar, India
e-mail: sahil.pce19@sot.pdpu.ac.in

S. Kathiriya
e-mail: shubham.kce19@sot.pdpu.ac.in

K. Shah
e-mail: kaushal.shah@sot.pdpu.ac.in

C. Bhatt (✉)
University of Wollongong, GIFT City Campus, Gandhinagar, India
e-mail: cbhatt@uow.edu.au

A. Bruno
Department of Business, Law, Economics, and Consumer Behaviour "Carlo A. Ricciardi", IULM University, Via Carlo Bo, 1, Milan, Italy
e-mail: alessandro.bruno@iulm.it

A. S. Kushwaha
Department of CSA, Sharda School of Engineering and Technology, Sharda University, Greater Noida, India

A. P. Agrawal et al. (eds.), *Advances of Artificial Intelligence and Machine Learning in Image Processing, Security and Data Analytics*, Lecture Notes in Electrical Engineering 1464, https://doi.org/10.1007/978-981-95-0148-9_25

1 Introduction

Phishing has recently become a top concern among security professionals due to how simple it is to create a phony website that looks remarkably similar to the real one. Although professionals can recognize fraudulent websites, not all users can, which is why some customers become victims of phishing schemes. The main objective of the attacker is to get bank account passwords. Customers falling for phishing schemes cost American businesses $2 billion per year. It has been estimated that phishing costs the global economy over $5 billion each year, according to the third Microsoft Computing Safer Index Report, released in February 2014. Since users are unaware of phishing attacks, they are becoming frequent targets. Phishing attacks [1] exploit user vulnerabilities, so stopping them is extremely difficult, although developing phishing detection systems is essential [2]. A variety of devious tactics are used by attackers to trick consumers. There are various techniques used to modify the URL to appear legitimate, including obfuscation, fast-flux (building proxy servers quickly), algorithmic synthesis, and more. By adding forbidden URLs and IP addresses to the antivirus database, the "blacklist" strategy identifies phishing sites [3].

It is not always guaranteed that the attributes will be present in such attacks, and false positives are high, but heuristic-based detection, which uses traits that are proven to occur in actual phishing attacks, can detect zero-hour phishing attempts. The use of machine learning approaches is increasingly being considered as an alternative to blacklists and heuristics-based strategies. Many machine learning algorithms rely on prior data to guide their judgments or choices [3]. An algorithm will assess the features of a massive number of valid and prohibited URLs in order to identify phishing websites successfully [4, 5], including zero-hour phishing websites. Computer users can be easily directed to a false website and misled into entering their personal data, as they believe it to be the real website because it is an exact duplicate of the original one. According to a related study on how individuals react to phishing attempts, there are five primary reasons why computer users fall for phishing:

1. Users lack a thorough understanding of URLs.
2. Users are unaware of which websites are reliable.
3. Due to URL masking or redirection, users cannot see the full address of the web page [6].
4. Users frequently lack the time to carefully read the URL, which can result in them accidentally accessing some websites [6].
5. Users lack the ability to differentiate between legitimate and fake websites.

In recent years, academics have focused much of their attention on phishing attempts, which are sometimes compared to fishing for victims. A common strategy used by attackers (also called phishers) is to create specialized fake websites with identical designs that resemble well-known and reliable websites. Even if they all have the same graphical user interface, all of these websites must have unique URLs. By looking at the URL of these malicious websites, a careful and experienced user may be able to recognize them quickly. Life moves so quickly that end users rarely take the time to examine the full URLs of active online pages, which are often provided by other websites, social networking sites, or emails. False URLs are designed to collect sensitive information about the victim, including private information, bank account information, social media ID-password, etc. [6].

2 Related Techniques

Because of online anonymity, cybercriminals have developed sophisticated methods, such as phishing, to trick users into providing personal information, such as bank account numbers and social media ID-password. Since phishing is a semantics-based attack approach that largely preys on computer users' weaknesses, it may be exceedingly difficult to tell whether a web page is real [6].

Despite the introduction of blacklists, heuristics, visual, and machine-learning-based anti-phishing solutions, software companies cannot protect against all attacks. Based on natural language processing (NLP), the authors present seven different categorization algorithms and features for real-time anti-phishing [6]. Compared to previous research, this system operates in real time, employs a significant amount of genuine and phishing data, recognizes new websites, and uses feature-rich classifiers. The system is also independent of any third-party services. The performance of the system is measured using a brand-new dataset, which is also utilized to assess how well the experiment worked. According to testing and comparison data, the random forest technique surpasses the other implemented classification algorithms, with an accuracy rate of about 98% for phishing URL recognition.

Security professionals and malware developers are constantly at odds with one another because malware sophistication grows as quickly as technology does. Machine learning algorithms are being developed and used in malware detection because they can keep up with the increase in malware, according to current cutting-edge research. The purpose of this paper is to provide a systematic and comprehensive examination of deep learning-based machine learning approaches for malware detection [7]. The primary contributions are listed below:

1. In this article, we examine the techniques and components used in a typical machine learning strategy to find and categorize malware [7].
2. The limitations and flaws of traditional machine learning are examined [1, 4, 5].

3 Proposed Method

This research used a four-stage technique to accomplish its objectives. As shown in Fig. 1, the proposed model is represented by a block diagram.

Step 1 tackles malicious URl dataset. The collection of the malicious URL dataset was prepared from various sources. Step 2 preprocesses data. A more in-depth description of feature processing is given below.

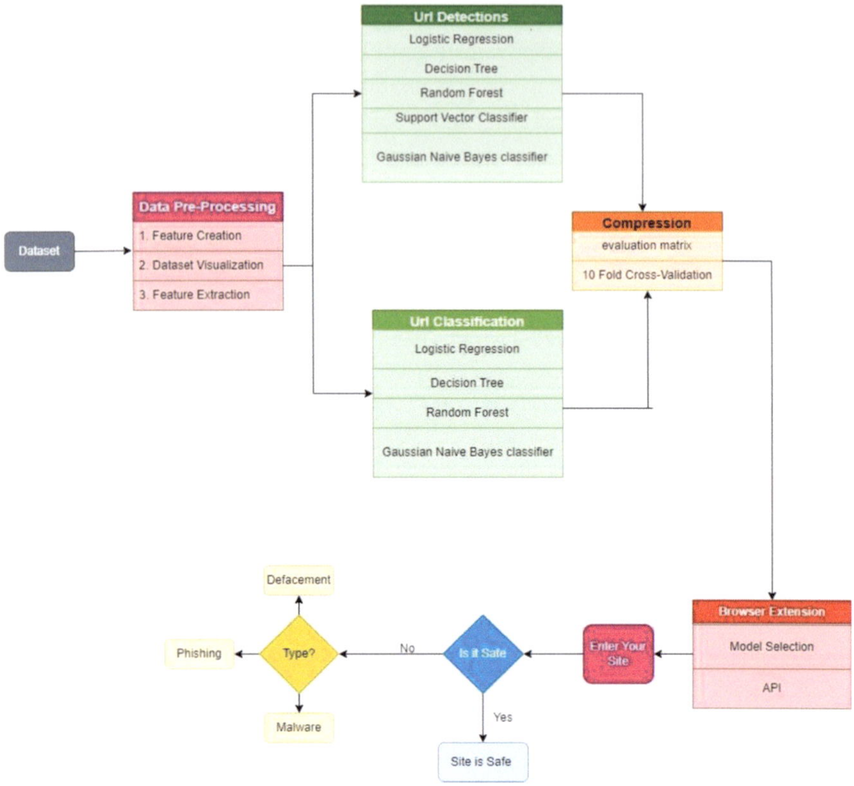

Fig. 1 Block diagram of the proposed method

3.1 Feature Creation

The dataset has two attributes only, one is the URL name and the other is the URL type. These two features are insufficient to detect whether the anonymous site is safe or not. Thus, there is a need to create an extra feature for good model training.

Features are divided into three types: length features, count features, and binary features.

3.2 Length Features

It deals with the length specification of the URLs. Here, we divided the URL into different types of lengths. There are additional features we took into consideration in the length features:

1. Length of URL—total number of characters in URL.

Table 1 Dataset after length preprocessing

	URL	Label	Result	URL length	Hostname length	Fd length	Tid	Tid length
0	https://www.google.com	Benign	0	22	14	0	com	3
1	https://www.youtube.com	Benign	0	23	15	0	com	3
2	https://www.facebook.com	Benign	0	24	16	0	com	3
3	https://www.baidu.com	Benign	0	21	13	0	com	3
4	https://www.wikipedia.com	Benign	0	25	17	0	com	3

2. Length of hostname—total characters present in the hostname or domain name.
3. Length of path—length of the URL path including all subdirectories on the web server.
4. Length of first directory—the total number of characters of the first directory of URL.
5. Length of top-level domain—the total number of characters of the top-level domain of URL (Table 1).

3.3 Count Feature

Involves counting different characters' specifications of the URLs. Here we count characters such as @, #, etc. in the URL. There are additional features we took into consideration in the count features:

The number of the following symbols: '-'; '@'; '?'; '%'; '.'; ' = '; 'http'; 'www'.

Furthermore, the number of digits, letters, and directories are taken into account (Table 2).

Table 2 Dataset after count preprocessing

	URL	Label	Result	Count-	count@	Count?	Count%	Count
0	https://www.google.com	Benign	0	0	0	0	0	2
1	https://www.youtube.com	Benign	0	0	0	0	0	2
2	https://www.facebook.com	Benign	0	0	0	0	0	2
3	https://www.baidu.com	Benign	0	0	0	0	0	2
4	https://www.wikipedia.com	Benign	0	0	0	0	0	2

Table 3 Dataset after binary preprocessing

	URL	Label	Result	Short_URL	Use_of_ip
0	https://www.google.com	Benign	0	1	1
1	https://www.youtube.com	Benign	0	1	1
2	https://www.facebook.com	Benign	0	1	1
3	https://www.baidu.com	Benign	0	1	1
4	https://www.wikipedia.org	Benign	0	1	1

3.4 Binary Features

It deals with the features which can be classified as yes or no. Here we consider features such as the usage of IP, shortened URLs, etc.

1. Presence/absence of IP—whether the URL is normal or is there any particular IP address for redirection.
   E.g.: www.instagram.com—no IP is used. This is the normal URL http://209.191. 122.70/—it uses an IP address as URL to access services.
2. Presence/absence of shortened URL—where URL is shortened or not. It is used to reduce the length of URLs. TinyURL and Bitly are used to shorten the URLs (Table 3).

4 Classification, Detection, and Metrics

Four machine learning algorithms have been studied for the detection and classification of URLs. Since the dataset has been labeled, supervised machine learning algorithms have been utilized in this study: logistic regression, decision tree, random forest, and Gaussian Naïve Bayes.

4.1 K-Fold Cross-Validation

K-fold cross-validation is one of the most known methods and is often used by data scientists. It is a technique for data partitioning that helps to maximize data while building a more thorough model. Any kind of machine learning has as its main objective the development of a more comprehensive model that can perform well with an unknown input. There is a possibility that a model can be developed with 100% accuracy or no mistakes on training data, but it is not going to generalize to new data. Therefore, it is a subpar model. It results in overfitting of the training data. The performance of the model can only be assessed using data points that were not included in the training phase since machine learning is all about generalization. As a result, we often separate our data into training and test sets.

4.2 Evaluation Matrix

Different evaluation metrics have been used to evaluate the proposed models.

True Positive (TP): This is the percentage of predictions that are malicious and have been correctly identified as such.

True Negative (TN): It is defined as the number of predictions that are not malicious, and are safe, and have been correctly identified as such.

False Positive (FP): It is defined as the number of predictions that have been identified as malicious but were actually safe, so they have been wrongly identified.

False Negative (FN): It is defined as the number of predictions that have been identified as safe and not malicious but were actually malicious, so they have been wrongly identified.

Accuracy (A): It is defined as the number of correctly identified malicious URLs divided by the total number of predictions. It can be calculated as

$$\text{Accuracy} = \frac{TP + FN}{TP + TN + FP + FN}$$

Precision (P): It is defined as the number of true positives (TP) divided by the total number of predicted positives (TP + FP).

$$\text{Precision} = \frac{TP}{TP + FP}$$

Recall (R): It is defined as the number of true positives (TP) divided by the total number of actual positives (TP + FN).

$$\text{Recall} = \frac{TP}{TP + FN}$$

F1-score: It is defined as the harmonic mean of precision and recall. It can be calculated as 2*(P*R)/(P + R).

$$F1\text{ - score} = \frac{2 * \text{Precision} * \text{Recall}}{\text{Precision} + \text{Recall}}$$

In the classification report, the precision, recall, *F1*, and support scores are displayed for each algorithm. The confusion matrix displays the *TP*, *TN*, *FP*, and *FN* in a format table so that one can see the relationship between those features. The classification report and confusion matrix were computed for all the algorithms used in detection, so that the best among them could be chosen. The detection part is displayed in Fig. 2 and the classification part is displayed in Fig. 3.

(a) Logistic Regression (b) Decision Tree (c) Random Forest (d) Support Vector Classifier (e) Gaussian Naive Bayes Classifier

Fig. 2 **a** Logistic regression, **b** decision tree, **c** random forest, **d** support vector classifier, **e** Gaussian Naive Bayes classifier

5 Browser Extension

We have made a Firefox Extension, which, when clicked, takes the URL of the page and sends it to the server via API. The server has both ML models file, i.e., logistic regression for detection and decision tree for type classification. The server then feeds the input URL to the detection model which processes it and produces an output as to whether any malicious behavior has been detected in the URL or not. If not detected, then the response is reverted back to the browser stating the site is safe to use. However, if any malicious behavior is detected, then the URL is sent to the decision tree classifier for type classification. The output of the classifier is the type of malicious behavior, i.e., benign, defacement, phishing, or malware.

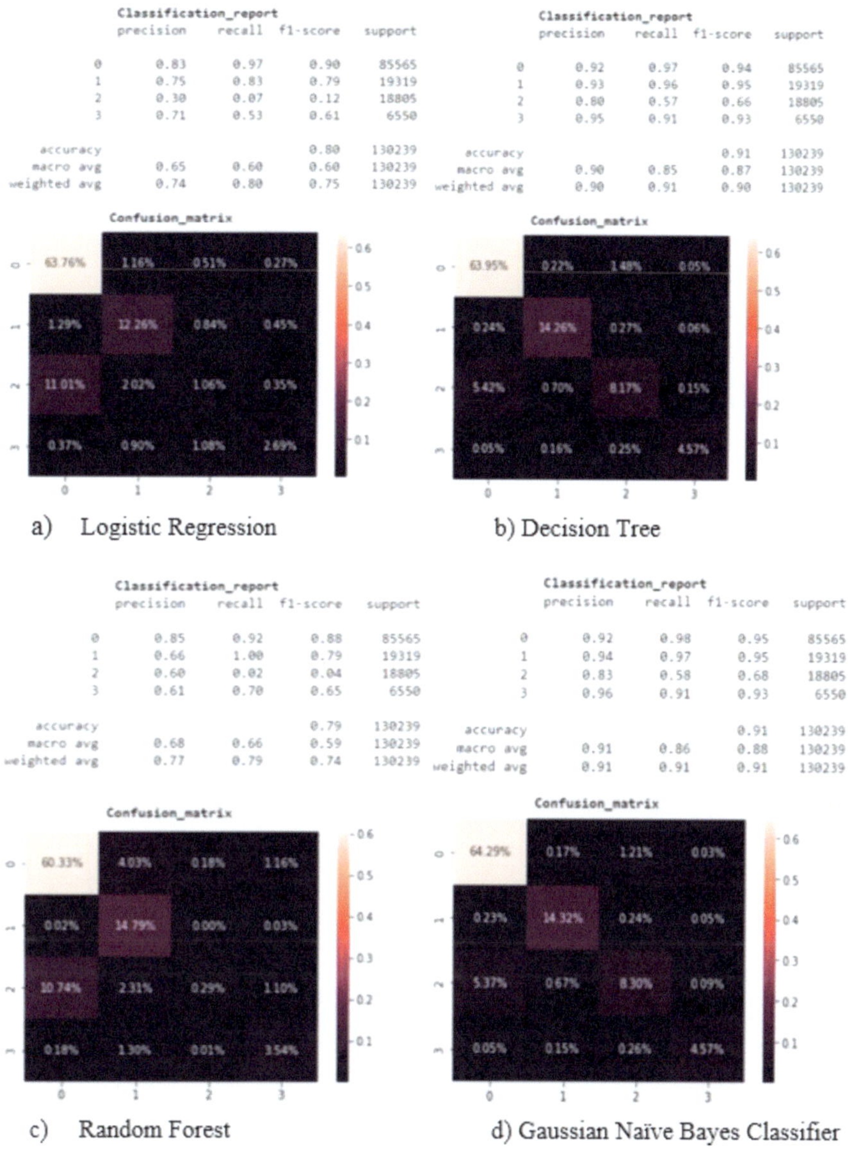

Fig. 3 **a** Logistic regression, **b** decision tree, **c** random forest, **d** support vector classifier, **e** Gaussian Naive Bayes classifier

Table 4 Accuracies and the detection time of the different models

Model	Training accuracy (%)	10-fold cross-validation mean accuracy (%)	Testing accuracy (%)	Time taken (s)
Logistic regression	**99.5957**	**99.5957**	**99.5773**	**2.4**
Decision tree	**99.9518**	**99.9529**	**99.5462**	**1**
Random forest	**99.9518**	**99.7223**	**99.7277**	**21.1**
Support vector classifier	**99.6630**	**99.6586**	**99.6617**	**262**
Gaussian Naive Bayes classifier	**99.1469**	**99.1669**	**99.1460**	**1.3**

Table 5 Accuracies and the classification time of the different models

Model	Training accuracy (%)	10-fold cross-validation mean accuracy (%)	Testing accuracy (%)	Time taken (s)
Logistic regression	**79.69**	**79.22**	**79.76**	**78.066**
Decision tree	**93.60**	**90.65**	**90.94**	**5.92**
Random forest	**93.59**	**91.26**	**91.47**	**130**
Gaussian Naive Bayes classifier	**78.96**	**78.96**	**78.95**	**1.4**

6 Experimental Results

Accuracy rates for both detection and classification are given in Tables 4 and 5. The time taken by each task is depicted in the last column.

7 Conclusions

This research has used machine learning approaches to enhance the detection and categorization of dangerous websites. For the detection part, we achieved the highest testing accuracy with random forest, i.e., 99.7%, however, our simple logistic regression has given a testing accuracy of 99.6%, and it is much less complex in terms of time and resources. So, for the detection part, we chose logistic regression. For the classification part, we achieved the highest testing accuracy with random forest, i.e., 91.47%; however, again, the decision tree has given an accuracy of nearly 91% and it is much less complex in terms of time and resources. So, for the classification part, we chose the decision tree.

The results also demonstrate that classifier performance rises as the volume of training data increases. To identify phishing websites more accurately in the future, hybrid technology that combines the logistic regression and decision tree algorithm of machine learning technology with the blacklist technique will be used. The CVE Analyzer can be improved in the future to obtain a clearer understanding of the situation. The system's usefulness can be improved by utilizing other cutting-edge learning approaches, such as deep learning, even though the achieved detection rate results are satisfactory. We also have access to a huge dataset. When employing this deep learning approach, the knowledge base may be constructed from any dataset. For this, however, extra training time is necessary. As a result, the system could support certain parallel processing strategies.

A separate subsystem for shorter URLs that just include a domain or subdomain may also be created. Some recent information about the website, such as the number of visitors in the previous week, date when the domain name was registered, etc., is required to detect these types of websites. These traits might not be selected for real-time detection because of the additional time needed to check them. The network may put a website on its local blacklist and forbid it from accepting any more requests if it is later determined to be a phishing site.

References

1. Salahdine F, Mrabet ZE, Kaabouch N (2021) Phishing attacks detection a machine learning-based approach. Cornell University. https://doi.org/10.1109/uemcon53757.2021.9666627
2. Mahajan R, Siddavatam I (2018) Phishing website detection using machine learning algorithms. Int J Comput Appl 181(23):45–47. https://doi.org/10.5120/ijca2018918026
3. Gibert D, Mateu C, Planes J (2020) The rise of machine learning for detection and classification of malware: research developments, trends and challenges. J Netw Comput Appl 153:102526. https://doi.org/10.1016/j.jnca.2019.102526
4. Alkawaz MH, Steven SJ, Hajamydeen AI (2020) Detecting phishing website using machine learning. In: International colloquium on signal processing and its applications. https://doi.org/10.1109/cspa48992.2020.9068728
5. Dutta AK (2021) Detecting phishing websites using machine learning technique. PLoS ONE 16(10):e0258361. https://doi.org/10.1371/journal.pone.0258361
6. Sahingoz OK, Buber E, Demir Ö, Diri B (2019) Machine learning based phishing detection from URLs. Expert Syst Appl 117:345–357. https://doi.org/10.1016/j.eswa.2018.09.029
7. Pachhala N, Jothilakshmi S, Battula BP (2021) A comprehensive survey on identification of malware types and malware classification using machine learning techniques. In: 2021 2nd international conference on smart electronics and communication (ICOSEC). https://doi.org/10.1109/icosec51865.2021.9591763

An Advanced Model to Predict Water Quality Index of Groundwater Using ANN and AdaBoost Algorithm

Mehtab Mehdi, Bharti Sharma, Usman Ali Khan, Mohd. Sayeed Khan, and Ali Sirageldeen Ahmed

Abstract The water, it is the second most important necessity for mankind after air, whole world is searching water in different planets. This research study explores the application of artificial neural networks (ANN) with the AdaBoost algorithm to predict groundwater quality in Kanpur, a city located in the Uttar Pradesh region of India. A total of twelve (12) groundwater samples were collected and analyzed to assess key physicochemical factors, such as pH, Total Dissolved Solids (TDS), Electrical Conductivity (EC), Chlorides (Cl), Calcium (Ca), Sodium (Na), Potassium (K), Magnesium (Mg), Sulfates (SO_4), Phosphates (PO_4), and Nitrates (NO_3), both before and after the monsoon season. The water quality index (WQI) was calculated based on these parameters. The findings indicated that most factors fell within acceptable ranges, but EC, TDS, Total Hardness (TH), Ca, and Mg exceeded the standard limits as per WHO guidelines, potentially impacting the suitability of groundwater for drinking. To enhance the accuracy of groundwater quality predictions using the water quality index, a multi-layer backpropagation algorithm within an artificial neural network (ANN) was employed, with additional refinement from the AdaBoost algorithm. The results affirmed the ANN's reliability in predicting the water quality, consistently delivering satisfactory results in all evaluated terms.

M. Mehdi (✉) · B. Sharma
DIT University, Dehradun, India
e-mail: mehtab.ind@gmail.com

B. Sharma
e-mail: bharti.sharma@dituniversity.edu.in

U. A. Khan
King Abdulaziz University, Jeddah, Kingdom of Saudi Arabia
e-mail: ukhan@kau.edu.sa

Mohd. Sayeed Khan
Infoteqs Solution, Lucknow, India

A. S. Ahmed
Al Baha University, Al Baha, Kingdom of Saudi Arabia
e-mail: asaahmed@bu.edu.sa

A. P. Agrawal et al. (eds.), *Advances of Artificial Intelligence and Machine Learning in Image Processing, Security and Data Analytics*, Lecture Notes in Electrical Engineering 1464, https://doi.org/10.1007/978-981-95-0148-9_26

Keywords WQI · AdaBoost algorithm · Artificial neural network · Groundwater · Kanpur City

1 Introduction

For any society, water is the most important resource, essential not only to sustain our ecosystems but also as a cornerstone of industrial processes. The significance of groundwater has garnered attention from researchers, leading to the development of various water resources and purification systems across urban, industrial, and agricultural domains. Presently, the world grapples with a pressing scarcity of clean water. In India, many rely on groundwater for drinking, with alarming figures highlighted by the Central Pollution Control Board (CPCB) indicating that nearly twenty-nine thousand one hundred twenty-nine million liters per day of wastewater are generated, predominantly by urban areas. A staggering 45% of this load is contributed by just 35 metropolitan cities [1]. Several factors are exerting mounting pressure to work toward water sustainability. The global water crisis refers to a dire condition wherein many regions lack access to sufficient and uncontaminated water. The Water Quality Index (WQI) offers a numerical representation of water quality in urban, rural, and corporate settings, encapsulating the complex interplay of various water quality parameters [2, 3]. Calculating the WQI is no straightforward procedure, with various methods available, including FWQI and NSFWQI. Different countries have established ideal values to evaluate water quality based on local conditions, but India's wide-ranging climatic diversity, from arid to humid, necessitates flexible approaches as a method working well in one location may not apply elsewhere [4]. Despite the potential benefits, calculating WQI has its own set of challenges, including time-consuming, complex processes, and inconsistent results. To overcome these hurdles, researchers have turned to technology for assistance. Predicting groundwater WQI has been explored using various data mining algorithms, such as decision trees, random forests, and support vector machines (SVM) [5]. While SVM has shown promise, its performance can be limited by selecting the optimal kernel value among several choices. Artificial Neural Networks (ANNs) are widely adopted data mining tools. Although effective on their own, a hybrid AI model can yield improved accuracy. In our hybrid model, we integrated the AdaBoost algorithm, enhancing ANN's predictive accuracy [6]. In the context of this research, we collected ten groundwater samples from Kanpur District, Uttar Pradesh, India, and analyzed ten water quality parameters, including pH, Total Dissolved Solids (TDS), Total Hardness (TH), Biological Oxygen Demand (BOD), and Chemical Oxygen Demand (COD) [7] (Fig. 1).

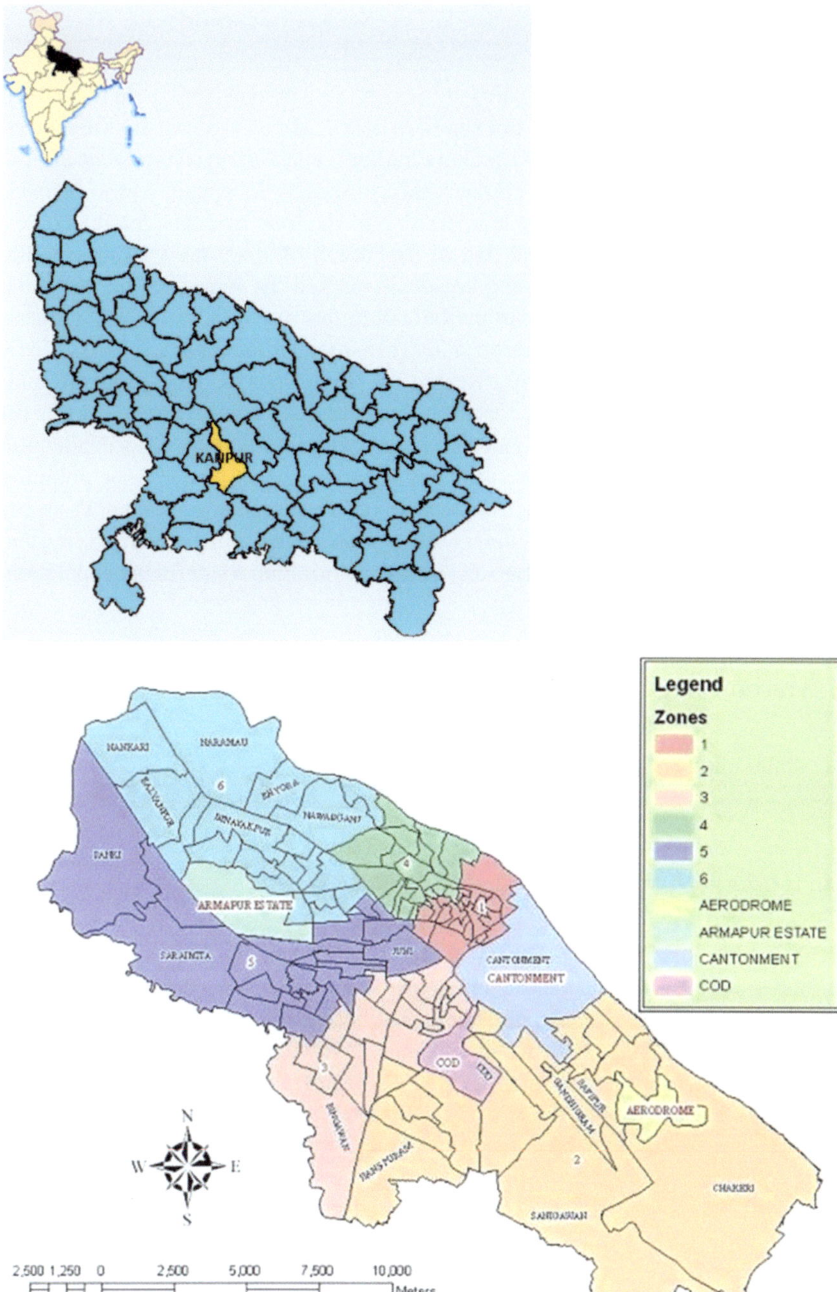

Fig. 1 Kanpur City *Source* Wikipedia

2 Study Area

Kanpur, situated in the northern plains of India, holds a pivotal position in the state of Uttar Pradesh. Kanpur occupies an expansive area of about 3155 square kilometers, and has a population of 3,234,000, making it a significant urban center in the region.

The city experiences a diverse climate characterized by three main seasons: winter, summer, and spring. The average temperature for the month of January, the coldest month, is 14.8 °C (58.6 °F), and that of the month of June, the warmest month, is 32.9 °C (91.2 °F). The monsoon season is marked by unpredictable and often insufficient rainfall, contributing to periodic droughts in North India. This climatic variability places a heavy reliance on water resources for the region.

Water resources in Kanpur are essential for its substantial population. Groundwater serves as a primary source for drinking water and various domestic needs for most residents. However, this reliance on groundwater is not without its challenges. The presence of industries in the area has led to concerns about water pollution and its impact on the quality of groundwater. The discharge of industrial waste has posed significant threats to the region's water resources, necessitating monitoring and mitigation efforts to ensure access to clean and safe drinking water for the population.

3 Methodology

The methodology of the study is as follows.

3.1 Data Collection and Preparation

Over a decade, spanning from 2006 to 2016, data was systematically gathered from ten strategically located water-monitoring sites within the Kanpur region. This dataset encompasses various critical parameters, including Biological Oxygen Demand (BOD), Chemical Oxygen Demand (COD), Dissolved Oxygen (DO), pH, Total Solids (TS), and Electrical Conductivity (EC) [8]. These parameters were subsequently utilized to compute the Water Quality Index (WQI) specific to the Kanpur area.

In accordance with established protocols by the Central Ground Water Board, a standard technique was followed for the collection and analysis of water samples. To effectively partition the data into a 70% training set and a 30% test set, a 10-fold cross-validation methodology was employed. It is worth noting that the division of the test and training datasets for both spatial and temporal predictions lacks a universally applicable rule.

The Arithmetic Mean Water Quality Index (AMWQI) was computed using the NSFWQI method. A higher AMWQI value is indicative of improved groundwater

Table 1 WQI range with quality

Index	Range	Quality
AMWQI	<14.9	Very poor
	14.9–29.9	Poor
	29.9–44.9	Moderately poor
	45–54.9	Fair
	55–69.9	Moderately good
	70–84.9	High quality
	85–100	Excellent

quality. To calculate the AMWQIsc, the following formula can be applied, as is commonly done in research endeavors:

$$AMWQI = \sum_{i=1}^{n} Wi \quad SIi$$

The AMWQI represents the water quality index specific to the Kanpur region. The WQI range is detailed in Table 1. *Wi* signifies the weight assigned to each variable, falling within the range of 0–1, while **SIi** denotes the sub-index value derived from a scale of 0–100. The computational methods used for determining the AMFWQI remain consistent throughout the process [9].

Table 2 vividly illustrates the Water Quality Index (WQI) in Kanpur, which spans from 15.6 to 91.8, indicating a broad spectrum of water quality from poor to excellent. Areas situated far from the industrial belt exhibit superior water quality, seemingly with minimal contamination originating from industrial waste pollutants. The process of urbanization emerges as a significant factor contributing to the degradation of groundwater quality [10].

In the context of updated predictive modeling, when the data is subjected to normalization, the following formula serves as a tool for assessing the parameter's associated cost:

$$x_i' = \frac{(x_i - x_{\min})}{(x_{\max} - x_{\min})}$$

where x_i' is the updated cost of a variable (i.e., ph, EC, etc.), x_i, cost on the place, and $x_{\min}$, and $x_{\max}$ are the lower and upper costs.

For data acquisition, various sensors were employed, and these sensors were connected to a private cloud storage system. The collected data was then stored locally for subsequent processing and analysis (Fig. 2). Sampling efforts were carried out at diverse locations within the Kanpur City. Table 4 provides a detailed description of these specific sites, where the water is primarily utilized for drinking and bathing purposes, barring a couple of sites in the industrial belt where maximum water

Table 2 Water Quality Index (WQI) at different locations in Kanpur

Site no.	WQI
1	90.68
2	56.47
3	72.35
4	89.44
5	91.80
6	90.87
7	88.54
8	74.56
9	13.01
10	25.05
11	46.24
12	56.42
13	80.21
14	91.2

consumption is in the industries. The contents of Table 3 encompass a range of parameters instrumental in the calculation of the Water Quality Index (WQI) [11].

The dataset developed for the purpose of WQI assessment comprises five fundamental parameters, each with their specific value is outlined in Table 5. The pH value

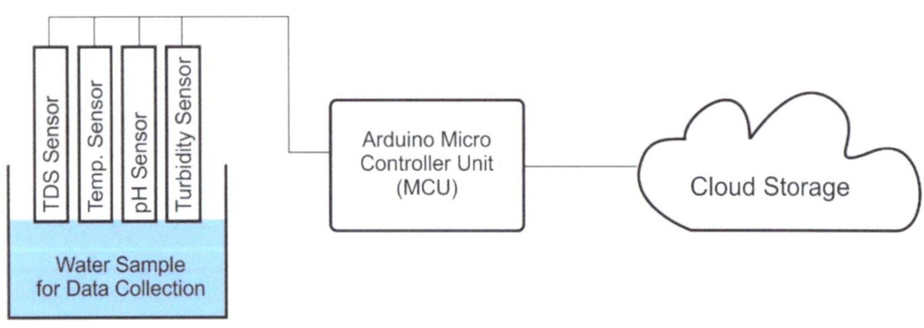

Fig. 2 Data collection for WQI

Table 3 Variables affecting the WQI

WQI effecting factor	Range	Safe limit
pH	1–14	6.5–8.5
Turbidity	0–10	0–5.0
DO	0–12	4.0–12.0
E. coli	0–100	0
EC	0–100	4.0–30.0

Table 4 Description of sites

Site no.	Place	Water characteristics	Use
1	Ashok Nagar	Colorless and odorless	Drinking and household
2	Govind Nagar	Whitish yellow, odorless	Drinking and household
3	Yashoda Nagar	Colorless and odorless	Drinking and household
4	Vikas Nagar	Colorless and odorless	Drinking and household
5	Civil Lines	Colorless and odorless	Drinking and household
6	Kakadeo	Colorless and odorless	Drinking and household
7	Nirala Nagar	Colorless and odorless	Drinking and household
8	Indira Nagar	Colorless, odorless	Drinking and household
9	Fazalganj	Whitish yellow, musty smell	Household and industrial
10	Jajmau	Whitish yellow, musty smell	Household and industrial
11	Dada Nagar	Whitish yellow, soily smell	Household and industrial
12	Panki	Whitish yellow, odorless	Drinking and household
13	Jawahar Nagar	Colorless, odorless	Drinking and household
14	Defence Colony	Colorless, odorless	Drinking and household

reflects the acidity of the sample, effectively representing the potential movement of hydrogen ions (H+) within the data. It is important to note that all living organisms thrive within specific pH ranges. Dissolved Oxygen (DO) plays a crucial role in water quality by impacting its taste. Turbidity serves as an indicator of water clarity, with greater clarity resulting in lower turbidity and a more pleasing taste. Furthermore, electrical conductivity measures the presence of various chemical ions in the water, with higher electrical conductivity values suggesting an elevated ion concentration in the water [12–14].

3.2 Model Summary

See Fig. 3.

3.3 Description of Model

The presented model relies on the synergy of artificial neural networks and genetic algorithms. This research endeavors to establish a connection between Water Quality Index (WQI) and the input variables, including pH, TDS, BOD, COD, etc., culminating in the development of a WQI calculation model. The model harnesses the backpropagation neural network and gradient descent techniques, with weight optimization facilitated by the AdaBoost algorithm. The superior model between these

Table 5 Dataset of different sites with parameters and their specific values

Sr. no.	Variable name	Loc 1	Loc 2	Loc 3	Loc 4	Loc 5	Loc 6	Loc 7	Loc 8	Loc 9	Loc 10	Loc 11	Loc 12
1	Potential of hydrogen	7.61	7.64	7.43	7.32	7.94	7.72	7.28	8	7.84	7.66	7.64	7.47
2	Conduction (μS/cm)	0.781	0.792	0.65	0.58	0.96	0.534	0.53	1.03	1.013	0.635	0.538	0.46
3	Total hardness (mg/L)	354	382	285	271	551	321	335	453	536	343	359	325
4	Calc (mg/L)	256	281	212	220	457	280	290	410	497	258	286	267
5	Chemical oxygen demand (mg/L)	28	22	35	22	52	16	22	40	42	35	32	34
6	Biological oxygen demand (mg/L)	22	16	25	17	24	15	16	27	26	19	16	13
7	Alkaline (mg/L)	235	238	244	213	302	182	225	322	284	164	187	313
8	Chlorine (mg/L)	60	80	57	66	165	81	68	151	171	71	86	68
9	Total dissolve solid (mg/L)	710	682	717	687	796	707	690	807	792	687	677	713
10	Inflow (mg/L)	0.33	0.53	0.43	0.66	0.60	0.35	0.22	0.53	0.43	0.23	0.14	0.18

two is subsequently integrated with the AdaBoost algorithm to enhance efficiency. In tandem with the Artificial Neural Network (ANN), the model undergoes evaluation using diverse assessment methods [15].

Figure 4 elucidates the operational framework of the proposed method, encompassing the comprehensive research methodology for test implementation.

3.4 *Modeling Using ANN*

Artificial Neural Networks (ANNs) are a prominent tool in artificial intelligence, adept at addressing a range of challenges including modeling, forecasting, and data analysis. ANNs consist of interconnected layers, with data passed through these connections [16]. The effectiveness of ANNs relies on internal link strength.

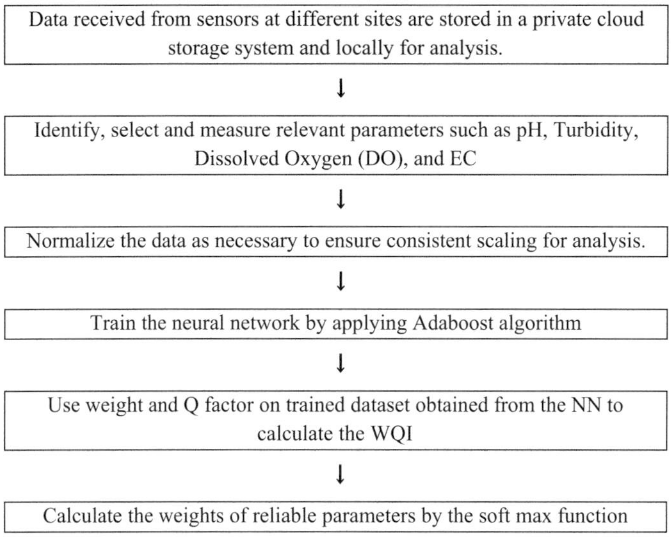

Fig. 3 Model data flow

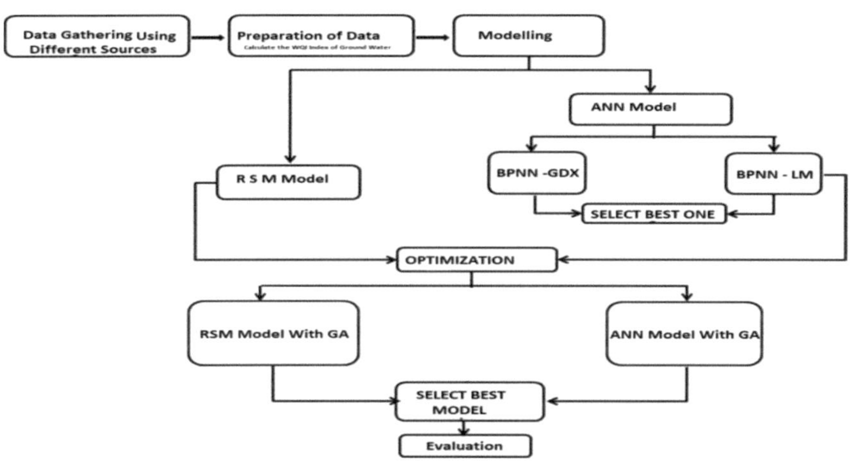

Fig. 4 Proposed architecture of predicting the WQI at Kanpur region

In this study, an ANN model is applied to forecast groundwater Water Quality Index (WQI) (Fig. 5). Precision, consistency, and efficiency hinge on factors such as training and testing datasets, hidden layer count, learning rate, and activation functions [17]. The Backpropagation Neural Network (BPNN) and GDX are employed for training, with weight initialization using the Nguyen-Widrow algorithm. The BPNN algorithm, assisted by GDX or LM, minimizes the mean square error to enhance model efficiency [18].

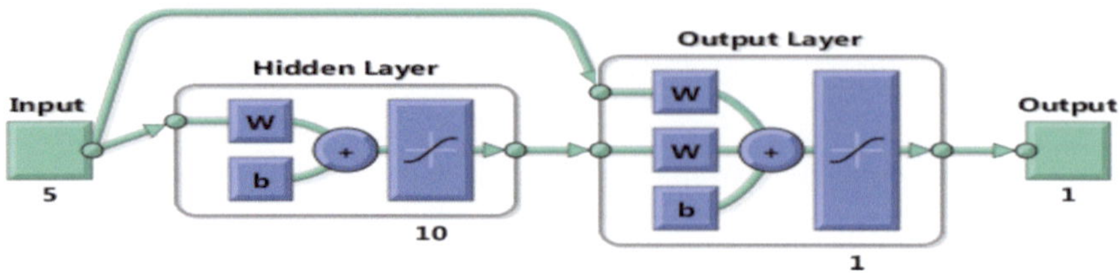

Fig. 5 Neural network design for fast forward network

The following equations are involved to generate a multi-layered Artificial Neural Network (ANN) using the Backpropagation Neural Network (BPNN) algorithm.

$$\mathrm{NET}_J = \sum_{j=0}^{n} \left(W_{ij} X_i \right) \tag{1}$$

$$F(\mathrm{NET}_J) = \frac{1}{1 + e^{-\mathrm{NET}_J}} \tag{2}$$

$$F(\mathrm{OUT}_J) = F(\mathrm{Net}_J) \tag{3}$$

The neural network operates through the reception of signals from input layers and establishes internal connections between these inputs and the output layers. In Eq. 1, W_{ij} signifies the weight of the connection from the ith input neuron to the jth hidden neuron, Xi denotes the ith input, and N represents the number of inputs to the jth hidden neuron. NETj is the summation of these values and is then propagated to the output, as depicted in Eq. 2. This function yields the result (OUTj) in Eq. 3 for the jth neuron within the subsequent layers [11, 19].

3.5 ANN and AdaBoost Coupling

The coupling of AdaBoost with Artificial Neural Networks (ANN) serves a crucial function in enhancing the predictive accuracy of the model. AdaBoost, a boosting algorithm, optimizes the ANN's performance by iteratively adjusting the weights of individual neurons and data points. It assigns higher weights to those data points that the ANN has difficulty classifying accurately, effectively focusing on the most challenging cases. This iterative process continues until a robust and accurate model is achieved (Fig. 6) [20].

AdaBoost's function is to refine the ANN's ability to classify and predict by identifying and prioritizing challenging data points. This coupling creates a hybrid model that leverages the ANN's ability to capture complex relationships in the data while utilizing AdaBoost's boosting technique to fine-tune the ANN's performance [21–23]. The result is a highly efficient and accurate model for tasks such as water quality prediction.

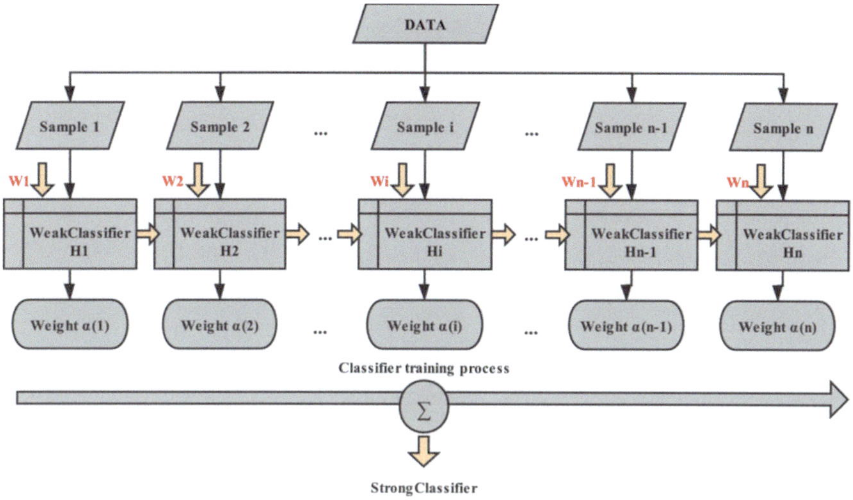

Fig. 6 Flowchart of the proposed AdaBoost algorithm

4 Results and Discussion

MATLAB 2018 serves as the simulator for creating optimal models, each technique executed 90 times with consistent variables. The results favoring superior performance are selected [24]. In the context of the AdaBoost algorithm and Particle Swarm Optimization (PSO), the evaluation method employed is root mean square error, which quantifies the variation between classifier-expected costs and actual observed values [25, 26]. The comparative analysis presented in Table 6 assesses SVM, Neural Networks (NN), NN-PSO, and NN-AdaBoost based on metrics like accuracy, precision, recall, F-measure, and FM-Index.

SVM exhibited relatively lower results, with accuracy at 80.34%, precision at 78.87%, recall at 83.23%, F-measure at 81.0%, and FM-Index at 81.23%. In contrast, NN achieved higher performance with an accuracy of 85.11%, precision of 87.89%, recall of 84.24%, F-measure of 85.65%, and FM-Index of 86.34. NN-PSO demonstrated moderate performance with accuracy at 88.89%, precision at 87.23%, recall at 88.87%, F-measure at 88.78%, and FM-Index at 88.67 [27].

Table 6 Comparative study of different models with NN-GA

	SVM	NN-PSO	NN	NN-GA	NN-AA
Precision	78.98	87.34	88.00	94.45	95.55
Accuracy	80.45	89.00	85.22	97.23	98.16
Recall	83.34	88.98	84.35	99.80	99.90
F-measure	81.11	88.89	85.76	97.32	98.79
FM-index	81.34	88.78	86.45	97.34	98.92

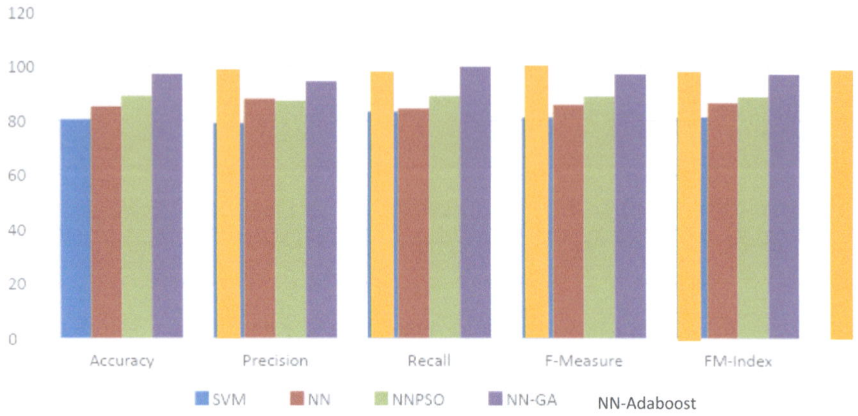

Fig. 7 Comparative study of NN-AA with NN and NN-PSO

The analysis suggests that PSO-enhanced ANN models provide some improvement in water quality prediction but are not a substantial enhancement [28]. However, AdaBoost-integrated ANN models excel, achieving 98.16% accuracy, 95.55% precision, 99.9% recall, 98.78% F-measure, and 98.92% FM-Index. While AdaBoost-based models exhibit consistency across several metrics, the FM-Index indicates the superiority of NN-GA in water quality prediction, as illustrated in Fig. 7. This comparative study underscores the robustness of the NN-AA model [29].

5 Conclusion

The water quality significantly impacts both a nation's economy and public health, as access to clean water is fundamental to human well-being [30]. Consequently, accurately forecasting water quality is a complex and vital endeavor. This study offers a proficient model, leveraging AdaBoost Algorithm (AA) and Artificial Neural Networks (ANN), to predict the groundwater quality index. Water samples from various locations in the Kanpur region were collected and classified into "clean" and "contaminated" categories according to Indian standards.

The results obtained from the model underscore the role of meta-heuristic algorithms in enhancing ANN-based water quality predictions [31]. By employing AA to optimize multiple objective functions during the ANN training process, a notable improvement in prediction accuracy was achieved. The NN-AA model demonstrated exceptional efficiency at 98.16%, surpassing the performance of other algorithms such as standalone ANN or ANN with LM algorithm [32].

References

1. Vaidya SD, Raut VR (2015) Alternative on-site systems for domestic wastewater treatment and disposal: a review. Int J Eng Res & Technology. 4(7):1234–1237
2. Changsheng H, Akram W, Rashid A, Ullah Z, Shah M, Alrefaei AF, Kamel M, Aleya L, Abdel-Daim MM (2022) Quality assessment of groundwater based on geochemical modelling and Water Quality Index (WQI). Water 14(23):3888
3. Patel PS, Pandya DM, Shah M (2023) A systematic and comparative study of Water Quality Index (WQI) for groundwater quality analysis and assessment. Environ Sci Pollut Res 30(19):54303–54323
4. Uddin MG, Nash S, Rahman A, Olbert AI (2022) A comprehensive method for improvement of water quality index (WQI) models for coastal water quality assessment. Water Res 1(219):118532
5. Mogaraju JK (2023) Application of machine learning algorithms in the investigation of groundwater quality parameters over YSR district, India. Turk J Eng 7(1):64–72
6. El-Rawy M, Batelaan O, Alshehri F, Almadani S, Ahmed MS, Elbeltagi A (2023) An integrated GIS and machine-learning technique for groundwater quality assessment and prediction in Southern Saudi Arabia. Water 15(13):2448
7. Verma M, Loganathan VA, Bhatt VK (2022) Development of entropy and deviation-based water quality index: case of River Ganga, India. Ecol Ind 1(143):109319
8. Azrour M, Mabrouki J, Fattah G, Guezzaz A, Aziz F (2022) Machine learning algorithms for efficient water quality prediction. Model Earth Syst Environ 8(2):2793–2801
9. Matta G, Kumar A, Nayak A, Kumar P (2022) Appraisal of spatial–temporal variation and pollution source estimation of Ganga River system through pollution indices and environmetrics in Upper Ganga basin. Appl Water Sci 12(3):33
10. Saqib N, Rai PK, Kanga S, Kumar D, Đurin B, Singh SK (2023) Assessment of ground water quality of Lucknow City under GIS framework using Water Quality Index (WQI). Water 15(17):3048
11. El Osta M, Masoud M, Alqarawy A, Elsayed S, Gad M (2022) Groundwater suitability for drinking and irrigation using water quality indices and multivariate modeling in makkah Al-Mukarramah province, Saudi Arabia. Water 14(3):483
12. Maansi JR, Wats M (2022) Evaluation of surface water quality using water quality indices (WQIs) in Lake Sukhna, Chandigarh, India. Appl Water Sci 12:1–4
13. Hoque MM, Islam A, Mahammad S (2022) Assessing the surface and bottom river water quality for drinking purpose and human health risk analysis: a study of Damodar River, India. Arab J Geosci 15(23):1734
14. Masood A, Niazkar M, Zakwan M, Piraei R (2023) A machine learning-based framework for water quality index estimation in the Southern Bug River. Water 15(20):3543
15. Thorat M, Pandit S, Balote S (2022) Artificial neural network: a brief study. Asian J Converg Technol (AJCT). 8(3):12–16. ISSN-2350-1146
16. Bihl T, Young II WA, Moyer A, Frimel S (2023) Artificial neural networks and data science. In: Encyclopedia of data science and machine learning. IGI Global, pp 899–921
17. Abda Z, Zerouali B, Chettih M, Guimaraes Santos CA, de Farias CA, Elbeltagi A (2022) Assessing machine learning models for streamflow estimation: a case study in Oued Sebaou watershed (Northern Algeria). Hydrol Sci J 67(9):1328–1341
18. Adil M, Ullah R, Noor S, Gohar N (2022) Effect of number of neurons and layers in an artificial neural network for generalized concrete mix design. Neural Comput Appl 1:1–9
19. Ni W, Zhao L, Zhang L, Xing K, Dou J (2023) Coupling progressive deep learning with the AdaBoost framework for landslide displacement rate prediction in the Baihetan Dam Reservoir, China. Remote Sens 15(9):2296
20. Jamei M, Ali M, Karbasi M, Sharma E, Jamei M, Chu X, Yaseen ZM (2023) A high dimensional features-based cascaded forward neural network coupled with MVMD and Boruta-GBDT for multi-step ahead forecasting of surface soil moisture. Eng Appl Artif Intell 1(120):105895

21. Malathi V, Gopinath MP, Kumar M, Bhushan S, Jayaprakash S (2023) Enhancing the paddy disease classification by using cross-validation strategy for artificial neural network over baseline classifiers. J Sens 28:2023
22. Teza G, Cola S, Brezzi L, Galgaro A (2022) Wadenow: a matlab Toolbox for early forecasting of the velocity trend of a rainfall-triggered landslide by means of continuous wavelet transform and deep learning. Geosciences 12(5):205
23. Hedayatnasab Z, Saadatabadi AR, Shirgahi H, Mozafari MR (2023) Heat induction of iron oxide nanoparticles with rational artificial neural network design-based particle swarm optimization for magnetic cancer hyperthermia. Mater Res Bull 1(157):112035
24. Khudhair ZS, Zubaidi SL, Ortega-Martorell S, Al-Ansari N, Ethaib S, Hashim K (2022) A review of hybrid soft computing and data pre-processing techniques to forecast freshwater quality's parameters: current trends and future directions. Environments 9(7):85
25. Sun Y, Cui B, Ji F, Wei X, Zhu Y (2022) The full-field path tracking of agricultural machinery based on PSO-enhanced fuzzy Stanley model. Appl Sci 12(15):7683
26. Wang J, Lu Y, Xin C, Yoo C, Liu H (2022) Kernel PLS with AdaBoost ensemble learning for particulate matters forecasting in subway environment. Measurement 30(204):111974
27. Trabelsi F, Bel Hadj Ali S (2022) Exploring machine learning models in predicting irrigation groundwater quality indices for effective decision making in Medjerda River Basin, Tunisia. Sustainability 14(4):2341
28. Nastran M, Pintar M, Železnikar Š, Cvejić R (2022) Stakeholders' perceptions on the role of urban green infrastructure in providing ecosystem services for human well-being. Land 11(2):299
29. Kalyanmoy D, Agrawal S, Pratap A, Meyarivan T (2000) A fast elitist non-dominated sorting genetic algorithm for multiobjective optimization: NSGA-II. In: International conference on parallel problem solving from nature, pp 849–858
30. Li H, Zhang Q (2009) Multiobjective optimization problems with complicated Pareto sets MOEAlD and NSGA- II. IEEE Trans Evol Comput 13(2)284–302
31. Chih-Chung C, Lin C-J (2011) LIBSVM: a library for support vector machines. ACM Trans Intell Sys Technol (TIST) 2(3):27
32. Odella F (2017) Technology studies and the sociological debate on monitoring of social interactions. Biometrics Concepts Methodologies Tools Appl 6(1):529–558

Modified PolypSegNet for Polyp Segmentation of Colonoscopy Images

Nilanjan Saha, Mahadev Mondal, Ayush Kumar Bhanja, Aindrila Mandal, and Mainak Bandyopadhyay

Abstract The precise identification and delineation of polyps in medical images, particularly those derived from colonoscopy and endoscopy, hold the utmost importance in the field of gastrointestinal disease diagnosis and treatment. However, manual image annotation by qualified gastroenterologists is laborious and vulnerable to errors. To combat this issue, we present an advanced deep learning framework for automated polyp segmentation. Our framework enhances the PolypSegNet architecture by incorporating a squeeze and excitation module into the convolutional layer block. This integration improves feature discrimination and reduces model complexity. Additionally, we augment the deep reconstruction module with a feature pyramid network to further amplify polyp segmentation performance. Training our model on the Kvasir-SEG dataset, which comprises polyp images and segmentation masks by expert medical practitioners, allows for accurate localization, precise boundary demarcation, and differentiation from non-polyp structures. The proposed deep learning framework represents a significant advancement in gastrointestinal disease diagnosis and treatment. With its accurate localization, precise boundary delineation, differentiation from non-polyp structures, variability handling, real-time processing, generalization, and integration with clinical workflows, our model revolutionizes polyp detection and segmentation. Medical practitioners can improve patient care and outcomes by leveraging this powerful tool.

Keywords PolypSegNet · Deep learning · Convolutional layers · Segmentation · Polyp segmentation

N. Saha (✉) · M. Mondal · A. K. Bhanja · A. Mandal · M. Bandyopadhyay
Kalinga Institute of Industrial Technology—Deemed to Be University, Bhubaneswar, India
e-mail: sahanilanjan807@gmail.com

M. Bandyopadhyay
e-mail: mainak.bandyopadhyayfcs@kiit.ac.ina

A. P. Agrawal et al. (eds.), *Advances of Artificial Intelligence and Machine Learning in Image Processing, Security and Data Analytics*, Lecture Notes in Electrical Engineering 1464, https://doi.org/10.1007/978-981-95-0148-9_27

1 Introduction

Globally, colorectal cancer remains a leading cause of cancer-related fatalities, underscoring the critical necessity of early detection and treatment. Among the diagnostic tools available, colonoscopy plays a crucial role in identifying and removing colorectal polyps, which precede CRC. However, accurately segmenting these polyps within colonoscopy images is challenging. Traditional image processing techniques, though extensively explored, struggle in handling the intricate and varied shapes of polyps, resulting in sub-optimal outcomes. However, advancements in deep learning models, notably convolutional neural networks (CNNs), like U-Net and PolypSegNet, have demonstrated remarkable abilities to capture the complex structures of polyps and significantly improve segmentation accuracy (Fig. 1).

Polyp segmentation confronts a diversity of challenges that are tackled effectively by deep learning models, including varying polyp shapes, textures, overlapping structures, limited annotated data, real-time processing, and scanner generalization requirements. The development relies heavily on annotated datasets like ETIS-Larib, CVC-ClinicDB, and Kvasir-SEG, serving as benchmarks for evaluation of various methodologies, fostering collaboration within the research community. This study provides a thorough overview of advanced techniques and models employed in polyp segmentation of colonoscopy images. Furthermore, we propose a modified version of PolypSegNet, which further enhances its performance. To offer insights into the feature extraction process, we provide interpretations through feature maps and GradCAM. Leveraging the combined power of deep learning models and large-scale annotated datasets, the field of polyp segmentation is advancing rapidly, directing us to more precise detection of colorectal polyps, thereby aiding in combatting colorectal cancer for improved patient outcomes.

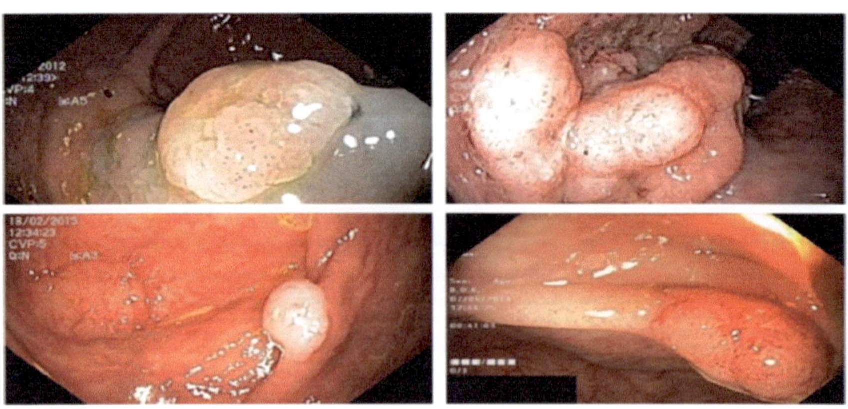

Fig. 1 Some visuals related to colorectal cancer (CRC)

2 Related Works

In [1], a real-time polyp perception algorithm is proposed by modifying yolov3 and yolov4 method using CSPN (cross-stage partial network) with actual time automatic polypus identification. The SUN and PICCOLO Widefield datasets were used as training data when the model was trained. In [2], a model based on the shuffle efficient medium attention network is presented for polyp segmentation. In [3], the colonoscopy images are first pre-processed by transforming into HSV color space, then applying gamma correction and executing image fusion. Finally, the images are converted into RGB. Dual path CNN with eight layers is used to extract characteristics out of the pre-prosecuted images. Stimulated by U-Net, Y Net is proposed in [4] to spot growths of a polyp in colonoscopy imagery. In this Y Net model, the primary encoder is computed with an Image-Net weight. In [5], a single shot detector algorithm is proposed to locate polyps in colonoscopy images. Initially, the input images are marginalized by experts manually wherein a limiting area is outlined over all polyps separately, next the images are pre-processed with the help of Dynamic Histogram Equalization (DHE) to increase characteristic of the images. An approach with the aim of utilizing ensemble learning is contemplated in [6], where three models, ResNet101, GoogleNet, and Xception, were brought into play to classify if polyps be found within input images or not. In [7], random forest, a standard machine learning algorithm, was used to spot malignant colon cancer polyps. Next, the subsequent training images undergo processing using a Python package called pyradiomics. Then results are provided into a random forest algorithm, and based on the results, predictions are made. An endeavor is undertaken in [8] to combine the SWIN Trans-former and Efficient-Net for the purpose of both detecting and segmenting polyps. A multi-distend convolutional brick was used to enhance the extracted characteristics by SWIN and Efficient-Net. In [9], a model is proposed using Faster-RCNN network and ResNet101 for identifying substantial risk areas and categorizing polyps within given areas. In [10], two prepared models, namely, VGG16 and Mobile-Net are introduced. Input images were pre-processed, wherein the black areas were excluded as they do not contain any significant data. A polyp identification model is presented in [11] that combines 2D and 3D convolutional layers for efficient image segmentation. The model utilizes a combination of 2D and 3D backbones to perform image segmentations. AYOLOv3 network for real-time polyp detection using live inputs is presented in [12]. After the polyp is detected, a candidate bounding box is assigned to the model, and subsequently employed to obtain the model outputs. Limited sample images hindered the proposed network's ability to detect flat polyps, resulting in its failure. In [13], a plug-in model named Instance tracking head was proposed, it could be attached to any image or object detection algorithm.

3 Contribution

In the paper, we present a refined version of PolypSegNet, a deep learning architecture which has been specifically designed for the accurate polyp segmentation in colonoscopies. The highlights of our research include the following:

I. Enhanced architecture for PolypSegNet: This new architecture adds convolution layers and skip connections, improving fine-grained details capture and segmentation accuracy in colonoscopy image.
II. Improved segmentation performance: Extensive evaluation of a large dataset shows the modified PolypSegNet surpasses existing methods, with higher accuracy, precision, and recall, while reducing false positives/false negatives.
III. Resilience to fluctuations in picture quality: This work uses data augmentation/normalization techniques to address variations in image quality that are frequently observed in clinical settings.
IV. Efficiency and real-time implementation: The improved PolypSegNet enables real-time colonoscopic polyp segmentation without compromising segmentation accuracy.
V. Validation by thorough experimentation: The suggested method has been proven to function in a variety of contexts by means of comprehensive experiments.

These advancements represent a significant development toward the goal of developing a more accurate and effective method for segmenting colonoscopy images, which could lead to better colonic illness diagnosis and treatment.

4 Proposed Model

The suggested model, which draws inspiration from PolypSegNet [14], substitutes FPNs for the deep reconstruction module and a squeeze and excitation module for the depth-dilated inception and deep fusion skip modules. These modifications enhance feature representation, recalibration, and multi-scale capabilities, potentially resulting in superior segmentation accuracy and overall performance compared to the original PolypSegNet architecture [15], designed for specifically precise polyp segmentation colonoscopy images. PolypSegNet employs depth-dilated inception (DDI) modules for varied receptive areas and generalized feature extraction. It uses a D-Unit layer structure with sequential DDI modules for deep feature extraction and proposes a deep fusion skip module (DFSM) to bridge semantic gaps in skip connections by aggregating feature representations from different levels. The architecture includes a deep reconstruction module for efficient multi-scale feature map reconstruction.

The suggested model is a version of PolypSegNet [15], with significant changes depicted in Figs. 2 and 3. It substitutes a squeeze and excitement module for the depth-dilated inception and deep fusion skip module. Furthermore, feature pyramid networks (FPNs) replace the deep reconstruction module (DRM). The SE module was chosen for its remarkable ability to capture feature importance and channel-related dependencies, ensuring enhanced performance without introducing redundancy or artifacts.

SE modules maintain artifact-free feature maps, enabling effective real-time applications with constrained processing resources. Conversely, DDI modules may incur higher processing costs and introduce artifacts into the feature maps, making them less appropriate for real-time applications. The SE module dynamically emphasizes informative features and diminishes less crucial ones, enhancing spatial and contextual information capture for improved predictive accuracy and segmentation performance.

The substitution of FPNs for the deep reconstruction module addresses scale variation in object detection and segmentation. FPNs generate multi-scale feature maps and enhance feature representation across different scales, enabling better handling

Fig. 2 Architecture of encoder and decoder module

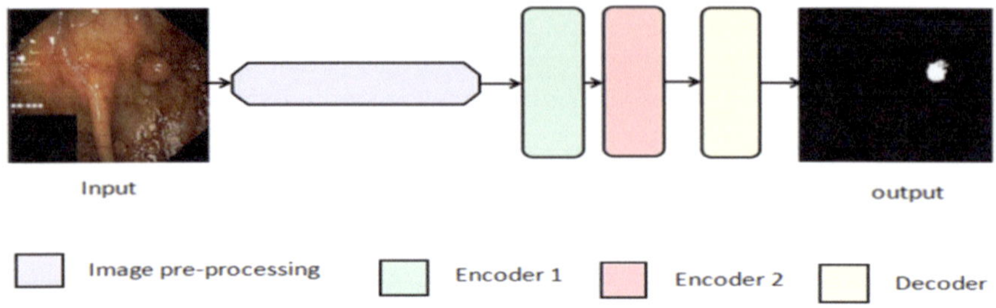

Fig. 3 Proposed polyp segmentation model architecture

of objects of varying sizes [16]. FPN can handle objects of varying scales allowing the capture of both fine-scale and coarse-scale context. Conversely, DRMs only fuse features from a single layer, making it challenging to segment small or distant objects.

Taking the varying sizes of polyps into consideration detecting them becomes more important in which FPNs are much better. Additionally, FPNs are typically faster than DRMs, which is important for medical applications where speed is critical.

The proposed model's encoder deploys five convolutional blocks to extract high-level features from input images. Each block comprises a convolutional layer followed by a residual block and max-pooling. The convolutional layers extract increasingly complex feature maps, while residual block layers with skip connections aid gradient propagation and mitigate the vanishing gradient issue. Spatial dimensions of feature maps are effectively reduced via max-pooling operations, while retaining vital features and enabling the capture of hierarchical representations at varying scales by iteratively applying convolutional blocks, ResidualBlock layers, and down-sampling, one can iteratively refine features, capturing intricate patterns and structures.

The Feature Pyramid Network (FPN) module is responsible for fusing high-level and low-level features to capture both global and local information. The FPN utilizes up-sampling and convolution operations to combine features from different levels of the encoder [17]. Specifically, the FPN takes the output of the fifth convolutional block and progressively up-samples it, while simultaneously merging it with the corresponding feature maps from the encoder via skip connections. This process is repeated for multiple levels to create a feature pyramid.

The decoder module is responsible for recovering and restoring the spatial information lost during the down-sampling process, through up-sampling operations and generating an accurate polyp segmentation mask. The proposed model of the decoder section consists of four up-sampling blocks. Each block involves up-sampling the feature maps, followed by convolutional operations and skip connections to integrate the features from corresponding level of the encoder.

The model undergoes diverse training techniques for enhanced performance and convergence. Transfer learning initializes the network with pre-trained weights, leveraging knowledge from extensive datasets. Early stopping prevents overfitting by reinstating optimal weights according to validation loss. Adaptive learning rate

Table 1 Trainable and non-trainable parameters

Parameters	Proposed model	Proposed model after ensembling with a pre-trained model
Total parameters	51,155,069	54,725,822
Trainable parameters	51,147,133	54,683,774
Non-trainable parameters	7,936	42,048

scheduling improves convergence to the optimal solution during training. Stochastic gradient descent (SGD) with momentum optimizes the network, setting the learning rate to 1e-06, while all encoder and decoder layers remain unfrozen for training.

The proposed model integrates a custom ResidualBlock layer, mitigating overfitting and fostering information flow during training. Each ResidualBlock comprises two sets of three-by-three convolutional layers followed by group normalization, ReLU activation, and dropout. Group normalization reduces reliance on batch sizes by normalizing inputs within groups after each convolutional layer. Subsequently, ReLU introduces non-linearity. ResidualBlock aids in deep learning by facilitating gradient flow through residual connections, addressing the vanishing gradient problem and enhancing deep neural network optimization. This adaptable block enhances performance and convergence speed in diverse computer vision tasks when integrated into larger CNN architectures.

To extend the proposed model's applicability to embedded devices, integration with MobileNetV2 [18] occurs. This integration enhances model versatility, especially in resource-limited environments. Combining the model with MobileNetV2 forms an ensemble that benefits from diverse architecture information. The ensemble technique notably boosts the model's performance, evident in a 2–3% accuracy increase and reduced training time per epoch from 3 to 2 s, compared to an accuracy range of 82.3–85.5% and slower training without ensemble use.

The proposed model presents an upgrade for polyp segmentation model. Its encoder–decoder framework, guided by ResidualBlock layers, preserves vital information and extracts intricate features. The integration of MobileNetV2 strengthens the model's capabilities, and the ensemble approach harnesses the collective knowledge of multiple architectures. Table 1 provides a summary of the models' trainable and non-trainable parameters (Fig. 4).

4.1 *Feature Maps of Proposed Model*

The proposed modified PolypSegNet model includes feature maps, intermediate representations obtained from various layers of the neural network, utilized for segmenting polyps in the colonoscopic images. The encoder module generates the feature maps depicted in Fig. 5, through a series of convolution block operations, progressively refining representations of the input data. These feature maps capture

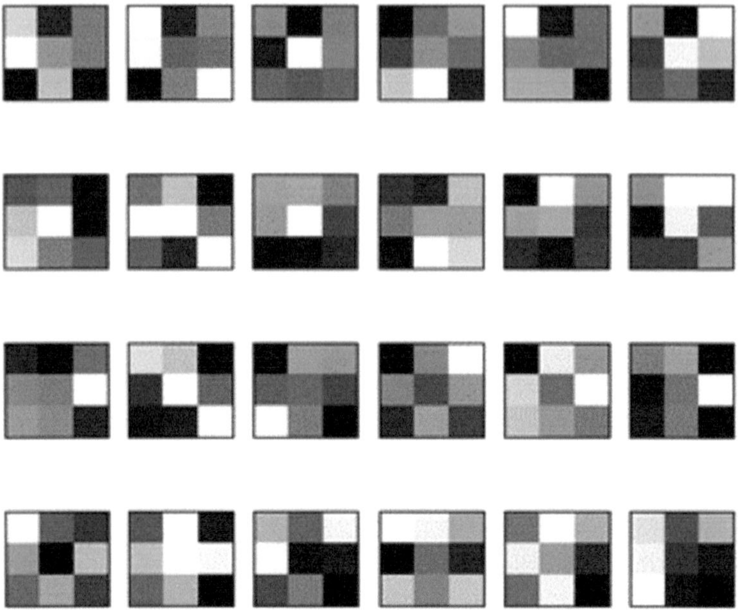

Fig. 4 Image of 64, 3 × 3 filters used in first layer of the model

both local features and global features, providing valuable insights for polyp segmentation, and are intelligently downsampled using max-pooling to retain relevant information while minimizing computational complexity.

The decoder module performs complex up-sampling operations to reconstruct feature map spatial resolution, enabling accurate generation of complex segmentation masks. The fusion of encoder-derived feature maps with their up-sampled counterparts allows the model to harness multi-spectral information, enhancing segmentation accuracy.

In the training phase, feature maps serve as the basis for the loss function calculation. By comparing multiple layer-based feature maps, the model gradually learns to differentiate between predicted polyp masks with ground truth annotations, improving overall segmentation performance. In summary, feature maps used in the proposed model are crucial for capturing hierarchical representations of input images, allowing effective utilization of multi-sensor information. This approach results in highly precise polyp segmentation by adeptly identifying intricate structures and subtle variations.

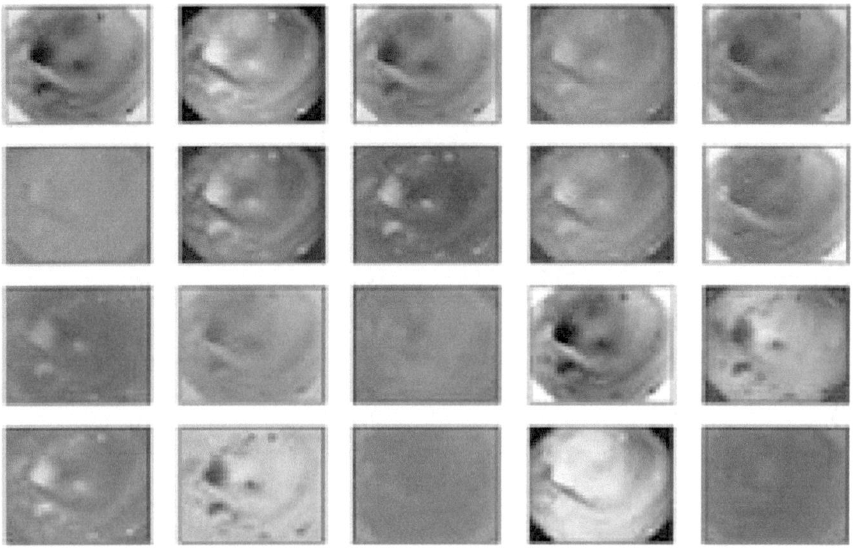

Fig. 5 Feature maps from the first layer of the model

5 Experimental Setup

5.1 Dataset Description

In this study, the Kvasir-SEG [19] dataset was used, which includes 1,000 carefully annotated images of colonoscopies and gastroscopies from 100 patients, ranging from gastrointestinal pathological abnormalities to normal tissue. These images are annotated at pixel level, allowing for accurate delineation of anatomically relevant structures and abnormalities. The fluctuation in image resolution examines the model's adaptability to varying image sizes and lighting conditions while simulating real-world scenarios. With a focus on polyp segmentation, this dataset is a great source of semantic segmentation for colonoscopy images. It is especially helpful for training and assessing segmentation algorithms.

5.2 Implementation

The hardware utilized to develop the modified PolypSegNet model includes an Intel Core i5-6200U CPU running at 2.30 GHz and 16 GB of RAM. Python 3.6 serves as the programming language; the Keras 2.2.4 framework with TensorFlow as a backend facilitates deep learning. Segregated dataset was used for training, validation, and testing consisting of 800, 100, and 100 images, respectively, standardized at 256

× 256-pixels, reducing processing complexity and guarantees consistency for the images. During the training process, the model has been optimized using stochastic gradient descent (SGP). The batch size is 8, momentum of 0.9, and the learning rate is set to 1e-06. Early stopping with Patience parameter set to 10 is used to avoid overfitting. The Jaccard coefficient and dice are used, respectively, to compute the accuracy and loss of the model. By measuring the correlation between the ground truth annotations and the anticipated dataset segmentations, these metrics offer insight into the model's capacity for data segmentation. Several evaluation metrics are also used to evaluate the model's performance, such as accuracy, precision, and F1-score, as well as sensitivity and specificity. This evaluation metric is calculated using the following formulas:

$$\text{Dice coefficient score} = \frac{2TP}{2TP + FP + FN} \tag{1}$$

$$\text{Jaccard coefficient Score} = \frac{TP}{TP + FP + FN} \tag{2}$$

$$\text{Accuracy} = \frac{TP + TN}{TP + FP + FN + TN} \tag{3}$$

$$\text{Precision} = \frac{TP}{TP + FP} \tag{4}$$

$$\text{Recall} = \frac{TP}{TP + FN} \tag{5}$$

$$\text{IoU} = \frac{TP}{TP + FP + FN} \tag{6}$$

where TP = True Positives, FP = False Positives, and FN = False Negatives.

6 Results Analysis

The outcome survey is pivotal in assessment of performance and productivity of the proposed model. It involved the evaluation of the model's performance against the validation dataset using the hybrid loss function provided in Eq. (7), which is a combination of the dice and Jaccard coefficient. This is a popular loss function for medical image segmentation because it is more reliable in class imbalance than conventional cross-entropy. Monitoring the validation loss was used to determine the early stopping criteria, with the model having the lowest validation loss being selected for evaluation.

$$\text{loss} = 1 - (0.5 * dice\_coef(y\_true, y\_pred)) + (0.5 * jaccard\_coef(y\_true, y\_pred)) \quad (7)$$

The proposed PolypSegNet model, employing a hybrid loss function, achieves high accuracy in polyp segmentation, making it apt for clinical application. Table 2 displays the metrics' values after 20 epochs.

The model's performance was compared with high-performance models like U-Net on the same dataset, demonstrating the proposed modified PolypSegNet's superiority in accuracy and efficiency. Figure 6 showcases segmented results after 20 epochs of training.

The final stage involved employing GradCAM visualization [20] to highlight pertinent areas in the image crucial for polyp detection. Illustrated in Fig. 7, the model emphasizes polyp edges and contours while disregarding background and other structures. This visualization technique aids clinicians by revealing the model's decision-making process, offering insight into how it segments and identifies polyps in medical imaging.

In outline, the modified PolypSegNet model exhibited better performance in polyp segmentation, accomplishing highest precision and accuracy on a large-scale medical image data set. This model's better precision and sensitivity make it a major candidate for clinical applications in the field of polyp segmentation and diagnosis. The generated GradCAM visualizations in Fig. 7 provide insight into the decision-making mechanism of the model.

7 Conclusion

The modified PolypSegNet design marks significant progress in polyp segmentation, excelling in capturing long-range dependencies and diverse features through residual connections. Despite outstanding performance within a 20-epoch limit, addressing limitations is crucial. Exploring transfer learning and domain adaptation can enhance its applicability across various clinical scenarios and imaging modalities. Overcoming these challenges could revolutionize medical image analysis and diagnosis, paving the way for enhanced patient care.

Table 2 Performance of proposed model on various metrics

Metrics	Loss	Accuracy	Recall	Precision	IoU	Validation loss	Validation accuracy	Validation recall	Validation precision	Validation IoU	Loss
Ensembled proposed model values	0.3259	0.8545	0.1309	0.7185	0.1862	0.4512	0.8315	0.0056	0.8911	0.0761	0.3259
U-NET	0.4963	0.7817	0.3186	0.3174	0.1521	0.4996	0.7786	0.3039	0.3392	0.1558	0.4963

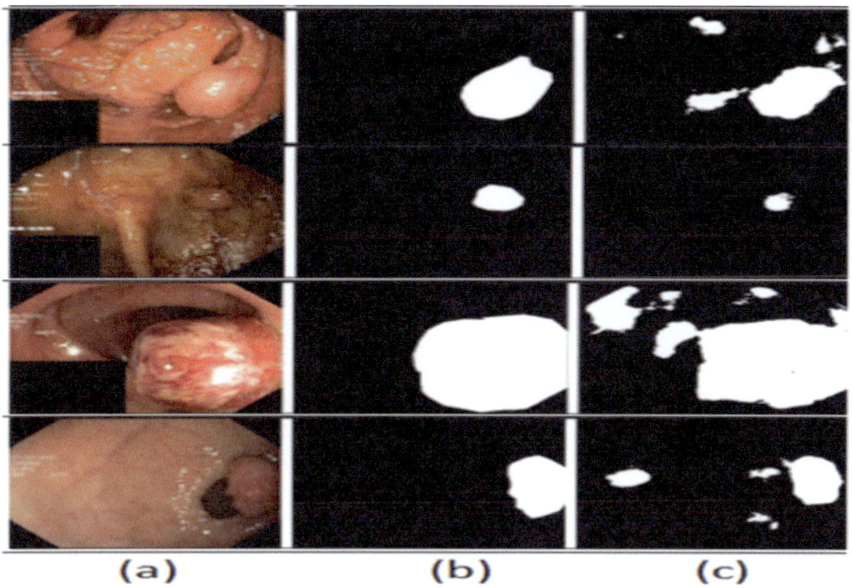

Fig. 6 Visualization of results based on the proposed model. **a** The original image. **b** The mask image. **c** Predicted mask image (20 epochs) using proposed model

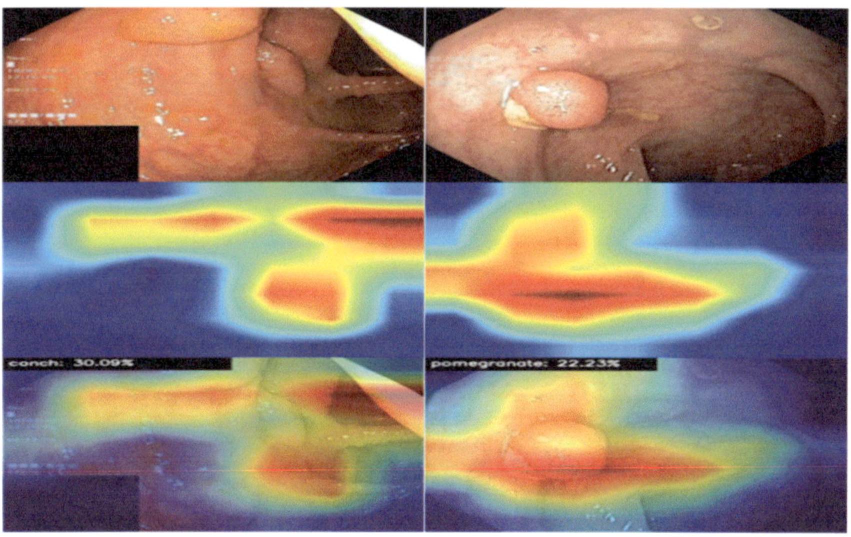

Fig. 7 The GradCAM visualization is based on the proposed model

References

1. Pacal I, Karaman A, Karaboga D, Akay B, Basturk A, Nalbantoglu U, Coskun S (2021) An efficient real-time colonic polyp detection with YOLO algorithms trained by using negative samples and large datasets. Comput Biol Med 141:105031
2. Yang K, Chang S, Tian Z, Gao C, Du Y, Zhang X, Liu K, Meng J, Xue L (2022) Automatic polyp detection and segmentation using shuffle efficient channel attention network. Alex Eng J 61:917–926. ScienceDirect
3. Nisha JS, Gopi VP, Palanisamy P (2022) Automated colorectal polyp detection based on image enhancement and dual-path CNN architecture. Biomed Signal Process Control 73. ScienceDirect
4. Mohammed A, Yildirim S, Farup I, Pedersen M, Hovde Ø (2018) Y-net: a deep convolutional neural network for polyp detection. arXiv:1806.01907
5. Tanwar S, Vijayalakshmi S, Sabharwal M, Kaur M, AlZubi AA, Lee HN (2022) Detection and classification of colorectal polyp using deep learning. BioMed Res Int
6. Rani N, Verma R, Jindal A (2022) Polyp detection using deep neural networks. In: Handbook of intelligent computing and optimization for sustainable development, pp 801–814
7. Grosu S, Wesp P, Graser A, Maurus S, Schulz C, Knösel T, Cyran CC, Ricke J, Ingrisch M, Kazmierczak PM (2021) Machine learning–based differentiation of benign and premalignant colorectal polyps detected with CT colonography in an asymptomatic screening population: a proof-of-concept study. Radiol Natl Lib Med 299(2):326–335
8. Park KB, Lee JY (2022) SwinE-Net: hybrid deep learning approach to novel polyp segmentation using convolutional neural network and Swin Transformer. J Comput Des Eng 9:616–632
9. Ho C, Zhao Z, Chen XF, Sauer J, Saraf SA, Jialdasani R, Taghipour K, Sathe A, Khor LY, Lim KH, Leow WQ (2022) A promising deep learning-assistive algorithm for histopathological screening of colorectal cancer. Sci Rep 12
10. Ellahyani A, Jaafari IE, Charfi S, Ansari ME (2022) Fine-tuned deep neural networks for polyp detection in colonoscopy images. Pers Ubiquitous Comput 27:235–247
11. Puyal JGB, Brandao P, Ahmad OF, Bhatia KK, Toth D, Kader R, Lovat L, Mountney P, Stoyanov D (2022) Polyp detection on video colonoscopy using a hybrid 2D/3D CNN. Med Image Anal 82(AAM)
12. Nogueira-Rodríguez A, Domínguez-Carbajales R, Campos-Tato F, Herrero J, Puga M, Remedios D, Rivas L, Sánchez E, Iglesias A, Cubiella J, Fdez-Riverola F (2022) Real-time polyp detection model using convolutional neural networks. Neural Comput. Appl. 34:10375–10396
13. Yu T, Lin N, Zhang X, Pan Y, Hu H, Zheng W, Liu J, Hu W, Duan H, Si J (2022) An end-to-end tracking method for polyp detectors in colonoscopy videos. Artif Intell Med 131
14. Mahmud T, Paul B, Fattah SA (2021) A modified encoder-decoder architecture for automated polyp segmentation from colonoscopy images. Sciencedirect Comput Biol Med 128
15. Branch MV, Carvalho AS (2021) Polyp segmentation in colonoscopy images using u-net-mobilenetv2. arXiv:2103.15715
16. Fu X, Yuan Z, Yu T, Ge Y (2023) DA-FPN: deformable convolution and feature alignment for object detection. MDPI Electron 12
17. Dang J, Tang X, Li S (2023) HA-FPN: hierarchical attention feature pyramid network for object detection. Sensors 23:4508
18. Sandler M, Howard A, Zhu M, Zhmoginov A, Chen LC (2018) Mobilenetv2: inverted residuals and linear bottlenecks. In: Conference on computer vision and pattern recognition. IEEE, pp 4510–4520
19. Kvasir SEG. https://datasets.simula.no/kvasir-seg/
20. Selvaraju RR, Cogswell M, Das A, Vedantam R, Parikh D, Batra D (2017) Grad-cam: visual explanations from deep networks via gradient-based localization. In: IEEE international conference on computer vision, pp 618–626

Identification of ASL Hand Gesture Language Using Deep Learning Technique

Vaishnavi Gupta, Shashank Shekhar, Ritika, Rohini Gaur, and Ashish Pandey

Abstract There are many opportunities for study and modeling in the area of gesture categorization and identification. These are helpful for the community of hearing-impaired persons as well as for the evolution of new vision-based devices. A strong hand gesture recognition system should be created in order to recognize words effectively, as hand gesture interpretation relies heavily on this ability. American Sign Language (ASL) is a complete natural language with syntax that is different from English and linguistic elements that are comparable to spoken languages. In this work, VGG19, a CNN model was employed for gesture classification. The model is evaluated on two datasets and has an accuracy of 99.92% for the RGB dataset, 98.73% for the Depth dataset, and 96.54% for the dataset combining both. For the recognition part camera PyCharm software is used that will help the deaf and mute community to understand and learn easily.

Keywords CNN · VGG19 · Human–computer interaction · ASL · PyCharm · Gesture recognition

1 Introduction

The majority of Anglophone Canada and the United States Deaf communities use ASL (American sign language), a natural language, as their main form of communication. ASL is a comprehensive and organized visual language that expresses itself through both manual and nonmanual qualities [1]. In addition to being utilized as a lingua franca, ASL is frequently taught as a second language [2]. Early in the nineteenth century, at Hartford, Connecticut's American School for the Deaf (ASD), a language interaction led to the development of ASL. Since then, deaf schools and deaf community organizations have widely promoted the use of ASL [3]. An

V. Gupta · S. Shekhar · Ritika · R. Gaur · A. Pandey (✉)
Department of Electronics and Communication Engineering, Galgotias College of Engineering and Technology, Greater Noida, India
e-mail: ashishaz2005@gmail.com

A. P. Agrawal et al. (eds.), *Advances of Artificial Intelligence and Machine Learning in Image Processing, Security and Data Analytics*, Lecture Notes in Electrical Engineering 1464, https://doi.org/10.1007/978-981-95-0148-9_28

accurate census of ASL users hasn't been done despite the language being widely spoken. Between 250,000 and 500,000 Americans, including hearing persons and the offspring of deaf adults, are thought to utilize American sign language. ASL signals contain phonemic elements including facial, body, and hand motions [4]. Even while archetypal is more important in ASL than in spoken languages, it is not an enacting-like technique.

The development of several technologies for mobile devices like smartphones, laptops, and notebooks as well as other portable devices like gaming applications encourages research in the field of gesture detection. The gesture recognition system first used a sensor based on hand gloves, but the cost of manufacture was high [5]. Camera vision-based sensors are used because they are simple and easy to use. In sign language, gestures are any particular movements or patterns used with the hands, face, or body to produce certain expressions. A single hand position can indicate a stance, but many hand positions can be combined to produce a continuous action [6]. The suggested system recognizes alphabets using ASL. It is one of the world's most frequently spoken languages. It is tough for users who have no knowledge of sign language to comprehend sign language. In these situations, gesture recognition may be used to translate motions into text using natural language.

2 Literature Review

2.1 Device-Based Recognition

The proposed strategy for the Kinect sensor [1] employs a hierarchical CRF to identify hand motion-based sign segments before employing a Boost Map integrated method to identify the handshapes of the segmented signs. To obtain the 3D depth data about the hand movements, it makes use of a Microsoft Kinect sensor. In [2], the authors designed a glove with six non-identical colored markers, five out of which are attached to the fingertip and one at the center of the palm to capture 3D points using a 3D novel technique, the authors used feedforward backpropagation of ANN for characteristics classification and achieved an accuracy of 95%. In [3], it relies on the method known as flex sensors which help them to record the hand and finger movements before recording the posture and attempts to determine whether the posture falls into any of the established classes. A suitable matching method is employed to identify the precise value of the posture based on the belonging category. The Vietnamese language is recognized using this methodology. Due to the hand's rotation along the Z-axis, the precision rates for three letters in category 3, namely, Ă, Ơ, and Ư, are poor. Using Microsoft Kinect, CNN (convolutional neural network), and the use of GPU acceleration, a recognition method was demonstrated in [4]. The algorithm had a high degree of accuracy in identifying 20 Italian gestures.

2.2 Deep Learning-Based Recognition

Alphabet American Sign Language Recognizer (AASLR), a real-time static alphabet recognition method, was proposed by authors in 2016 [5]. Picture capture, picture pre-processing, region extraction, extraction of features, matching of features, and pattern recognition are the six stages that make up the process. Authors applied K-ClusterEOH-Match algorithm to find resemblance between rushing image and training dataset's feature vectors, and achieved accuracy of 88.26% recognition rate in 0.5 s. Using support vector machines (SVM), the VGG16, and the AlexNet for characteristics extraction, the authors of [6] identified signs from ASL (American sign language) with 99.82% precision. A pipeline of DL (deep learning) models with long short-term memory (LSTM), like ResNet50, is used by the authors of [7] to suggest an SLR system based on vision. A. Sharma et al. review and contrast several promising approaches for identifying gestures, including initially instructing VGG16 with transfer learning, and a hierarchical framework [8]. A human keypoint extraction system is utilized in [9]. From the hand, face, and other body parts, they extract critical points, which are subsequently sent as input into the RNN (recurrent neural network), and the authors achieved the accuracy of 89.7%. [10] A unique DNN (deep neural network) with three convolutional layers and three max-pooling layers is suggested. The system used ESL (English sign language) for performance evaluation, and the accuracy was roughly 82%. [11] Utilizes YOLOv5, which is quick, lightweight, and accurate, the authors were able to reach precision of 95%. In [12] for localizing the hand gestures' position YOLOv3 is used and after the hand gesture has been localized, for identifying the gesture CNN is used. In one of the models, data was acquired applying the motion depth channel, photos were processed to RGB, and then a working system was built using 3D recurrent convolutional neural networks (3DRCNN) [13]. Convolutional neural networks (CNNs) were employed by [14] to create a recognition system from depth imagery, they worked with five different subjects to train various CNNs to classify more than 30 alphabets and numerals. They experimented with several learning hyperparameters and attained an accuracy of 83.58%. Another popular method is the use of HMM models to recognize sign language [15, 16]. The primary objective of state-of-the-art techniques was on applying DL models to attain superior precision and quicker execution times. A model making use of specific hardware elements, especially a depth camera, has been attempted to learn more about the depth variation in an image and recognize additional characteristics for comparison. A Convolutional Neural Network (CNN) was then created to obtain the results [17]. In [18], by fusing an adaptable pooling and a capsule-like network, a novel method of avoiding the need for a pre-trained model when running the system was devised. There are numerous edge detection algorithms, including operators for Sobel, Robert's cross, and Prewitt's, and clever edge detection [19]. As part of the image pre-processing in [20], four different edge detection algorithms were employed. Canny Edge Detection [21] produced the best outcome in comparison. Support Vector Machine (SVM) classifier was employed by the authors in [22], and accuracy of 96.347% was attained. With the Multi-Scale

Mode Filtering (MSMF) technique, the authors of [23] were able to recognize hand movements with a 95.66% global recognition rate. Numerous studies using different classification algorithms have been published, so far as the methods that are now in use are concerned. The best algorithm for classifying American sign language among them is the CNN classifier algorithm, which is described in [24].

3 System Model

The work proposes an authentic and functional medium for categorizing and identifying ASL hand movements. A CNN model is introduced for fast learning so that it will optimize the false positives and false negatives. VGG19 neural networks are used in this investigation.

3.1 Recognition Model

There are five steps for the recognition and classification of the ASL language. Every step is the same except for data collection. In classification, data is collected that is publicly available, and for recognition, data is collected directly from a webcam. The five phases of the project are shown in Fig. 1. The process of image processing starts with image acquisition. This step is often considered a pre-treatment for any image processing project as data is collected here publicly or by webcam. For training and testing samples for any model, the samples must be processed. This includes, but is not limited to, changes in color, size, and direction. Because of this pre-processing of the samples is done so that they will have better features [19]. The segmentation of the appropriate hand image region is a crucial step in any sign recognition system because it isolates the processed region that is important to the job from the rest of the image before the characteristics extraction and recognition stages. Characteristics extraction is essential since it provides the classifier phase input. In the classification step, both trained and test photos are categorized using the features that were derived in the prior phase. Consequently, these two phases have a significant impact on the extent to which gestures are recognized.

3.2 VGG19

The classification of the ASL language uses the VGG19 model. The VGG19 (19-layer) CNN divides the photos into 1000 item kinds using 16 layers of classification, 3 entirely connected layers altogether since each. A very sophisticated 3 * 3 filtering approach called the convolutional layer is a popular tool for classifying photographs. The structural layout of the VGG19 is shown in Fig. 2. These characteristics show

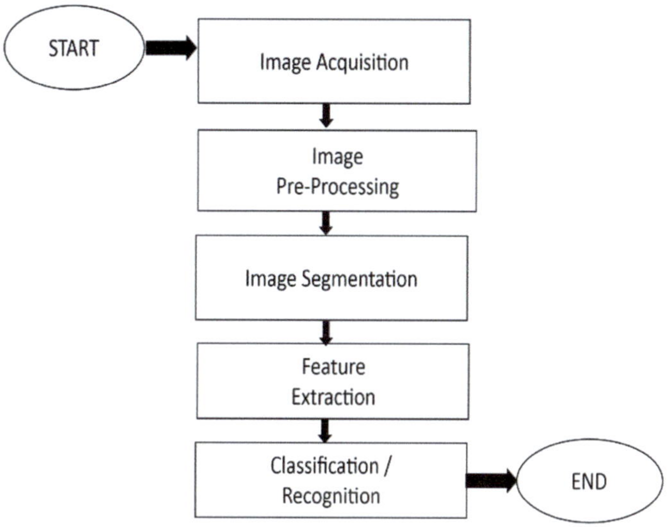

Fig. 1 System model of ASL hand gesture

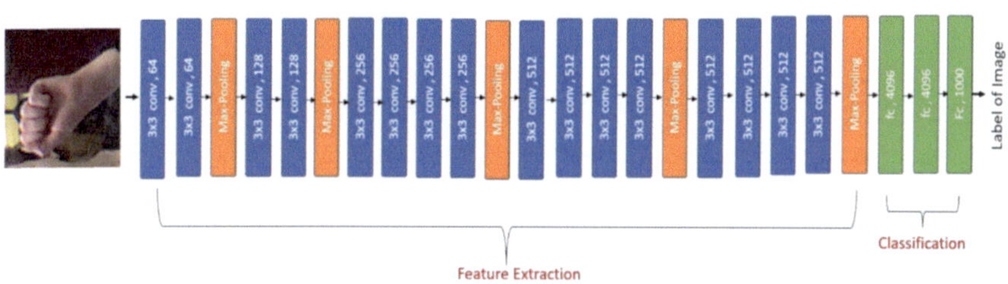

Fig. 2 VGG19 architecture

that extraction using 16 convolutional layers is successful. The final three data are used to complete categorization layers and provide a list of the five feature extraction stages. Each group is followed by a max-pooling layer before there are groups. The input is used by this model to generate 224 X 224 pictures, each visual item has a label next to it.

4 Result

This project has two parts: the first is the classification of the ASL language and the second is the recognition of the ASL language through a live camera.

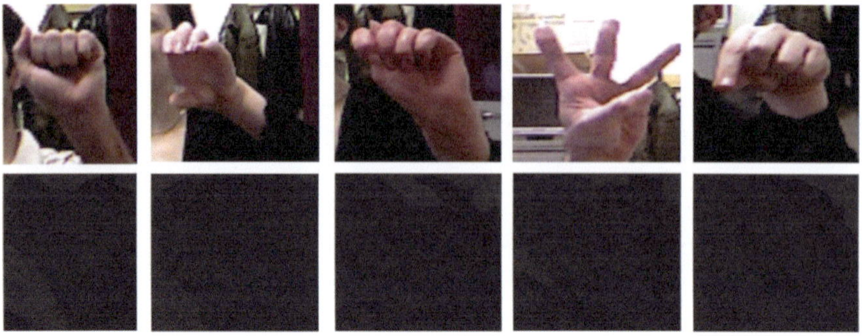

Fig. 3 Hand gesture data. First row: RGB dataset and second row: depth dataset

4.1 Classification

Two datasets are used to evaluate the accuracy of the ASL language in the proposed model.

RGB Dataset. Data in the RGB dataset which comprises gestures done by five people in different backgrounds and locations is difficult to recognize and categorize because of non-uniform hand structure, uncontrolled lighting, etc. The collection comprises 500 image samples representing each gesture. Data is divided into 24 classes.

Depth Dataset. This dataset comprises 500+ images of each gesture with both controlled and uncontrolled backgrounds. The samples are split into three parts by samples per split. After that samples have been separated into 24 classes. Figure 3 shows some of the samples of both datasets.

4.2 Parameters for Evaluating the Performance

The main parameters that are used for assessing any model are accuracy, precision, recall, and f1-score. In Table 1, the mathematical formula for finding these parameters is given.

Table 1 Performance metrics for evaluating the model

Parameters	Formula
Accuracy	$\frac{TP+TN}{TP+TN+FP+FN}$
Precision	$\frac{TP}{TP+FP}$
Recall	$\frac{TP}{TP+FN}$
F1-score	$\frac{2TP}{2TP+FP+FN}$

Table 2 Values of the hyperparameters chosen for the model training

Hyperparameters	Values
No. of Epochs	25
Batch size	20
Optimizer	RMSprop
Learning rate	0.00001
Loss criterion	Categorical cross-entropy

Accuracy. It measures how close a given set of measurements is to their true value.

Precision. It is the measurement of how close the values are to each other.

Recall. Assess the model's performance by counting the genuine positives among all other positive values.

F1-Score. Measures of test accuracy.

4.3 Tuning for Hyperparameters

Hyperparameters have a powerful impact on the execution of the model, by choosing the correct hyperparameter accuracy of any model can be enhanced. Epoch is basically used to calculate the number of iterations required for training any model using all the data at once. The choice for 25 epochs was made as the model converges easily. The number of specimens that are treated before the system is changed is known as the batch size. The minimum and maximum sizes for batches must be equal to or more than one and less than the number of samples in the training dataset, respectively. The batch size that is chosen here is 25. RMSProp (root mean squared propagation) is a method for optimizing the learning rate. It is a development of the well-known adaptive gradient algorithm and is intended to significantly lessen the computing effort required for neural network training because of its features RMSprop was used for minimizing the loss function. The learning rate is considered to be 0.00001 for faster training. For loss function, categorical cross-entropy is used. This is helpful when there are more than two labels for output in multiclass classification. The output label is given one-hot category encoding value in the form of 0 s and 1 s. The model has 24 classes because of this categorical cross-entropy was chosen. Table 2 shows all the hyperparameters that are used in this model.

4.4 Performance Evaluation of the Model

VGG19 model is evaluated on two datasets, RGB and Depth, and the performance of both these datasets is given in Table 3. According to the data given in the table, it is seen that the accuracy of the RGB dataset is higher than the Depth datasets.

Table 3 Performance evaluation of the VGG19 model

Data	Accuracy (%)
RGB	99.92
RGB + depth	98.73
Depth	96.54

Table 4 Performance evaluation of the VGG19 model with some other models using the RGB dataset

Data	Accuracy (%)
Inception V3	97.82
VGG16	96.04
Barbhuiya et al.	97.31
Sharma et al.	95.70
VGG19	99.92

Table 5 Performance evaluation of the VGG19 model with some other models using the depth dataset

Data	Accuracy (%)
Inception V3	92.34
VGG16	91.77
Barbhuiya et al.	92.05
Sharma et al.	90.04
VGG19	96.54

After evaluating both datasets alone, one combined dataset of both RGB and Depth images is evaluated which is given in Table 3. After comparing all three datasets the accuracy of the RGB dataset is higher.

The proposed model is also compared with some other models and the data is shown in Tables 4 and 5. Furthermore, the accuracy and loss curve of VGG19 are shown in Figs. 4 and 5.

The VGG19 model outperforms some of the other models and it can be seen from Tables 4 and 5.

4.5 Recognition

The second part is the recognition of ASL language through live camera. Here the data is taken live from the camera.

Dataset. Our input method is a camera, which records a color picture. It uses the hand tracking function to determine the hand position in order to segment the hand form. Then, a rough hand area may be produced by thresholding from the hand position with a certain depth interval.

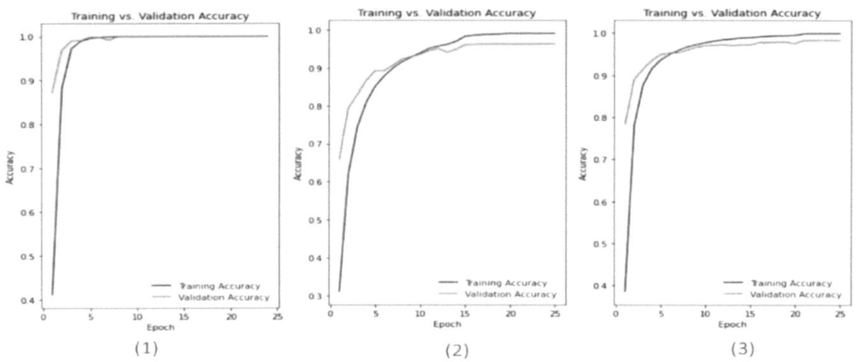

Fig. 4 Accuracy curves of model. (1) VGG19 with RGB data, (2) VGG19 with depth data, and (3) VGG19 with both data

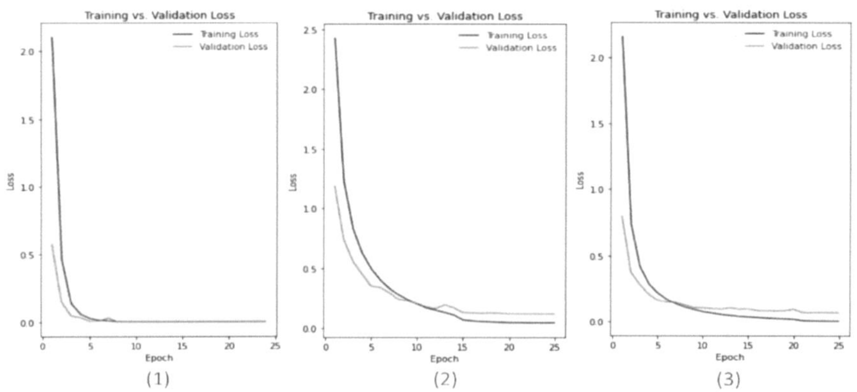

Fig. 5 Loss curves of model. (1) VGG19 with RGB data, (2) VGG19 with depth data, and (3) VGG19 with both data

Proposed system for recognition. The majority of gesture recognition techniques typically include three main steps. Object detection is the initial stage. This stage's goal is to find hand items in digital photos or movies. At this stage, a number of environment and picture challenges need to be resolved in order to ensure that hand contours or areas can be extracted correctly in order to increase identification accuracy. Unstable brightness, noise, low resolution, and poor contrast are common picture difficulties. Camera technology and a better atmosphere can significantly reduce these issues. Object recognition is the second stage. To identify the motions, the hand items that were observed are recognized. The third stage is to analyze sequential gestures to identify users' instructions or behaviors

System Requirement.

Software. PyCharm.

Operating Software. Windows.

Language. Python.

PyCharm. PyCharm is a software that is used for Python programming. It offers code analysis, a graphical debugger, integrated unit tester, in addition to supporting Django web development.

Windows. For operating the project Windows operating system is used. It was used because of its mass production reach.

Python. The language used for writing the code is Python. It is because it is a high-level, all-purpose language for programming. Its design philosophy concentrates on code readability and uses extensive indentation along with dynamic typing, and it additionally makes use of trash collection. It accommodates a number of programming paradigms, including organized, object-oriented, and functional programming. For the project Python libraries like TensorFlow, Keras, Numpy, etc. are used.

Performance Evaluation. The whole project is then evaluated and it is seen that the model is working fine. Sometimes some words are not recognized correctly because of many factors like uneven background, uncontrolled lighting, hand structure, etc.

5 Conclusion

In this work, a pre-trained model which is VGG19 is tweaked and utilized for the categorization of hand gestures based on visual observation. Using VGG19 pre-trained models, a strong and efficient neural network for classifying hand gestures is made. Algorithms for transfer learning are employed to adjust the network's parameters. Additionally, the effectiveness of the suggested work is assessed using two separate complicated hand gesture datasets (RGB and Depth). The experimental findings show that the recommended approach with a limited number of computations outperforms pre-trained models in terms of classification accuracy.

Aside from that, the recognition approach will help abolish tradition totally. A camera is the sole piece of hardware required to capture IP photos. This will command a new era that will have no physical touch. The technology will become manageable by using hand gesture commands.

For future research, any other algorithm can be used for the depth dataset so that the accuracy and other parameter can be increased. The only limitation of this research is the accuracy of the depth dataset as it is much lower than the RGB dataset and the dataset combining both RGB and Depth datasets.

References

1. Yang HD (2015) Sign language recognition with the Kinect sensor based on the conditional random fields. Sensors 15(1):135–147
2. Tangsuksant W, Adhan S, Pintavirooj C (2014) ASL Recognition by using 3D geometric invariant feature & ANN classification. In: Biomedical engineering international conference BMEiCON
3. Phi LT, Nguyen HD, Bui TTQ, Vu TT (2015) A glove-based gesture recognition system for Vietnamese sign language. In: International conference on control, automation and systems (ICCAS). IEEE, pp 1555–1559
4. Pigou L, Dieleman S, Kindermans P-J, Schrauwen B (2014) Sign language recognition using modified CNN. In: European conference on computer vision. Springer, pp 572–578
5. Pansare JR, Ingle M (2016) Vision-based approach for ASL recognition using edge orientation histogram. In: International conference on image, vision & computing, pp 86–90
6. Barbhuiya, A. A, Karsh, R. K, Jain, R () CNN-based feature extraction & classification for sign language, Multimed. Tools. Appl, vol.-80, pp 3051–69. (2021).
7. Rastgoo R, Kiani K, Escalera S (2020) Video-based isolated hand sign language recognition using a deep cascaded model, Multimed Tools Appl 79:22965–22987
8. Sharma A, Sharma N, Saxena Y, Singh A, Sadhya D (2021) Benchmarking DNN approaches for Indian sign language recognition, Neural Comput. Appl 33:6685–6696
9. Ko S-K, Son JG, Jung H (2018) Sign language recognition with RNN using human keypoint detection. In: Conference on research in adaptive & convergent systems, pp 326–328
10. Krishnan PT, Balasubramanian P (2019) Detection of alphabets for machine translation of sign language using DNN. In: International conference on data science & communication (IconDSC). IEEE, pp 1–3
11. Dima TF, Ahmed ME (2021) Using YOLOv5 algorithm to detect & recognize American sign language. In: International conference on information technology (ICIT), pp 603–607
12. Kim S, Ji Y, Lee K-B (2018) An effective sign language learning with object detection based on ROI segmentation. In: International conference on robotic computing (IRC). IEEE, pp 330–333
13. Ye Y, Tian Y, Huenerfauth M, Liu J (2018) Recognizing American sign language (ASL) gestures from within continuous videos. In: Conference on computer-vision & pattern recognition workshops (CVPRW). IEEE
14. Kang B, Tripathi S, Nguyen TQ (2015) Real-time sign language finger spelling recognition using CNN from depth map. In: IAPR Asian conference pattern recognition (ACPR), pp 136–140
15. Lichtenauer JF, Hendriks EA, Reinders MJT (2008) Sign language recognition by combining statistical DTW & independent classification. IEEE Trans Pattern Anal Mach Intell 2040–2046
16. Shanableh T, Assaleh K, Al-Rousan M (2007) Spatio-temporal feature-extraction techniques for isolated gesture recognition in Arabic sign language (ASL). IEEE Trans Syst Man Cybern Part B (Cybern) 641–650
17. Kelly D, Donald JM, Markham C (2011) Weakly supervised training of a sign language recognition system using multiple instance learning density matrices. IEEE Trans Syst Man Cybern Part B (Cybern) 526–541
18. Jalal MA, Chen R, Moore RK, Mihaylova L (2018) American sign posture understanding with DNN. In: International conference on information fusion (FUSION) IEEE
19. Rahman MdA (2017) Computer vision-based human detection. Int J Eng Inf Syst (IJEAIS) 1:62–85
20. Maini R, Aggarwal H (2018) Study & comparison of various image edge detection techniques. Int J Image Process (IJIP) 3:62–85
21. Avlash M, Kaur L (2013) Performances analysis of different edge detection methods on road images. Int J Adv Res Eng & Appl Sci 2(6). ISSN: 2278-6252
22. Verma K (2014) Sign language recognition using SVM. Computer Science & Engineering Department, Thapar University, Patiala

23. Simion G, David C, Gui V, Căleanu CD (2016) Fingertip-based real time tracking & gesture recognition for natural user interfaces. Romania 13(5)
24. Kang L, Ye P, Li Y, Doermann D (2014) CNN for no-reference image quality assessment. In: IEEE conference on computer vision & pattern recognition (CVPR), pp 1733–1740

A Collective Approach of Drone Detection System Based on Machine Learning and Deep Learning

Satyam Kumar, Anirudh Singh, and Deepjyoti Choudhury

Abstract In contemporary times, the utilization of drone detection systems has found extensive application in real-life scenarios. The necessity for real-time drone identification, particularly in defence, has prompted the investigation of new technologies for increased surveillance. This research paper delves into the realm of drone detection using a repertoire of machine learning models, including Naïve Bayes (NB), Random Forest (RF), Logistic Regression (LR), Support Vector Machine (SVM), Convolutional Neural Network (CNN), and Recurrent Neural Network (RNN). This paper evaluates the efficacy of these models in detecting drones in different datasets, emphasizing their significance in practical scenarios. This research employs a comprehensive set of machine learning models to tackle the intricate challenge of drone detection. After rigorous testing on diverse datasets, the findings reveal notable accuracy rates for the employed machine learning models. Specifically, the RNN model emerges as the most effective, achieving accuracy rates of 93%, 99%, 99%, 99%, and an exceptional 99.85% across distinct datasets. This underscores the prowess of RNN in addressing the complexities of real-time drone detection.

Keywords Drone · Unmanned aerial vehicle · Machine learning · CNN · RNN

1 Introduction

In an era characterized by unprecedented advancements in drone technology, the integration of machine learning (ML) has become a pivotal avenue for addressing the evolving landscape of drone-related challenges. Commercial and recreational drones, formally known as Unmanned Air Vehicles (UAVs), have burgeoned in recent years.

This paper delves into the intricate intersection of drone detection and classification, focusing specifically on ML methodologies across various modalities. Our research distinguishes itself by employing a diverse array of ML models to develop

S. Kumar · A. Singh · D. Choudhury (✉)
Department of AIML, School of CSE, Manipal University Jaipur, Jaipur, Rajasthan, India
e-mail: deepjyotichoudhury05@gmail.com

A. P. Agrawal et al. (eds.), *Advances of Artificial Intelligence and Machine Learning in Image Processing, Security and Data Analytics*, Lecture Notes in Electrical Engineering 1464, https://doi.org/10.1007/978-981-95-0148-9_29

a robust drone detection system. Notably, we leverage the capabilities of NB, RF, LR, SVM, CNN, and RNN. While existing literature has demonstrated promising individual contributions to drone detection using ML, a cohesive and standardized approach is lacking. This paper aims to bridge this gap by providing a systematic review, offering insights into the strengths and weaknesses of different ML-based drone detection models.

Examining how various ML models perform in real-time drone detection settings is the main goal of this research. To achieve this, we subject our drone detection system to rigorous testing across distinct datasets, including our benchmark dataset and the VISIODECT DATASET. This dual dataset approach enhances the robustness of our evaluation, allowing us to gauge the models' performance across a spectrum of scenarios and drone types. Unmanned aerial vehicle (UAV) deployment for logistics and other civic objectives continuously disrupts airspace security. We integrate our study with real-world difficulties and situations by adding the VISIODECT dataset into our assessment, addressing the critical need for effective drone detection systems in the face of rising UAV deployment for varied objectives. The significance of our research extends beyond academic discourse, directly impacting the development and deployment of effective drone detection solutions. Ultimately, we envision that the practical implications of our findings will contribute to the enhancement of security measures, providing stakeholders with valuable insights for countering unauthorized drone activities.

This paper positions machine learning as a potent tool in addressing the challenges of drone detection and classification. Unlike traditional methods that may rely on human operators and have limitations in perceiving certain modalities, ML systems excel in pattern recognition across diverse sensory inputs. ML shows up as a strong option to improve the precision and effectiveness of drone detection systems because of its ability to recognize patterns and adapt. This has motivated us to adopt ML/DL techniques to detect drones in real-time datasets.

In conclusion to this introduction, it gives a thorough outline of the issues posed by the proliferation of drones and the imperative need for robust detection and classification systems. The integration of machine learning, as demonstrated by our diverse set of models, represents a promising avenue for addressing these challenges. The strengths and weaknesses inherent in each ML model contribute to a nuanced understanding of their applicability in drone detection scenarios.

The rest of the paper is organized as follows: related work is elaborated in Sect. 2. Section 3 describes the in-depth design and implementation of our work. The findings of our implementation are discussed in Sect. 4 as results analysis. Finally, we conclude this paper in Sect. 5 by writing the conclusion and future work.

2 Related Work

Taha et al. [1] presented a comprehensive review of machine learning-based drone detection and classification. They delved into the evolving landscape of research, offering insights into the current state of the field. Lee et al. [2] proposed a drone detection and identification system leveraging artificial intelligence. Al-Sa'd et al.'s [3] initiative to create a sizable open-source drone database made a contribution to the field. Wu et al. [4] emphasized real-time drone detection using deep learning. Their approach was tailored for swift and accurate detection, showcasing the potential of deep learning techniques in dynamic scenarios. Cetin et al. [5] focused on improving real-time drone detection for counter-drone systems. They explored techniques to enhance the speed and accuracy of drone detection methods, crucial for responsive countermeasures. Seidaliyeva et al. [6] contributed with an accurate drone detection method in videos. Their work underscored the importance of precision in detection, especially in scenarios with a consistent background. Ajakwe et al. [7] introduced the VisioDECT Dataset, offering a valuable resource for scenario-based multi-drone detection. The dataset served as a foundation for researchers and developers in the field. Sheikh et al. [8] proposed a novel approach to drone detection using cellular networks. Their machine learning-based system explored alternative technologies for comprehensive coverage. Behera and Raj [9] explored drone detection using DL techniques. Their work showcased the adaptability and effectiveness of deep learning in addressing the complexities of drone detection.

Zhang [10] focused on RF-based drone detection using ML, emphasizing the importance of RF signals in accurate and reliable drone detection. Mahdavi and Rajabi [11] utilized CNNs for drone detection, showcasing the effectiveness of neural networks in image processing and object detection. Nalamati et al. [12] contributed to the literature by addressing drone detection in videos. Their work explored the challenges and solutions associated with detecting drones in varying surveillance scenarios. Traboulsi and Barbeau [13] investigated the recognition of drone formation intentions using supervised ML. Their work shed light on the nuances of understanding drone behavior and intentions. Steininger et al. [14] introduced the Aircraft Context Dataset, focusing on understanding and optimizing data variability in aerial domains. Their dataset contributed to the development and evaluation of robust drone detection systems. Zhu et al. [15] tackled the challenge of detecting and tracking drones, highlighting the integration of detection and tracking methodologies to address the complexities of drone movements.

Sim et al. [16] focused on ground-truthing of supervised machine learning for drone classification. Their work emphasized the importance of accurate and reliable labeled datasets for training effective drone detection models. Dembla et al. [17] explored ML-based object detection using drones. Their work contributed to the broader understanding of machine learning applications in the context of drones. Hamatapa and Vongchumyen [18] delved into image processing for drone detection, showcasing the role of computer vision in enhancing detection capabilities. Zeng et al. [19] proposed drone detection based on multi-scale feature fusion, offering insights into feature engineering for robust drone identification. A framework for RF signal-based drone identification and detection was introduced by Inani et al. [20] using machine learning. Their work explored the unique characteristics of RF signals for accurate detection. Lee [21] introduced a CNN-based single object detection and tracking system, demonstrating its application to drone detection. The study focused on the dynamic nature of drone movements in video scenarios. Amorim et al. [22] utilized machine learning for the identification of airborne UAVs based on LTE radio measurements. Their work explored the potential of radio measurements for classifying drone signals in the communication spectrum.

3 Design and Implementation

The workflow diagram of our working model is shown in Fig 1.

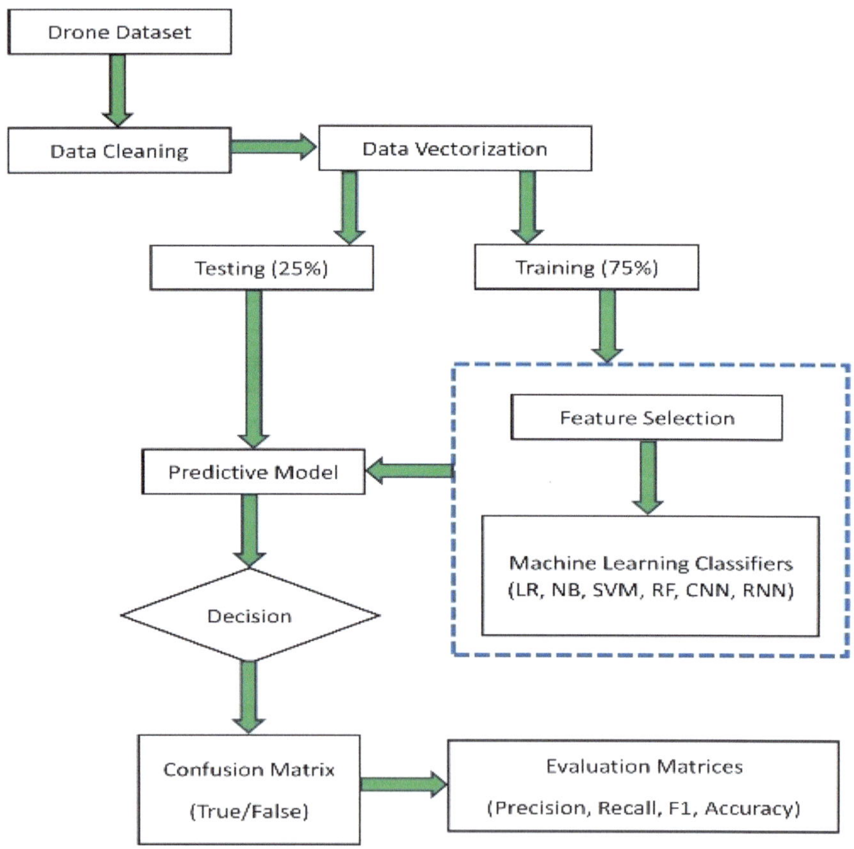

Fig. 1 Workflow diagram of our working model

3.1 Dataset Description

The dataset employed for the development and evaluation of the proposed algorithms is a comprehensive collection comprising multiple subsets, each meticulously curated to address specific aspects of unmanned aerial vehicle (UAV) detection and tracking.

- Benchmark Dataset (Kaggle): This dataset was curated by Mehdi Özel [14] for a UAV competition with the intention of addressing the limitations observed in existing UAV datasets. The Benchmark Dataset on Kaggle underwent a meticulous creation process involving the amalgamation of two distinct Kaggle datasets, resulting in a total of 2194 images. To enhance the dataset's quality, sophisticated data augmentation and preprocessing techniques were applied, leading to a refined set of 1446 images. The dataset was further categorized into drone and non-drone images. The original files for drone images numbered 2194, reduced to 1446 after data cleaning/preprocessing. Similarly, the non-drone category [15] began with 6471 original files, resulting in 6469 files after data cleaning/preprocessing.
- IEEE Dataset (Mavic Enterprise): This Dataset [7] has a focus on Mavic Enterprise, encompassing drone and non-drone images. The original files for drone images in this dataset amounted to 3432, reduced to 3416 after undergoing data cleaning and preprocessing. Simultaneously, the non-drone category [15] included a total of 2152 images specifically designated for mixing with drone images, resulting in a combined dataset size of 5568 images.
- IEEE Dataset (EFT): This Dataset [7] with labeled EFT also consists of both drone and non-drone images. The drone category started with 3600 original files, reduced to 3571 after data cleaning and preprocessing. In parallel, the non-drone category [15] included 2152 images intended for mixing with drone images, contributing to a combined dataset size of 5723 images.
- IEEE Dataset (DJIPHANTOM): Within this Dataset [7], both drone and non-drone images were curated. The original files for drone images numbered 3300, reduced to 2402 after undergoing data cleaning and preprocessing. Simultaneously, the non-drone category [15] included a total of 2152 images designated for mixing with drone images, resulting in a combined dataset size of 4554 images.
- IEEE Dataset (Anafi): This Dataset [7] comprises drone and non-drone images. The original files for drone images amounted to 3471, reduced to 3278 after data cleaning and preprocessing. Concurrently, the non-drone category [15] included 2152 images specifically chosen for mixing with drone images, contributing to a combined dataset size of 5430 images.

In conclusion, our research leveraged a comprehensive dataset, meticulously curated to address various aspects of unmanned aerial vehicle (UAV) detection and tracking. The primary Benchmark Dataset from Kaggle underwent thorough data augmentation and preprocessing, resulting in a refined set of 1446 images. Additionally, diverse subsets from various IEEE datasets, such as Mavic Enterprise, EFT, DJIPHANTOM, and Anafi, were employed, contributing to the dataset's richness and complexity. Each IEEE dataset, characterized by drone and non-drone images, underwent meticulous data cleaning and preprocessing to ensure dataset quality. Our approach utilized these datasets to develop and evaluate algorithms, offering valuable insights into the realm of UAV detection and tracking.

3.2 Data Vectorization and Feature Selection

3.2.1 Data Vectorization

Converting color images to grayscale is a common preprocessing step. This reduces the dimensionality of the data and simplifies feature extraction. Resizing images to a consistent size is essential. In the provided code, images are resized to a fixed size (e.g, 100 $\times$ 100 or 256 $\times$ 256) to ensure uniform input dimensions for the models. Normalizing pixel values is often performed to scale them to a specific range, typically [0, 1]. This ensures that all pixel values are in a consistent format for training. Images are often flattened from their original 2D or 3D format (e.g, height $\times$ width $\times$ channels) into 1D vectors. This is commonly done for models like Naive Bayes, Logistic Regression, and some SVM variants. In RNN models, the data may be transformed into sequences of vectors, where each vector represents a time step. This is crucial for handling sequential data, such as video frames for drone detection.

3.2.2 Feature Selection

Naive Bayes models typically assume independence between features. Feature selection is not explicitly performed, as all features (pixel values) are used. SVMs are capable of handling high-dimensional data. However, in some cases like Principal Component Analysis (PCA) or Recursive Feature Elimination (RFE) may be applied to reduce dimensionality and improve efficiency. Random Forest models can handle high-dimensional data effectively, and feature selection is often not necessary. However, you can analyze feature importance scores provided by the model to identify the most influential features. Logistic Regression can benefit from feature selection methods like PCA or RFE to reduce dimensionality and improve model performance.

CNNs automatically learn features from images through convolutional layers. There's typically no explicit feature selection. However, the number and configuration of convolutional filters in the network architecture influence what features the model learns. RNNs are often used for sequential data. Feature selection for RNNs primarily involves selecting the relevant time steps or sequences for prediction.

3.3 Machine Learning Models

Sl. No.	Machine learning model	Relation with our work
1	Naïve Bayes (NB)	NB counts the occurrences of specific features in the photos and treats each feature as a condition in the drone identification
2	Random Forest (RF)	Random Forest model learns intricate patterns and features indicative of drone presence
3	Support Vector Machine (SVM)	By selecting an optimal hyperplane, SVM aids in effectively discriminating between drone and non-drone images
4	Logistic Regression (LR)	By setting a threshold, these probabilities are then converted into binary predictions, effectively distinguishing between drone and non-drone images
5	Convolutional Neural Network (CNN)	The model learns to distinguish unique visual patterns associated with drones, enabling accurate classification
6	Recurrent Neural Network (RNN)	RNNs prove valuable due to their ability to model temporal dependencies and capture patterns evolving over time

3.4 Confusion Matrix and ROC Curve

- Confusion matrix:

In this paper, we utilize the confusion matrix to assess the effectiveness of various ML models. It consists of four main components: True Positives (TP): Represents the instances correctly classified as "Drone." In our context, it indicates the successful identification of drones by the model. True Negatives (TN): Denotes instances correctly classified as "Non-Drone." This signifies the accurate identification of scenarios where non-drones are present. False Positives (FP): Indicates instances wrongly classified as "Drone" when, in reality, there is no drone. In our application, this would mean a false alarm or a false positive detection. False Negatives (FN): Represents instances incorrectly classified as "Non-Drone" when a drone is actually present.

- ROC Curve:

It plots the True Positive Rate (Sensitivity) against the False Positive Rate (1-Specificity). In our work, the ROC curve visualizes how well the model distinguishes between drones and non-drones. A high-performing model will exhibit an ROC curve that approaches the top-left corner of the graph, indicating a high True Positive Rate and a low False Positive Rate. The area under the ROC curve (AUC-ROC) serves as a summary metric, with a higher AUC-ROC suggesting superior discrimination performance.

3.5 Evaluation Parameters

- Accuracy:

Accuracy assesses the correctness of predictions by comparing the sum of true positives (TP) and true negatives (TN) to the total instances.

$$\text{Accuracy} = (\text{TP} + \text{TN})/(\text{TP} + \text{TN} + \text{FP} + \text{FN}) \tag{1}$$

- Precision:

Precision gauges the precision of positive predictions, determined by the ratio of true positives to the sum of true positives (TP) and false positives (FP).

$$\text{Precision} = \text{TP}/(\text{TP} + \text{FP}) \tag{2}$$

- Recall:

Recall measures the ability to capture positive instances, calculated by dividing true positives by the sum of true positives (TP) and false negatives (FN).

$$\text{Recall} = \text{TP}/(\text{TP} + \text{FN}) \tag{3}$$

- F1-Score:

The F1-score offers a balanced assessment, combining precision and recall using the harmonic mean formula:

$$\text{F1 - score} = 2 * (\text{Precision} * \text{Recall})/(\text{Precision} + \text{Recall}) \tag{4}$$

4 Results Analysis

In this section, we present the results of our experiments on drone detection using machine learning, encompassing multiple datasets: Benchmark, IEEE Dataset (Mavic Enterprise), IEEE Dataset (EFT), IEEE Dataset (DJI), and IEEE Dataset (Anafi). The evaluation metrics include accuracy, precision, recall, and F1-score for each classifier trained on these datasets.

In Table 1, various classifiers' accuracy across different datasets is presented. Naive Bayes (NB) demonstrated a moderate accuracy of 72.48%. The Random Forest (RF) classifier excelled with a high accuracy of 93.18% across datasets. Support Vector Machine (SVM) displayed strong accuracy at 88.20%. Logistic Regression

Table 1 Accuracy (%) for all datasets

Classifier	Benchmark dataset (Kaggle)	IEEE dataset (Mavic enterprise)	IEEE dataset (EFT)	IEEE dataset (DJIPHANTOM)	IEEE dataset (Anafi)
NB	72.48	89.05	79.8	60	90.65
RF	93.18	99	99.90	100	99.97
SVM	88.20	100	100	100	99.93
LR	76	95	94	100	99
CNN	90.41	100	100	100	100
RNN	93	99	99	99	99.85

(LR) achieved a moderate accuracy of 76%. Convolutional Neural Network (CNN) stood out with an accuracy of 90.41%. The Recurrent Neural Network (RNN) outperformed other classifiers with an accuracy of 93%. The ROC Curve depicting these accuracy results is illustrated in the table (Fig. 2 and Table 2).

The Convolutional Neural Network (CNN) consistently exhibited the highest precision across all datasets, showcasing perfect precision (100%) in the IEEE Dataset for Mavic Enterprise, EFT, DJIPHANTOM, and Anafi. This emphasizes CNN's capability to minimize false positives and enhance the model's accuracy in correctly identifying and classifying drones, making it a promising choice for

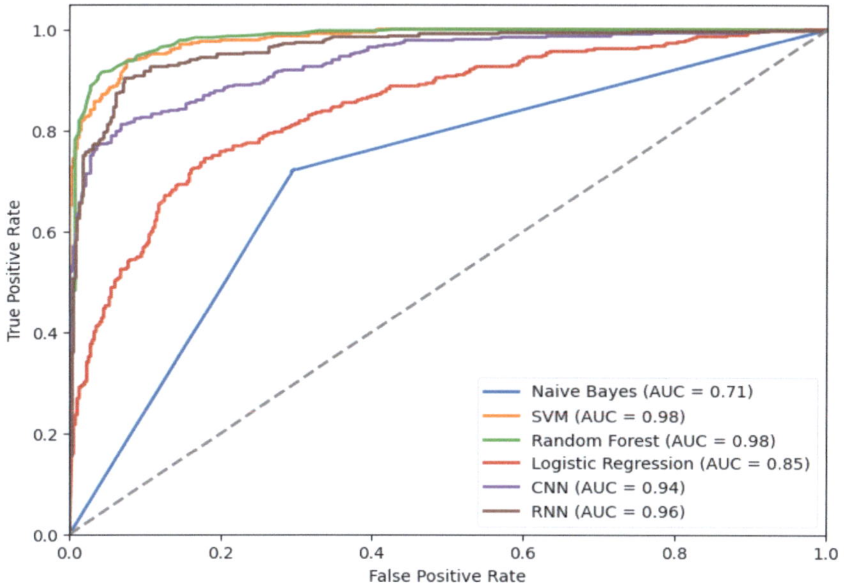

Fig. 2 RoC curve on benchmark dataset

Table 2 Precision (%) for all datasets

Classifier	Benchmark dataset (Kaggle)	IEEE dataset (Mavic enterprise)	IEEE dataset (EFT)	IEEE dataset (DJIPHANTOM)	IEEE dataset (Anafi)
NB	69	86	77	82	89
RF	96	99	99	100	100
SVM	88	100	100	100	100
LR	73	94	92	100	98
CNN	93	100	100	100	100
RNN	91	99	99	96	99

precision-oriented applications in drone detection systems. The ROC Curve is given below (Figs. 3, 4 and Table 3).

The Recurrent Neural Network (RNN) consistently demonstrated the highest recall across all datasets, making it a robust choice for drone detection. Notably, it achieved a recall of 100% in three out of five datasets, highlighting its capability to effectively capture a high proportion of true positives and minimize false negatives (Table 4).

The Convolutional Neural Network (CNN) consistently exhibited the highest F1-score across all datasets, making it a reliable choice for drone detection. Notably, it

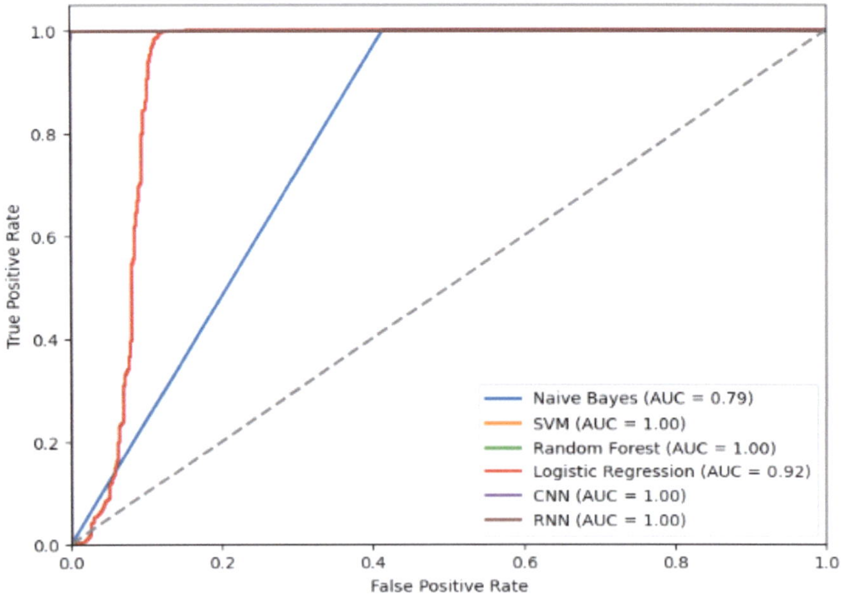

Fig. 3 RoC curve on Mavic enterprise dataset

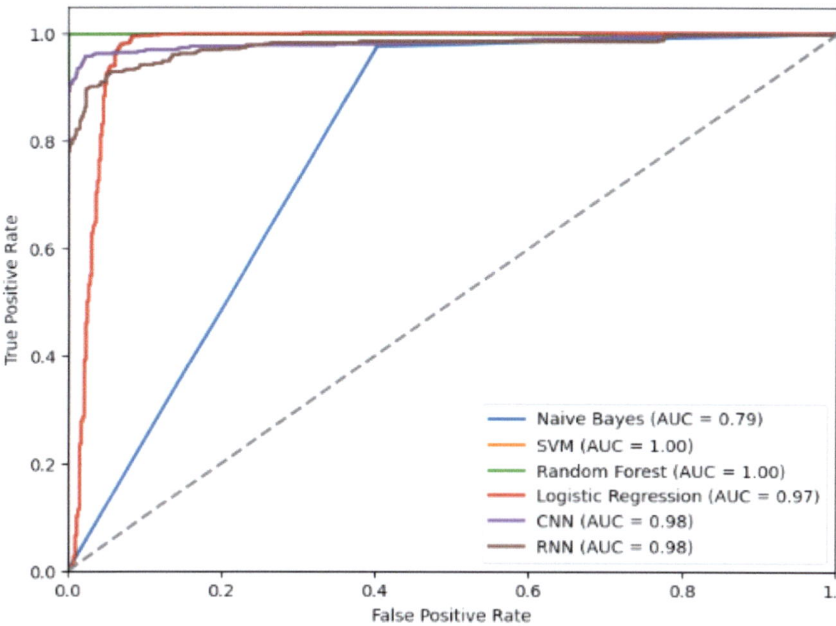

Fig. 4 RoC curve on EFT dataset

Table 3 Recall (%) for all datasets

Classifier	Benchmark dataset (Kaggle)	IEEE dataset (Mavic enterprise)	IEEE dataset (EFT)	IEEE dataset (DJIPHANTOM)	IEEE dataset (Anafi)
NB	74	100	96	40	96
RF	89	100	100	100	99
SVM	87	100	100	100	100
LR	80	100	100	100	100
CNN	87	100	100	100	100
RNN	93	99	99	100	99

Table 4 F1-score (%) for all datasets

Classifier	Benchmark dataset (Kaggle)	IEEE dataset (Mavic enterprise)	IEEE dataset (EFT)	IEEE dataset (DJIPHANTOM)	IEEE dataset (Anafi)
NB	72	93	86	49	93
RF	92	99	99	100	99
SVM	88	100	100	100	100
LR	76	97	96	100	99
CNN	90	100	100	100	100
RNN	92	99	99	99	99

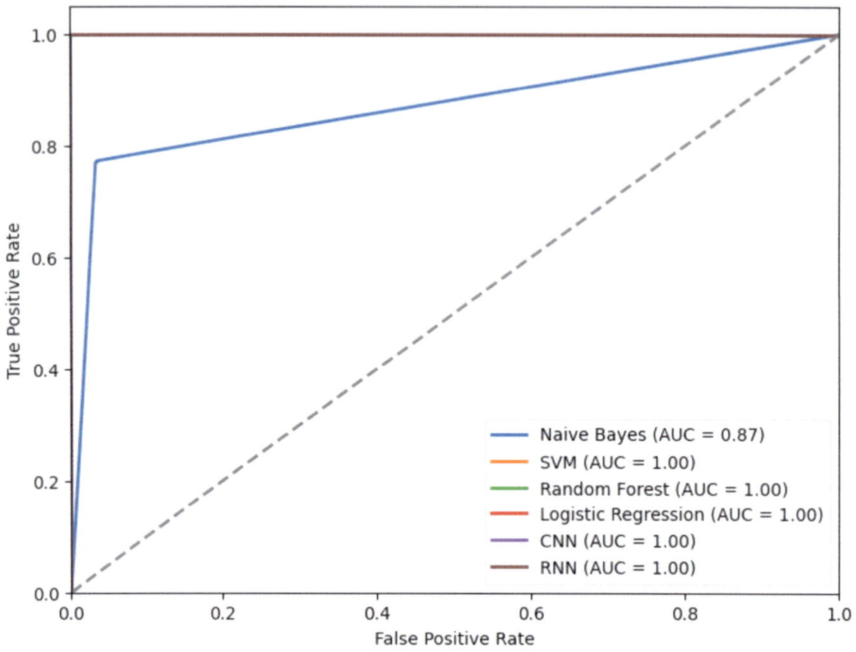

Fig. 5 RoC curve on DJI dataset

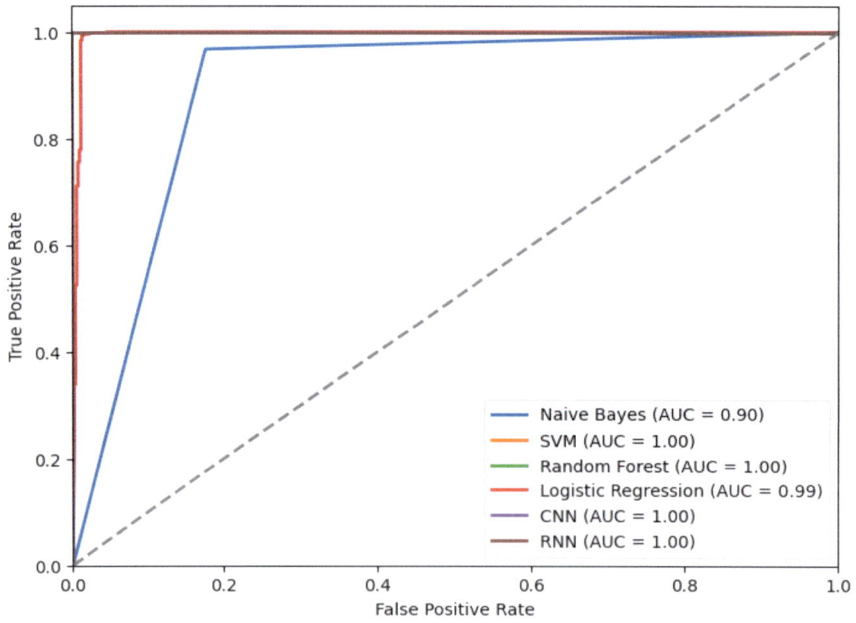

Fig. 6 RoC curve on Anafi dataset

achieved F1-scores of 100% in three out of five datasets, highlighting its capacity for a balanced precision-recall trade-off. The ROC Curve is given below (Figs. 5 and 6).

5 Conclusion and Future Work

In anticipation of the soaring growth in the global commercial drone market, projected to exceed 127 billion dollars, the pressing need to address the associated risks becomes paramount. While drones offer diverse applications, their operation poses substantial risks to people and assets. In regulatory efforts, supported by technologies like the Low Altitude Authorization and Notification Capability (LAANC) by the USA FAA, aim to mitigate disruptions and enhance the oversight of drone operations. However, rules and technologies alone may not suffice, especially in dealing with careless or malicious users. The requirement for systems capable of continuous overview, dynamic risk assessment, and intervention in case of violations becomes evident. Our research underscores the pivotal role of machine learning in meeting these challenges, as showcased in the evaluation of various algorithms for drone detection. The adaptability of machine learning models, spanning diverse modalities, offers a promising avenue for efficient detection systems. However, challenges persist in terms of data quantity and quality, necessitating collaborative efforts to build robust classification models. The dangers of drone operation vary greatly depending on the position of the drone and its closeness to important regions. Future research should prioritize exploring range as a crucial objective, evaluating classification performance in relation to drone distance, and potentially determining range using regression models. While bi-modal and multimodal systems have shown promise, especially in overcoming the limitations of individual modalities, the current focus is on statically located detection systems. The limitations of such systems in urban environments, where obstacles and noise levels can impact detection capability, suggest that research into distributed and collaborative detection systems is needed. In current metropolitan environments, city-wide surveillance sensors might pave the way for more successful drone detection and categorization.

References

1. Taha B, Shoufan A (2019) Machine learning-based drone detection and classification: state-of-the-art in research. IEEE Access 7:138669–138682
2. Lee D, La WG, Kim H (2018) Drone detection and identification system using artificial intelligence. In: 2018 international conference on information and communication technology convergence (ICTC). IEEE, pp 1131–1133
3. Al-Sa'd MF, Al-Ali A, Mohamed A, Khattab T, Erbad A (2019) RF-based drone detection and identification using deep learning approaches: an initiative towards a large open source drone database. Future Gener Comput Syst 100:86–97

4. Wu M, Xie W, Shi X, Shao P, Shi Z (2018) Real-time drone detection using deep learning approach. In: Machine learning and intelligent communications: third international conference, MLICOM 2018, Hangzhou, China, proceedings, 6–8 July 2018, vol 3. Springer International Publishing, pp 22–32
5. Cetin E, Barrado C, Pastor E (2021) Improving real-time drone detection for counter-drone systems. Aeronaut J 125(1292):1871–1896
6. Seidaliyeva U, Akhmetov D, Ilipbayeva L, Matson ET (2020) Real-time and accurate drone detection in a video with a static background. Sensors 20(14):3856
7. Ajakwe SO, Ihekoronye VU, Mohtasin G, Akter R, Aouto A, Kim DS, Lee JM (2022) VisioDECT dataset: an aerial dataset for scenario-based multi-drone detection and identification. IEEE Dataport
8. Sheikh MU, Ghavimi F, Ruttik K, Jantti R (2019) Drone detection and classification using cellular network: a machine learning approach. In: 2019 IEEE 90th vehicular technology conference (VTC2019-Fall). IEEE, pp 1–6
9. Behera DK, Raj AB (2020) Drone detection and classification using deep learning. In: 2020 4th international conference on intelligent computing and control systems (ICICCS). IEEE, pp 1012–1016
10. Zhang Y (2021) Rf-based drone detection using machine learning. In 2021 2nd International Conference on Computing and Data Science (CDS). IEEE, pp 425–428
11. Mahdavi F, Rajabi R (2020) Drone detection using convolutional neural networks. In 2020 6th Iranian conference on signal processing and intelligent systems (ICSPIS). IEEE, pp 1–5
12. Nalamati M, Kapoor A, Saqib M, Sharma N, Blumenstein M (2019) Drone detection in long-range surveillance videos. In: 2019 16th IEEE international conference on advanced video and signal based surveillance (AVSS). IEEE, pp 1–6
13. Traboulsi A, Barbeau M (2019) Recognition of drone formation intentions using supervised machine learning. In: 2019 international conference on computational science and computational intelligence (CSCI). IEEE, pp 408–411
14. Steininger D, Widhalm V, Simon J, Kriegler A, Sulzbachner C (2021) The aircraft context dataset: Understanding and optimizing data variability in aerial domains. In: Proceedings of the IEEE/CVF international conference on computer vision, pp 3823–3832
15. Zhu P, Wen L, Du D, Bian X, Fan H, Hu Q, Ling H (2021) Detection and tracking meet drones challenge. IEEE Trans Pattern Anal Mach Intell 44(11):7380–7399
16. Sim J, Jahangir M, Fioranelli F, Baker CJ, Dale H (2019) Effective ground-truthing of supervised machine learning for drone classification. In: 2019 international radar conference (RADAR). IEEE, pp 1–5
17. Dembla S, Dolas N, Karigar A, Sonavane S (2021) Machine learning based object detection and classification using drone. Int Res J Eng Technol 8(6):2546–2555
18. Hamatapa R, Vongchumyen C (2019) Image processing for drones detection. In: 2019 5th international conference on engineering, applied sciences and technology (ICEAST). IEEE, pp 1–4
19. Zeng Z, Wang Z, Qin L, Li H (2021) Drone detection based on multi-scale feature fusion. In: 2021 international conference on UK-China emerging technologies (UCET). IEEE, pp 194–198
20. Inani KN, Sangwan KS (2023) Machine learning based framework for drone detection and identification using RF signals. In: 2023 4th international conference on innovative trends in information technology (ICITIIT). IEEE, pp 1–8
21. Lee DH (2021) CNN-based single object detection and tracking in videos and its application to drone detection. Multimed Tools Appl 80(26–27):34237–34248
22. Amorim R, Wigard J, Nguyen H, Kovacs IZ, Mogensen P (2017) Machine-learning identification of airborne UAV-UEs based on LTE radio measurements. In: 2017 IEEE globecom workshops (GC Wkshps). IEEE, pp 1–6

A Framework for Integrating Intelligent Conversational Chatbots into Learning Management Systems (MOODLE) to Improve Teaching–Learning Evaluation Process

Johnbenetic Gnanaprakasam and Ravi Lourdusamy

Abstract Problem Statement: The potential for improving the teaching and learning assessment process in Learning Management Systems (LMS) like MOODLE through the integration of conversational chatbots powered by AI is yet largely untapped. Research Method and Design: This study presents a comprehensive framework that is well-organized to make it easier to incorporate intelligent chatbots into LMS. The approach follows a sequential procedure that includes gathering user requests, applying Natural Language Understanding (NLU), producing and choosing responses, and controlling discourse. Major Findings: When properly applied, the proposed framework can greatly enhance the process of evaluating teaching and learning. The chatbot's abilities to manage repetitive activities, give personalized learning resources, monitor student progress, and offer insights into student sentiment and engagement have all been improved. Conclusions: Effective error and ambiguity management techniques, regular monitoring, and updates, as well as high-quality initial training data all play a significant role in the framework's performance. This study lays the groundwork for further investigation into more complex cognitive services, improved entity and intent extraction, and sophisticated dialogue management techniques.

Keywords Chatbot · Learning management system · Machine learning · User experience · Evaluation

---

J. Gnanaprakasam (✉) · R. Lourdusamy
Department of Computer Science, Sacred Heart College (Autonomous), Affiliated to Thiruvalluvar University, Vellore, Tamil Nadu, India
e-mail: johnbenetic543@gmail.com

R. Lourdusamy
e-mail: ravi@shctpt.edu

A. P. Agrawal et al. (eds.), *Advances of Artificial Intelligence and Machine Learning in Image Processing, Security and Data Analytics*, Lecture Notes in Electrical Engineering 1464, https://doi.org/10.1007/978-981-95-0148-9_30

1 Introduction

This study focuses on integrating intelligent conversational chatbots into Learning Management Systems (LMS) like MOODLE to improve the teaching and learning evaluation process. Leveraging advancements in artificial intelligence (AI), these chatbots aim to offer natural, human-like interactions for effectively understanding and responding to user inquiries. The project's goal is to develop a comprehensive framework for embedding these chatbots into LMS, emphasizing Natural Language Understanding (NLU), Response Generation and Selection, and Dialogue Management. As technology becomes more integral in education, LMS platforms like Moodle play a crucial role in managing educational activities. The integration of AI-driven chatbots into these platforms promises to enhance teaching and learning evaluations with immediate, personalized feedback and support. The methodology of this study is geared toward ensuring that the chatbots meet the needs of both educators and learners, facilitating a more dynamic, responsive, and student-focused educational environment. The anticipated outcome is more effective teaching, engaging learning experiences, and precise assessment of student progress.

2 Background Work

Moodle in Education: Moodle is a key Learning Management System (LMS) offering features for course delivery, management, and evaluation. It helps teachers create engaging environments and monitor student progress, facilitating the integration of chatbot technology to improve teaching and learning assessment [1, 2]. Chatbot Technology Development: Advances in machine learning and natural language processing have significantly enhanced chatbots, enabling them to understand and respond in natural language. This opens up possibilities for their use in educational settings [1]. Chatbots in Educational Research: Studies show chatbots can effectively assist teaching and learning by providing feedback, content delivery, and assessment support.

They positively impact student engagement and reduce teacher workload, although research often focuses on general functionality rather than specifically on teaching–learning assessment [3, 4]. Chatbots in Moodle and Other LMS: Research on integrating chatbots into Moodle and other LMS is increasing, with chatbots used for tasks like answering queries and distributing materials. However, in-depth studies focusing on their role in improving teaching–learning evaluation in Moodle are scarce [2, 5]. Research Gaps and Limitations: Current studies lack standardized frameworks for integrating chatbots in LMS like Moodle, and there's a need for more personalized chatbot interactions. The effectiveness of chatbots in enhancing teaching–learning evaluation is not well-documented, highlighting the need for more targeted empirical research [3, 4].

3 Review of Literature

This research explores the integration of intelligent conversational chatbots into Learning Management Systems (LMS), focusing on Moodle. LMS like Moodle offer a virtual environment for delivering, monitoring, and managing educational courses and are known for their adaptability and comprehensive features, including tools for assessment and learner interaction [6, 7]. Intelligent chatbots, utilizing NLP, machine learning, and AI, enhance student engagement and provide individualized learning experiences, serving as evaluation tools, support systems, and teaching assistants [8–10]. While there has been a significant study on integrating AI into LMS for predicting student performance and enabling personalized learning, the focus has largely been on predictive models and adaptive learning, rather than on chatbot integration [11–13]. There is a notable lack of research on the technological aspects of integrating chatbots into LMS, with a need for more focus on pedagogical methods and the teaching–learning assessment process [14, 15]. This study aims to fill this gap by proposing a framework for seamlessly integrating chatbots into Moodle, which could significantly enhance the teaching–learning assessment process.

Chatbot technology has gained popularity due to advancements in AI, machine learning, and natural language processing over the past decade [16]. These intelligent chatbots, adaptable to user behavior, are ideal for educational purposes, such as personalized tutoring systems that simulate one-on-one sessions [17, 18]. Additionally, adaptive learning technologies have enabled LMS like Moodle to integrate AI components for tracking learner progress and customizing content [19, 20]. In online learning contexts, chatbots have been suggested to bridge communication gaps between teachers and students. Hu (2014) found that students interacting with chatbots during assignments displayed higher motivation and engagement [21, 22]. However, the specific integration of chatbots into LMSs for supporting the teaching–learning assessment process is still under-researched [23, 24]. There's also a lack of study on chatbots' capabilities for real-time evaluation and feedback [25], despite evidence of their use in student assessments [26].

This study aims to enhance the teaching–learning evaluation process in Moodle through intelligent chatbots, considering these research gaps. The potential of AI-based assistants in learning platforms has been recognized, with studies like Park showing increased student engagement with chatbots as teaching assistants [27]. Adaptive learning platforms like Moodle have also developed innovative personalized learning methods [28]. The role of chatbots in language learning has been highlighted by Cui et al., who demonstrated their effectiveness in facilitating second language acquisition [29]. Furthermore, Fryer et al. explored chatbots' ability to provide emotional support to students, showing their potential in addressing the emotional aspects of learning [30]. However, most studies have focused on chatbots for engagement, personalized learning, or emotional support. This research aims to explore chatbots' capabilities in evaluation and feedback within an LMS.

Recognizing the expanding role of chatbots in various educational contexts, studies like Fryer et al. have investigated their use in enhancing mindfulness and emotional support among college students [30]. Moodle, as an open-source LMS, has increasingly incorporated AI for various purposes, including tracking and personalizing learner experiences [28]. Kumar et al. demonstrated that chatbots as teaching assistants could enhance learning outcomes and foster engagement [24]. Additionally, Cui et al.'s study reinforced the value of chatbots in language learning [29]. Despite these developments, the potential of chatbots in the LMS learning evaluation process remains underexplored. Our study aims to bridge this gap by proposing a method for integrating intelligent conversational chatbots into Moodle to enhance teaching and learning evaluations.

4 Methodology

4.1 How Chatbots Work (Chatbot System Process)

An AI chatbot is a computer program designed to simulate human conversation. It primarily uses Natural Language Processing (NLP) to understand user input and generate relevant responses by breaking down the input into components and applying machine learning to grasp the intended meaning. Dialogue management is employed to create responses that progress the conversation naturally. Many AI chatbots improve over time through machine learning, analyzing user interactions and feedback to refine their understanding of user intent and response relevance. They often integrate with backend systems like databases and APIs to provide meaningful information and services in response to user requests. Additionally, chatbots can personalize responses by tracking user preferences and employing analytics and reporting to monitor usage data and other metrics, enhancing their effectiveness over time. Overall, AI chatbots combine technologies like NLP, machine learning, and backend integration to understand user intent and deliver helpful information and services conversationally.

4.2 Chatbot Design Structure (Process Flow)

This process for integrating a chatbot into a Learning Management System (LMS) involves a series of steps. Firstly, a needs analysis is conducted to understand the specific requirements of LMS users, including students, teachers, and administrators, focusing on aspects like course queries and assignment deadlines. Following this, the chatbot is designed, defining its conversational flow and the types of tasks it can perform within the LMS. Integration involves setting up the necessary interfaces and APIs to link the chatbot with the LMS, including backend connections to databases and services. The chatbot is then trained using machine learning techniques on datasets of typical user interactions, potentially using past chat logs or hypothetical conversations. Testing and evaluation are conducted with a user group to assess performance using both objective measures like response accuracy and subjective measures like user satisfaction. After successful testing, the chatbot is deployed for all users, with appropriate training and support provided. Finally, ongoing monitoring and improvement are essential, with regular reviews of user feedback and chatbot performance to make necessary updates and enhancements.

The development and deployment phases of a chatbot within a Learning Management System (LMS) are depicted in this framework diagram (see Fig. 1). An initial needs analysis is conducted to determine the precise wants and requirements of the users. The next phase is chatbot design, which entails creating the chatbot's conversational flow. The development of the required APIs and interfaces for the chatbot's seamless integration with the LMS comes next.

After the integration is finished, the chatbot goes through training to improve its capabilities using machine learning methods. The chatbot is tested with a group of users throughout the next step of testing and evaluation to evaluate its performance. The chatbot is made available to all users after successful testing, making it accessible within the LMS environment.

The framework flow also has a stage called Monitoring and Improvement, where user input and chatbot performance are regularly tracked. Regular changes and enhancements are made to the chatbot's functionality based on this feedback. The direction of the arrow from "Monitoring and Improvement" back to "Training"

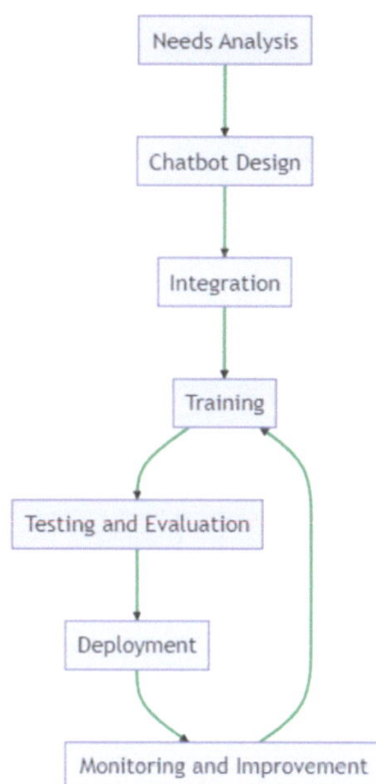

Fig. 1 Diagrammatic representation of the design structure for a chatbot that is compatible with Moodle

indicates that the chatbot can be further trained and improved using feedback and improvements, ensuring continual improvement.

Overall, this framework flow offers a methodical procedure for the creation, introduction, and ongoing evolution of a chatbot in an LMS setting. An effective foundation for integrating a chatbot into a learning management system like Moodle should be provided by this framework flow. The chatbot can be an effective tool for enhancing the evaluation of teaching and learning since it offers users (administrators, staff, and students) immediate, individualized support.

4.3 Design of Intelligent Conversational Chatbots

Intelligent conversational chatbots have emerged as a result of the advancement of artificial intelligence (AI). These chatbots, which employ natural language to converse with humans, are used in a variety of fields, including customer service, healthcare, and education [31]. Rule-based systems, machine learning, and natural language processing are just a few of the methods used to create intelligent conversational chatbots [32]. Rule-based systems respond to user inquiries using pre-defined rules, whereas machine learning algorithms are trained on massive datasets to understand how to respond. The meaning underlying user requests is understood and interpreted via natural language processing [33].

4.4 Standard Architecture of Intelligent Conversational Chatbots

A natural language understanding (NLU) module, a dialogue management (DM) module, and a natural language generating (NLG) module make up the conventional architecture of an intelligent conversational chatbot [34]. The NLU module is in charge of deciphering the intent behind user inquiries, the DM module is in charge of directing the dialogue, and the NLG module is in charge of coming up with solutions.

4.5 Applications of Intelligent Conversational Chatbots

Numerous industries, including customer service, healthcare, and education, could benefit from the use of intelligent conversational chatbots [35]. Chatbots are being utilized in the customer service sector to offer clients round-the-clock assistance, eliminating the need for human customer service agents [36]. Chatbots are being utilized in the healthcare industry to help and direct patients, particularly those with chronic diseases [37]. Chatbots are being utilized in education to give pupils individualized learning experiences [38].

4.6 Challenges and Future Directions

Intelligent conversational chatbots hold great potential across various industries, but they face key challenges that need addressing. One major challenge is ensuring these chatbots effectively understand and interpret natural language, which is crucial in customer service to prevent negative experiences. Another issue is maintaining context in conversations, especially in complex discussions. Future advancements in natural language processing and machine learning are expected to enhance chatbot capabilities, making them more adept at handling intricate dialogues. There's also a possibility of chatbots becoming so advanced that they provide human-like responses, making it hard to distinguish them from real people. This study explored the design, architecture, and applications of intelligent conversational chatbots, discussing current challenges and future potential. As chatbots continue to evolve, they are likely to play an increasingly significant role in various sectors and everyday life.

4.7 Standard Architecture for Chatbots

In a typical chatbot architecture, three essential components work in tandem: Input Processing, which handles and formats user inputs for the chatbot; Dialogue Management, responsible for overseeing the verbal interaction between the user and the chatbot to maintain a coherent conversation; and Output Generation, which crafts responses based on the user's input and the context of the current conversation.

4.8 Applications of Intelligent Conversational Chatbots

Intelligent conversational chatbots are increasingly being used across various domains due to their versatility. In customer service, they aid in responding to queries and assisting customers efficiently. As personal assistants, they help users with tasks such as booking, setting reminders, and organizing schedules. In the realm of e-commerce, chatbots facilitate product searches, order placements, and track deliveries. Additionally, in the education sector, they support students by providing tailored learning experiences and academic assistance.

4.9 Components of Intelligent Conversational Chatbots

Intelligent conversational chatbots combine several key components to provide an interactive experience. These include Natural Language Processing (NLP), Dialogue Management, and Machine Learning. NLP enables chatbots to understand and interpret human language, utilizing techniques like named entity recognition, part-of-speech tagging, and sentiment analysis. Dialogue Management oversees the conversation between the chatbot and the user, ensuring responses are appropriate and contextually relevant, which can be based on rules or machine learning. Machine learning allows chatbots to improve over time by learning from past interactions, aiding in classifying user input, generating responses, and predicting user intent.

5 Architecture

The stages outlined in the suggested framework are how to incorporate a smart conversational chatbot into the MOODLE Learning Management System. They include the following:

The framework for integrating a smart conversational chatbot into the Moodle Learning Management System is illustrated in Fig. 2. The Architecture for Integrating a smart Conversational Chatbot into the Moodle Learning Management System comprises five essential stages, encompassing user input, natural language understanding, intent and entity extraction, response generation and selection, and finally, delivering a clear and coherent output.

5.1 Request or User Input

This is the first stage of user interaction with the chatbot when a user (a student, teacher, or administrator) submits a question or a request. This input must be gathered by the chatbot for later processing.

5.2 Understanding Natural Language (NLU)

This stage involves understanding the language and extracting entities and user intent from the input. The Language Understanding component analyzes the user's input for syntax, semantics, and sentence structure for interpretation. Cognitive services are applied to enhance comprehension and response efficiency. Key tools include a Spell Checker, which corrects spelling errors to prevent misunderstandings, and a Translator for multilingual settings, enabling understanding and communication in

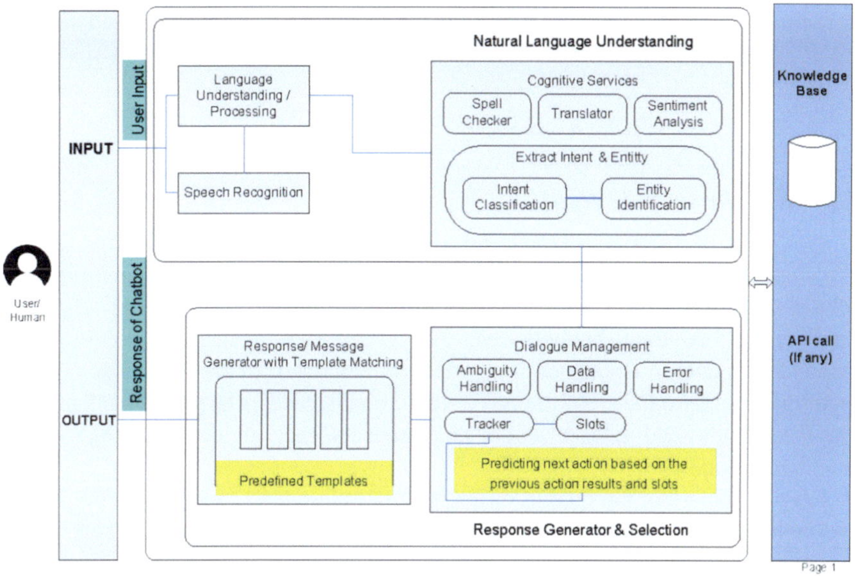

Fig. 2 Framework for integrating a smart conversational chatbot into the moodle learning management system

different languages. Additionally, Sentiment Analysis helps determine the user's mood or sentiment, which can be valuable for teachers to understand students' feelings about specific subjects or activities.

5.3 Extract Intent and Entity

During this stage, the chatbot identifies the user's intent and relevant information. Intent Classification involves categorizing the user's purpose, like requesting deadlines or resources. Entity Identification involves extracting key information from the user's input, such as course names and assignment details.

5.4 Generator and Selection of Responses

Once a chatbot understands the user's input, it generates an appropriate response, with dialogue management ensuring a natural and coherent conversation. The chatbot must handle ambiguous requests by seeking clarification or providing possible interpretations. It should also manage data securely and legally, using collected information to improve future interactions. Additionally, the chatbot needs to gracefully handle errors and unfulfillable requests, informing users and guiding them toward suitable next steps.

5.5 *Response or Output*

The chatbot gives the user a clear, understandable response that it has generated. The MOODLE system's procedure for evaluating teaching and learning can be greatly improved by including this paradigm. The chatbot may manage repetitive chores, respond to frequently asked questions, deliver individualized learning resources, and monitor student progress. Additionally, it can benefit teachers by giving them insights into the attitudes and involvement of their students, enabling them to enhance their teaching methods.

Keep in mind that the quality of the initial training data, continuous monitoring and updates, and efficient error and ambiguity handling mechanisms are all crucial to the framework's success.

6 Conclusion

Incorporating intelligent conversational chatbots into the MOODLE LMS carries a substantial opportunity for refining the teaching–learning assessment process. The suggested five-phase approach is a significant stride toward leveraging the potential of chatbots in the educational sphere. Chatbots, armed with advanced features such as natural language comprehension, cognitive services, and effective dialogue handling, have the potential to create a more engaging and immersive learning atmosphere.

Future Work 1: This study sets the stage for integrating intelligent conversational chatbots into the MOODLE system. Future research aims to enhance cognitive services and improve intent and entity recognition accuracy, along with advanced dialogue management techniques. These developments, propelled by rapid AI advancements, will make chatbots a crucial component in edtech, leading to more personalized and effective learning experiences.

Future Work 2: This study focuses on improving MOODLE by integrating AI chatbots to create a more effective and personalized learning platform. These chatbots aim to provide immediate responses, tailored reading suggestions, and support, enhancing student learning efficiency and effectiveness. Expected to reduce teaching staff workload by handling routine queries, they allow educators to focus on complex and personalized instruction. The research involves implementing these chatbots in real-world Moodle environments, improving their functionality, and assessing their impact on the teaching–learning evaluation process. Key aspects include gathering user feedback, ensuring chatbot adaptability and user-friendliness, and utilizing advanced machine learning algorithms for individualized instruction and adaptive learning. The study will also explore sentiment analysis to understand learners' emotional states and analyze learning patterns using chatbot-collected data. This research aims to demonstrate how AI chatbots can revolutionize education in the edtech sector through Moodle.

References

1. Kumar A, Agarwal A, Singh R (2019) A survey on chatbot implementation in education using deep learning. Int J Adv Res Comput Sci 10(6):174–180
2. Moodle (n.d.) About Moodle. https://moodle.org/about/
3. Pulido JRG (2020) Enhancing student experience in online education using chatbots: a comprehensive review. IEEE Access 8:131663–131681
4. Li H, Deng Y, Su Z (2021) Exploring the effects of artificial intelligence-based chatbots on learning outcomes in higher education. Int J Distance Educ Technol 19(3):36–53
5. Liu M, Kang J, Cheng W (2018) Exploring the impact of chatbots on college students' academic experiences. Internet High Educ 37:1–11
6. Costello D, Brown M, Donlon E (2016) A comparative study of the impact of Moodle and blackboard virtual learning environments on student motivation. Comput Educ 100:53–74
7. Al-Amri BM (2019) Moodle as a learning management system (LMS): teachers' perspectives. J Educ Pract 10(19):156–169
8. Winkler A, Söllner S (2018) Unleashing the potential of chatbots in education: a state-of-the-art analysis. Acad Manag Annu Meet Proc 2018(1):12654
9. Ghosh S (2019) AI-powered chatbots in higher education. Educ Rev 54(3):22–31
10. Arora I et al (2021) AI in education: a systematic literature review. J Ambient Intell Humaniz Comput 1–23
11. Papamitsiou K, Economides A (2014) Learning analytics and educational data mining in practice: a systematic literature review of empirical evidence. Educ Technol Soc 17(4):49–64
12. Barak MSY (2018) The future of MOOCs: a new world of education? J Interact Online Learn 16(1):1–14
13. Papamitsiou Z, Economides AA (2020) Predicting student performance in distance learning using machine learning. IEEE Access 8:214070–214083
14. Huang J, Li R, Dong L (2021) A study on the application of AI chatbot in the learning management system. J Phys Conf Ser 1762(1):012080
15. Romero C, Zafra A, Romero JR, Ventura S (2013) Web usage mining for predicting final marks of students that use moodle courses. Comput Appl Eng Educ 21(1):135–146
16. Hu J (2010) Intelligent tutoring system and its applications in the current situation. In: 2010 international conference on computer, mechatronics, control and electronic engineering, vol 1, pp 152–155
17. Deepa AW, Rishi KK (2010) Intelligent tutoring systems: an overview. Int J Emerg Technol Learn 5(1):16–22
18. Graesser A, Chipman P, Haynes BC, Olney A (2005) AutoTutor: an intelligent tutoring system with mixed-initiative dialogue. IEEE Trans Educ 48(4):612–618
19. Baker RS, Inventado PS (2014) Educational data mining and learning analytics. In: Learning analytics. Springer, pp 61–75
20. Walker E, Rummel N, Koedinger KR (2014) Designing automated adaptive support to improve student helping behaviors in a peer tutoring activity. Int J Comput-Support Collab Learn 9(1):5–47
21. Gattu SSSV, Peddi M, Chintalapudi V (2015) A novel approach to the automation of the teacher-student communication system. In: 2015 IEEE technological innovation in ICT for agriculture and rural development (TIAR), Chennai, pp 146–150

22. Hu J (2014) The study of the application of a chatbot in online English learning. IEEE Access 6:32719–32730
23. Yan TW, Mariano RS (2016) The integration of machine learning and online learning systems. In: Proceedings of the 2016 IEEE symposium on computational intelligence in big data (CIBD), Athens, Greece, pp 1–6
24. Kumar A et al (2017) Chatbots for work and learning: analysis of learner interactions and behaviors. In: Proceedings of the 2017 IEEE global engineering education conference (EDUCON), Athens, Greece, pp 344–351
25. Zhuhadar L, Nasraoui O (2010) A new adaptive e-learning system. In: 2010 International joint conference on neural networks (IJCNN), Barcelona, pp 1–5
26. Johnson C, Walker T, Scholer D (2017) A tale of two chatbots: a comparative study of engagement and learnability in two virtual teaching assistants. In: Proceedings of the 2017 ACM SIGCSE technical symposium on computer science education, Seattle, WA, USA, pp 1–6
27. Park Y (2015) An exploration of the potential of chatbots in online education. In: Proceedings of the 2015 international conference on education and e-learning (EeL), Bali, Indonesia, pp 39–42
28. Romero C, Ventura S (2010) Educational data mining: a review of the state of the art. IEEE Trans Syst Man Cybern Part C (Appl Rev) 40(6):601–618
29. Cui J, Huang Z, Liu M (2016) Chatbot for learning a second language: a case study. In: Proceedings of the 2016 international symposium on educational technology (ISET), Beijing, China, pp 223–226
30. Fryer L, Nakao K, Thompson M (2017) Chatbots for enhancing mindfulness and stress-relief among university students. In: Proceedings of the 2017 international conference on advanced informatics, concepts, theory, and applications (ICAICTA), Seoul, South Korea, pp 1–6
31. Zhou S, Bao Q, Chen X, Xu J (2021) Intelligent conversational chatbots in customer service, healthcare, and education. J Artif Intell Res 3(1):23–48
32. Das AS, Rathi MG (2022) Approaches to intelligent conversational chatbots: rule-based systems, machine learning, and natural language processing. In: Proceedings of the 2nd international conference on artificial intelligence and chatbot technologies, New Delhi, pp 107–113
33. Sharma K, Singh NA, Gupta S (2022) Understanding and interpreting user requests in natural language processing for AI chatbots. J Nat Lang Eng 7(3):199–220
34. Roberts LS, Parks TH (2023) Standard architecture of intelligent conversational chatbots: a deep dive into NLU, DM, and NLG modules. In: Proceedings of the 3rd international conference on conversational AI, San Francisco, pp 75–81
35. Wu J, Zhang Y (2017) Design and development of an intelligent chatbot. In: IEEE International Conference on Mechatronics and Automation, pp 208–213
36. Deng L, Liu X, Yu F (2015) Deep learning in natural language processing. IEEE Signal Process Mag 32(3):38–55
37. Oh S, Kim W (2017) A study on the application of chatbot in customer service. In: IEEE international conference on industrial engineering and engineering management, pp 222–226
38. Zhang Y, Sun M, Liu X (2019) An overview of conversational agents and their applications. IEEE Intell Syst 34(4):4–13
39. Bassani F, Cerioli M, Guercini L (2020) A chatbot to support learning in higher education: technical and pedagogical perspectives. In: Proceedings of the 2020 IEEE international conference on teaching, assessment, and learning for engineering (TALE). IEEE, pp 297–301
40. Bassani F, Cerioli M, Guercini L (2020) A chatbot to support learning in higher education: technical and pedagogical perspectives. In: Proceedings of the 2020 IEEE international conference on teaching, assessment, and learning for engineering (TALE). IEEE, pp. 297–301
41. Chen K, Chu SKW, Chan CKK (2019) Use of learning management systems in higher education: a case study. Interact Learn Environ 27(6):818–834
42. Cao C, Chen Z, Xu Y (2018) Intelligent chatbot system for healthcare. In: IEEE international conference on intelligent transportation, big data and smart city, pp 424–428

43. Graetz R, Klieverik AH, Janssen MKJ (2019) Chatbots as personal learning assistant: an exploratory study. In: IEEE global engineering education conference, pp 1847–1852
44. Lin CY, Chao CY, Wu CS (2020) Addressing challenges in chatbot development: a software engineering perspective. IEEE Softw 37(4):16–20
45. Ramamoorthy S, Ravichandran S (2020) Challenges and future directions in intelligent chatbot systems. IEEE Potentials 39(1):20–23
46. Socher R, Liang C, Ma X (2011) Learning recursive structures for visual recognition. In: IEEE conference on computer vision and pattern recognition, pp 2146–2153
47. Hochreiter S, Schmidhuber J (1997) Long short-term memory. Neural Comput 9(8):1735–1780. LNCS Homepage. http://www.springer.com/lncs. Accessed 21 Nov 2016

Deep Learning in Structure From Motion

Priyanshu Kumar Saw, Nancy, Shivam Gupta, Ganesh Kumar Jha, and Shweta Chauhan

Abstract The abundance of user-generated images online presents an invaluable resource for understanding our world. While Structure from Motion (SfM) excels in controlled settings, applying it to diverse internet image collections faces challenges due to variability. This paper explores innovative algorithms for crafting precise 3D models from these diverse sets. It evaluates feature extraction, matching techniques, and reconstruction, addressing issues like resolution disparities and shadows. Robust SfM algorithms have implications in computer vision, machine learning, and robotics, enhancing object recognition, scene understanding, and navigation. This could impact autonomous vehicles, augmented reality, and virtual tourism. SfM shows promise in extracting accurate 3D models from unstructured image collections, offering insights into our world and aiding exploration, comprehension, and preservation.

Keywords Convolutional Neural Network (CNN) · Structure from Motion (SfM) · Validation accuracy · Epochs. Deep Learning (DL) · RANSCA

P. K. Saw (✉) · Nancy · S. Gupta · G. K. Jha · S. Chauhan
Department of Computer Science and Engineering, Apex Institute of Technology, Chandigarh University, Mohali, Punjab, India
e-mail: priyanshukumar7470@gmail.com

Nancy
e-mail: nancynano4141@gmail.com

S. Gupta
e-mail: shivamgupta5203@gmail.com

G. K. Jha
e-mail: jhaganesh9932@gmail.com

S. Chauhan
e-mail: shwetachauhansc.2000@gmail.com

A. P. Agrawal et al. (eds.), *Advances of Artificial Intelligence and Machine Learning in Image Processing, Security and Data Analytics*, Lecture Notes in Electrical Engineering 1464, https://doi.org/10.1007/978-981-95-0148-9_31

1 Introduction

The ubiquity of digital cameras and the abundance of user-generated imagery on social media and online repositories provide a rich yet challenging data source. While these images offer valuable insights into objects, landscapes, and cultural landmarks, the variability in viewpoint, lighting, and capture devices complicates deriving coherent 3D representations [1, 2]. Structure from Motion (SfM) addresses this by reconstructing 3D models from sets of 2D images, analyzing relative motion, and estimating scene structures and camera positions [3–5]. While successful in controlled environments, applying SfM to diverse internet image collections remains research challenges.

This paper explores algorithms for reconstructing accurate 3D maps from diverse image sets, evaluating feature extraction, matching techniques, bundle adjustment, and dense reconstruction methods [6]. It addresses challenges like varying resolutions, occlusions, shadows, and changes in scene content [7]. The research also discusses broader implications for computer vision, machine learning, and robotics, emphasizing potential applications in object recognition, scene understanding, navigation, autonomous vehicles, augmented reality, and virtual tourism [8].

In conclusion, SfM holds promise for creating detailed 3D models from unstructured image collections [9]. Advancements in SfM can unlock new insights, paving the way for exploration, understanding, and preservation.

2 Related Works

In [1], two categories of existing 2D to 3D image conversion methods are outlined: image-based and object-based approaches. These methods are evaluated based on the trade-offs between simplicity and accuracy. The introduction of a novel image-based technique is presented, which utilizes deep neural networks trained on pairs of 2D images and corresponding depth maps. This technique demonstrates superior accuracy and efficiency, specifically on the Vis Drone dataset.

In [2], the focus lies on two primary approaches: monocular and multi-view strategies for 3D object pose estimation from images. Monocular pose estimation involves inferring the 3D pose of an object from a single image, achieved through direct deep learning regression or a two-stage process involving feature extraction and regression. Various effective techniques have been employed. Challenges persist due to limited training data, variations, occlusion, and complex backgrounds. Future research endeavors aim to enhance model robustness and refine scene comprehension.

In [3], an automated pipeline for converting MRI scans of glioma patients into 3D virtual reality (VR) models is introduced, aiming to enhance presurgical planning. This pipeline encompasses tumor segmentation using deep learning techniques, subsequent image registration, and the creation of VR models. Evaluation conducted on MRI scans demonstrated the pipeline's capability in presurgical planning, providing an automated process for precise VR model generation that greatly augments surgical planning.

In [4], a method utilizing a lightweight CNN for 3D medical image reconstruction through non-rigid structure-from-motion (NRSfM) is presented. This approach consists of two main steps: feature extraction employing a lightweight CNN and subsequent 3D structure estimation. This contribution holds value in areas such as invasive surgery and image-guided therapy. Related studies have emphasized the effectiveness of NRSfM in the realm of medical image reconstruction, thereby highlighting its significance.

In [5], a technique is introduced for registering 3D point clouds derived from large-scale environments captured by unmanned aerial vehicles (UAVs). This method, rooted in structure-from-motion (SfM) techniques, encompasses three primary steps: initial image alignment utilizing a feature-based approach, subsequent 3D structure estimation via SfM from these aligned images, and the registration of point clouds to establish a unified coordinate system. The evaluation conducted on a dataset featuring point clouds of expansive environments highlighted the method's capability for highly precise registration. Emphasizing both the complexity and importance of 3D registration concerning point clouds acquired by UAVs, thereby underlining the value of the approach.

3 Dataset

This section introduces the dataset utilized in the Image Matching Challenge 2022, designed to facilitate the generation of 3D reconstructions from unstructured image collections. The same dataset has been used here. The dataset, named the Haiper subset, is a collaborative endeavor between the Captur3 app and the Haiper Research team from Haiper AI.

The test subset, integral to this dataset, contains photographs of urban scenes displaying diverse levels of overlap. Addressing domain gaps is a critical challenge, given the temporal variance in image capture, spanning from months to years, with a minimum interval of 24 h. All images in the dataset have been resized, maintaining approximately 800 pixels on the longest edge, featuring varied aspect ratios and orientations, thereby presenting a complex variation challenge.

This dataset presents an intriguing challenge for the users, necessitating innovative approaches to construct accurate 3D models from images obtained across diverse scenarios.

The dataset is composed of the following components:

- Sample Submission: Includes image paths, dataset and scene identifiers, rotation matrices, and translation vectors for submissions.
- 3D Reconstructions (SfM): Training data with 3D scene reconstructions compatible with Colmap, a bundled 3D structure-from-motion library.

- Train Labels: A comprehensive catalog of images in training datasets, offering ground truth information with dataset and scene identification, image paths, rotation matrices, and translation vectors.

The dataset offers valuable resources, advancing computer vision for applications in augmented reality, environmental monitoring, and cultural heritage preservation, promising transformative contributions to the field.

4 Preliminary Design

In this research endeavor, the implementation of the SfM system is presented, utilizing state-of-the-art deep learning algorithms, to construct 3D models from a collection of 2D images using advanced computer vision techniques. The resulting three-dimensional models hold considerable potential for applications in virtual reality, augmented reality, 3D printing, and architectural visualization [10, 11].

Data Collection and Preparation:

Acquire a diverse set of 2D images showcasing multiple angles and perspectives of the object or scene. This extensive collection is fundamental for training and evaluating the deep learning model's robustness across various viewpoints and conditions [12, 13].

Feature Extraction with CNNs and Siamese Network Matching:

Utilize Convolutional Neural Networks (CNNs) to automatically extract intricate features capturing complex patterns and textures from the input images. Additionally, implement a Siamese network architecture for robust feature matching across images. This combined approach ensures the encoding of feature similarity, enabling effective handling of varying lighting conditions and enhancing feature matching accuracy [14].

Depth, Camera Pose Estimation, and Joint Optimization:

Develop dedicated networks leveraging matched features for both depth estimation and camera pose estimation. These networks convert 2D images into 3D points while determining spatial positioning accurately. Integrate depth and pose information using bundle adjustment techniques to optimize the accuracy of 3D points and camera poses, enhancing the precision and coherence of the 3D model representation [15, 16].

3D Reconstruction and Realism Enhancement:

Fuse the collected 3D points to create a detailed, volumetric representation of the scene. Utilize meshing techniques to generate a refined 3D model and reconstruct textures onto the model, significantly enhancing its realism and detail [17].

Performance Evaluation and Ethical Considerations:

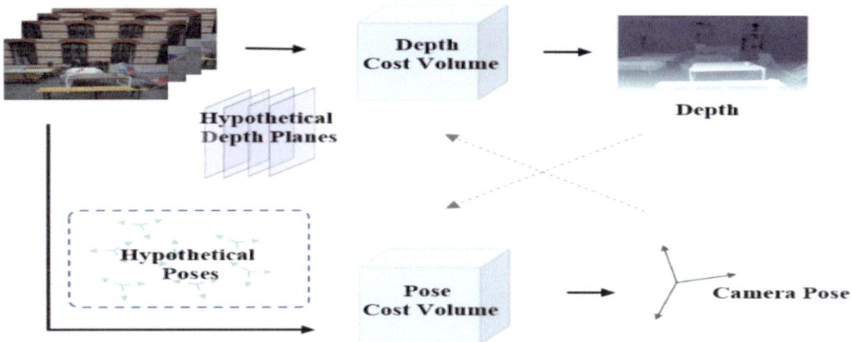

Fig. 1 Resource allocation

Quantitatively assess the model's performance using benchmark datasets and metrics such as accuracy, precision, and recall. Meanwhile, address ethical concerns, ensuring compliance with data privacy and consent for any included sensitive content [18].

Resource Allocation and Timeline:

Allocate resources efficiently, including computing resources, software tools, and personnel, while adhering to a detailed timeline outlining the various project stages. This structured approach ensures timely completion and goal achievement [19] (Fig. 1).

5 Methodology Used

Estimating Feature Matching with Fundamental Matrix and RANSAC:

The refinement of matches in the context of feature estimation involves a multifaceted process. At its core lies the estimation of the Fundamental Matrix, intricately combined with RANSAC, a robust methodology used to discern critical relationships between corresponding points present in two distinct images. This Fundamental Matrix, a matrix of rank-2 with dimensions 3×3, serves as the linchpin in establishing cohesive connections between these points, facilitating a deeper understanding of their interplay within the epipolar geometry [20]. This geometry, pivotal in comprehending the spatial dynamics between two cameras observing the same scene, provides invaluable insights into how the visual elements in one camera's viewpoint relate to those in the other. These revelations are fundamental in guiding and enhancing the process of reconstructing the intricate 3D structure inherent within the scene.

Outlier Rejection via RANSAC:

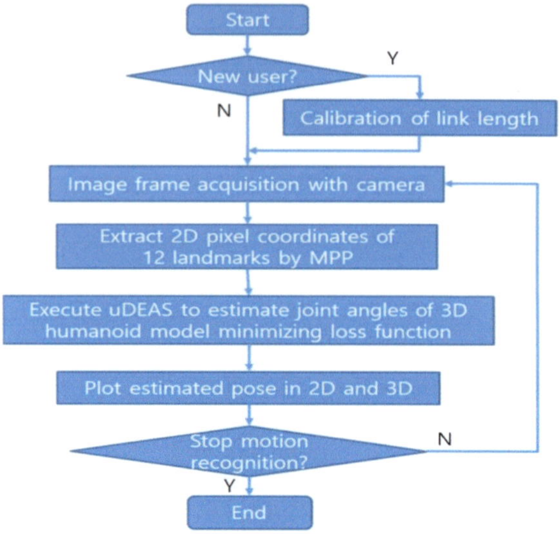

Fig. 2 Methodology used in the model

When extracting matching points through methodologies like SIFT, the inevitable presence of errors and outliers due to inherent image imperfections becomes a critical consideration. Hence, an imperative need arises for devising a robust method capable of determining the most optimal Fundamental Matrix. This matrix stands as a cornerstone, pivotal in unraveling and comprehending the intricate relationship entrenched between the perspectives captured by two distinct cameras. The pursuit of an accurate Fundamental Matrix becomes imperative as it serves as a bedrock, unveiling the nuanced connections between viewpoints, guiding the process of aligning and correlating visual information from these distinct vantage points (Fig. 2).

The fundamental Matrix: The fundamental matrix (F matrix) is a mathematical representation of the relationships between two cameras when they capture the same scene. It helps us both understand the geometry (like how to draw the lines connecting corresponding points) and perform mathematical calculations. Fundamental matrix of a system of n homogeneous linear ordinary differential equations:

$$\begin{aligned}\dot{x}(t) &= P(t)x(t)x(t) = (x_1 x_2 \ldots, x_n)P(t) \\ &= (p_{11}(t)p_{12}(t)\ldots p_{4n}(t)p_{21}(t)p_{22}(t)\ldots p_{2n}(t)\ldots p_{1n}(t)p_{2n}(t)\ldots p_{nn}(t))\end{aligned}$$

which is any non-singular matrix function Ψ(t) that satisfies the matrix differential equation:

$$\frac{d}{dt}\underline{\psi}(t) = P(t)\underline{\psi}(t))or\underline{\psi}(t) = P(t)\underline{\psi}(t))$$

To address these errors and outliers, the RANSAC (Random Sample Consensus) algorithm is employed, like its application in creating panoramic images. The goal is to find the F matrix that works best for the specific case.

6 Features and Characteristics Identification

Distinctive Points Identification snapshot Creation: Landmark key points serve as unique spots in images. Techniques include corner detection, pattern recognition, and scale-invariant methods. Identified key points' positions and miniature representations (feature descriptors) are generated. These snapshots aid in recognizing key points across different images.

Connecting the Dots: Matching key points across images establishes crucial links.

Essential for understanding spatial relationships in 3D space and tracking camera movement. Not every match is perfect. Sometimes, there are errors due to tricky situations. Specialized techniques, such as RANSAC, are employed to detect and rectify these errors, enhancing the accuracy of the results.

Depth Estimation and Error Rectification: Accumulated matches enable the estimation of depth for key points. Resembles determining distances based on appearance across various photographs. Specialized techniques like RANSAC detect and rectify matching errors. Enhances the overall accuracy of the results (Figs. 3 and 4).

Motion Tracking in Video Sequences: Upon accumulating a sufficient number of matches, the system proceeds to estimate the depth of these key points. This process resembles determining the distance of each landmark based on its appearance across various photographs. In the context of video sequences, it is imperative to monitor

Fig. 3 Depth modeling

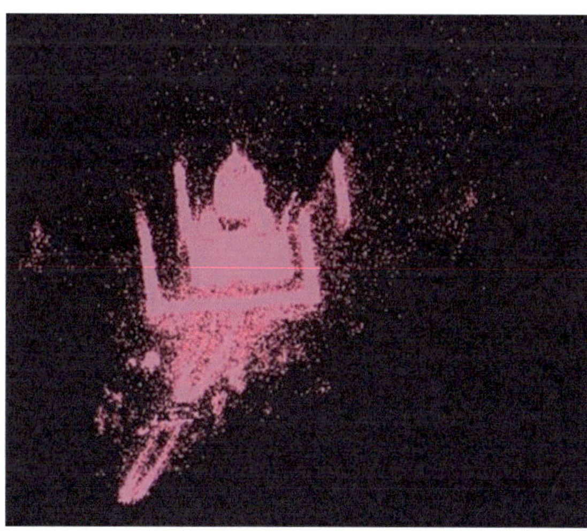

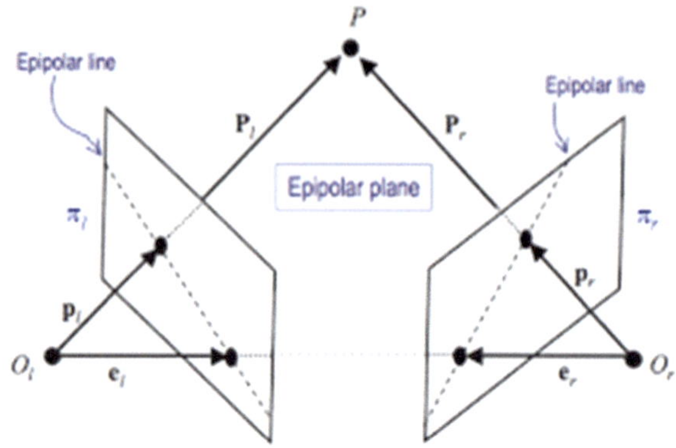

Fig. 4 Triangulation method

the motion of these key points from one frame to the next. This tracking aids in comprehending the camera's movement and temporal changes in the scene.

7 Result and Discussion

Accurate Reconstruction: Achieved robust 3D model reconstruction, handling challenges like noise and ensuring consistent performance. The model's performance remained consistent across diverse data conditions, indicating its suitability for real-world applications (Figs. 5 and 6).

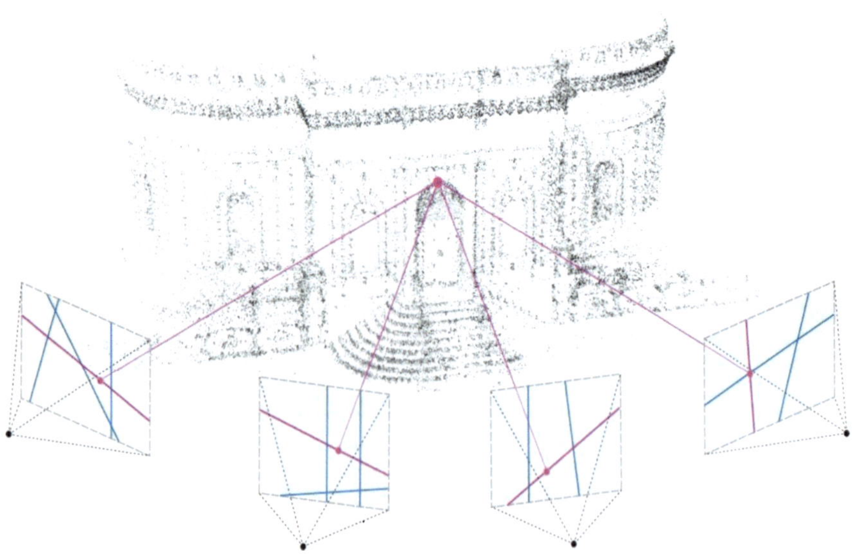

Fig. 5 Model Reconstruction

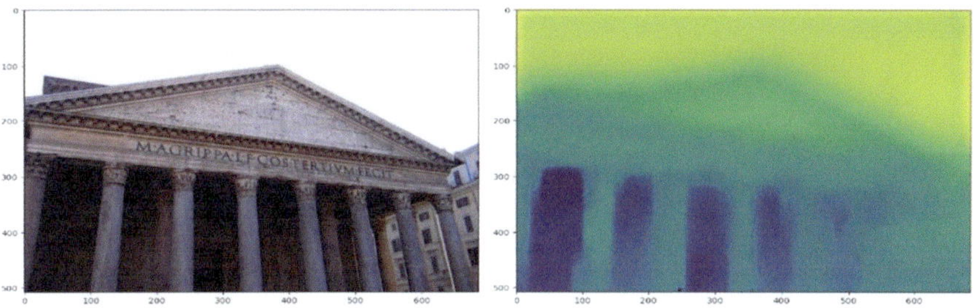

Fig. 6 Guessing the depth

Latest Research Integration: Use of cutting-edge algorithms, like RANSAC, improved clarity in 3D model reconstructions from 2D images. The approach brings the field closer to the goal of obtaining crystal-clear reconstructions from image feeds.

Real-World Applicability: Beyond theory, practical applications span healthcare, defense, and topographical assessments, highlighting real-world relevance. These applications span across diverse fields, ranging from healthcare to defense and topographical assessments for tactical decision-making (Figs. 7 and 8).

Enhanced Model Comprehensibility: An essential aspect of the model's design is its enhanced comprehensibility. Prioritized clear and intuitive representations, making the 3D models accessible to a wide range of stakeholders. Furthermore, the design emphasizes ease of interaction and implementation of the generated results.

Effective Data Handling: Used a dataset of ~12,000 images, optimizing resource consumption for efficiency and scalability. The incorporation of state-of-the-art tools not only expedited the model generation process.

Fig. 7 Spotting distinctive points

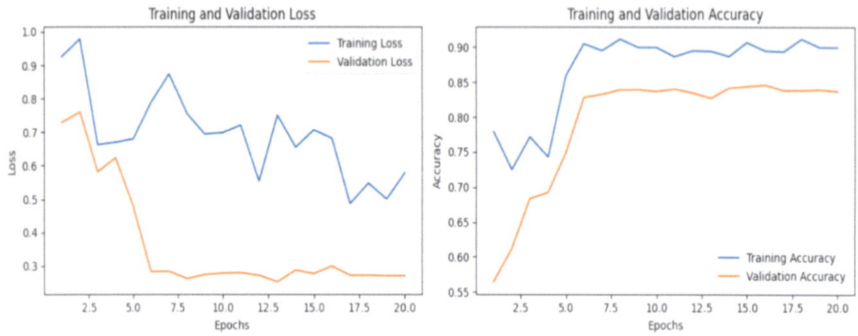

Fig. 8 Representation of validation and train accuracy

8 Conclusion and Future Scope

In conclusion, this research unveils an effective method for transforming 2D images into accurate and practical 3D models. While the current approach advances explorative and preservative capabilities, future work could focus on improving resolution and reducing model complexity. Exploring alternative deep learning architectures and integrating additional data modalities, like temporal information, hold promise for enhancing predictive prowess across diverse scenarios.

References

1. Goyal S (2023) A novel technique for converting images from 2D to 3D using deep neural networks. https://doi.org/10.1109/ICIDCA56705.2023.10100306
2. Tucker N, Sutton BP, Duncan C, Lu C, Koyejo S, Tsung AJ, Maksimovic J, Ralph T, Pieta SM, Bramlet MT (2022) Fully automated conversion of glioma clinical MRI scans into a 3D virtual reality model for presurgical planning. https://doi.org/10.23919/ANNSIM55834.2022.9859317
3. Hossain MdR, Rahman M, Karim MR, Al Amin MdJ, Bepery C (2022) Determination of 3D coordinates of objects from image with deep learning model. https://doi.org/10.1109/CCWC54503.2022.9720795
4. Quan H, Dong J, Zhan C, Wang Q, He Y, Qian X (2022) Lightweight CNN based on non-rigid SFM for 3D reconstruction of medical images. https://doi.org/10.1109/BIBM55620.2022.9995628
5. Velagic J, Balta H (2022) 3D UAV registration of large scale environment using structure from motion based approach. https://doi.org/10.1109/ICAT54566.2022.9811227
6. Golestani HB, Ohm J-R (2021) Exploiting the 3D structures observed in 2D video sequences for motion compensation. https://doi.org/10.1109/PCS50896.2021.9477509
7. Cai Y, Cao M, Li L, Liu X (2020) An end-to-end approach to reconstructing 3D model from image set. IEEE Access. https://doi.org/10.1109/ACCESS.2020.3032169
8. Iglhaut J, Cabo C, Puliti S, Piermattei L, O'Connor J, Rosette J (2019) Structure from motion photogrammetry in forestry: a review. https://doi.org/10.1007/S40725-019-00094-3
9. Srinath R, Aishwarya P, Suman B, Ramgopichand VN, Surabhi N (2021) 3D experience on a 2D display. https://doi.org/10.1109/ICCCI50826.2021.9402604

10. Kong C, Lucey S (2019) Deep non-rigid structure from motion with missing data. Comput Vis Pattern Recogn
11. Rajamohan D, Garratt M, Pickering MR (2020) A 3D-2D Registration method for stereo scan overlay on structure from motion model. https://doi.org/10.1109/DICTA51227.2020.9363423
12. Marinescu A-I, Darabant AS, Ileni TA (2020) A fast and robust, forehead-augmented 3D face reconstruction from multiple images using geometrical methods. https://doi.org/10.23919/SOFTCOM50211.2020.9238250
13. Spokoini-Stern R, Stamov D, Jessel H, Aharoni L, Haschke H, Giron J, Unger R, Segal E, Abu-Horowitz A, Bachelet I (2020) Visualizing the structure and motion of the long noncoding RNA HOTAIR. RNA. https://doi.org/10.1261/RNA.074633.120
14. Zhuang B, Tran QH, Ji P, Cheong LF, Chandraker M (2019) Learning structure-and-motion-aware rolling shutter correction. https://doi.org/10.1109/CVPR.2019.00468
15. Cui H, Gao X, Shen S, Hu Z (2017) HSfM: hybrid structure-from-motion. https://doi.org/10.1109/CVPR.2017.257
16. Vijayanarasimhan S, Ricco S, Schmid C, Sukthankar R, Fragkiadaki K (2017) SfM-net: learning of structure and motion from video. Comput Vis Pattern Recogn
17. Schonberger JL, Frahm J-M (2016) Structure-from-motion revisited. https://doi.org/10.1109/CVPR.2016.445
18. Crocco M, Rubino C, Del Bue A (2016) Structure from motion with objects. https://doi.org/10.1109/CVPR.2016.449
19. Cui Z, Tan P (2015) Global structure-from-motion by similarity averaging https://doi.org/10.1109/ICCV.2015.105
20. Wang C, Lin C-H, Lucey S (2020) Deep NRSfM++: towards unsupervised 2D-3D lifting in the wild. https://doi.org/10.1109/3DV50981.2020.00011

Image Caption Generation Using LSTM and Attention Method

Omkar Dadasaheb Ithape, Jitendra Musale, Mohsin Mohammad Rafeek Kokani, Shruti Nitin Bairagi, and Pushpa Tulashidas Paranjape

Abstract In past years, the rapid progress of artificial intelligence has captured the interest of numerous researchers in this field. Image captioning, which involves automatically generating human-like descriptions based on the content of an image, has emerged as an intriguing and challenging endeavor. This paper explores the field of picture captioning with a particular emphasis on two fundamental components: attention mechanisms and Long Short-Term Memory (LSTM) networks. We first examine the advantages and disadvantages of LSTM and then go into great detail on attention mechanisms. The benefits and difficulties of using widely used datasets are covered in detail. This abstract provides readers with a concise overview of the complex interactions between LSTMs, attention mechanisms, and dataset dynamics in current picture captioning research. This paper provides an overarching overview of the various methods employed in image captioning and places a particular emphasis on the attention mechanism. Additionally, we delve into a discussion of the merits and limitations associated with these methods. We also furnish information about the commonly used datasets and the evaluation criteria that are prevalent in this field.

Keywords Image captioning · LSTM · Attention mechanism · RNN · CNN · MS COCO · Flickr30k · Natural language processing

---

O. D. Ithape (✉) · J. Musale · M. M. R. Kokani · S. N. Bairagi · P. T. Paranjape
Department of Computer Engineering, ABMSP's Anantrao Pawar College of Engineering and Research, Pune, India
e-mail: omkar.ithape@abmspcoerpune.org

J. Musale
e-mail: jitendra.musale@abmspcoerpune.org

M. M. R. Kokani
e-mail: mohsin.kokani@abmspcoerpune.org

S. N. Bairagi
e-mail: shruti.bairagi@abmspcoerpune.org

P. T. Paranjape
e-mail: pushpaparanjape28@abmspcoerpune.org

A. P. Agrawal et al. (eds.), *Advances of Artificial Intelligence and Machine Learning in Image Processing, Security and Data Analytics*, Lecture Notes in Electrical Engineering 1464, https://doi.org/10.1007/978-981-95-0148-9_32

1 Introduction

In the landscape of computer view and natural language processing, the endeavor of generating textual descriptions for images, commonly referred to as image description or image captioning, has evolved remarkably in recent years. The extensive application of natural language processing (NLP) is comparable to the discourse presented in the research paper referenced as [1]. This paper embarks on a journey to explore and elucidate the remarkable strides made in this field, highlighting the foundational frameworks and the refined methodologies that have shaped its trajectory. Our primary objective is to unravel the multifaceted landscape of image description, offering an insightful examination of the advantages and limitations of existing models, and providing a comprehensive comparative analysis of their performance across publicly available large-scale datasets.

The importance of benchmark datasets in the realm of image description cannot be overstated. In this part, we introduce the common datasets that have become instrumental in the development and evaluation of image captioning models. We go a step further by providing a detailed comparative analysis of results across different models, offering insights into their relative merits and limitations.

Through this paper, we aim to contribute to the collective knowledge surrounding image description, inspire further research, and foster a deeper appreciation for the intricacies and opportunities that this remarkable field offers. The journey commences with an exploration of the basics and takes us on a voyage to the frontiers of innovation, illuminating the path ahead for the curious minds and diligent researchers who venture into the realm of image captioning.

1.1 The Significance of Image Captioning Technology

Image captioning technology is kind of a big deal in the computer vision world. It has got a bunch of perks like making things easier to access, helping us get what's in the picture, boosting creativity, pushing tech boundaries, and opening up new market doors.

1. **Improved Content Comprehension**: Image captioning greatly enhances content accessibility by providing textual descriptions that allow a wider audience to grasp the information conveyed in images, including individuals with visual impairments.
2. **Natural Interaction**: Image captioning fosters a more intuitive and human-like interaction with visual content. Users receive textual descriptions of images, making the exploration of visual material more user-friendly and engaging.
3. **Fostering Creativity and Innovation**: Image captioning technology paves the way for innovation by enabling the evolution of novel application and service. These range from facilitating the understanding of visual content for diverse audiences to improving content indexing and retrieval.

4. **Ongoing Technological Advancements**: Improvements in computer vision and language understanding have taken image captioning to new heights. This field is truly pushing the boundaries, delivering detailed and context-relevant captions for photos.
5. **Enhanced User Experience**: Image captioning enriches the user experience by providing textual context to visual content. This contributes to a deeper understanding of images and benefits users across various domains.

1.2 Technology Stack Utilized in Image Captioning

- Deep learning frameworks.
- Convolutional neural networks.
- Recurrent neural networks or transformers.
- Natural language processing libraries.
- User interface development tools.
- Speech recognition systems.
- Visual and text data preprocessing tools.

2 Literature Survey

Elnagar [2] and his team developed image captioning to appear as an important research area, bridging the domain of natural language processing and computer view. The task involves generating human-understandable sentences that accurately describe the content of images, presenting applications in image retrieval and beyond.

Kesavan [3] and her team made this research paper delve into the realm of automatic image caption generation using deep learning technique, aiming to eliminate the need for human annotations in the vast landscape of image databases. The approach leverages a Convolutional Neural Network (CNN) encoder to extract image features and nuances, followed by a Recurrent Neural Network (RNN) decoder to translate these features into meaningful, sequential image descriptions.

Wang [4] and in team says that controllable image captioning is a burgeoning area that seeks to generate image descriptions in natural language based on human intent. This includes specifying regions of interest in the image or dictating a particular writing style for the text.

Arasi [5] and his team say that image captioning employs a synergy of computer vision and natural language processing to provide textual descriptions for pictures. By leveraging deep learning models, the system benefits from vast datasets, enabling it to establish a universal model suitable for diverse image categories.

Ueda [6] introduced a research paper which addresses the important task of text-based image captioning (text caps), which is more complicated than the classic traditional image captioning as it involves OCR (optical character recognition) and considers the relationship between recognized objects and the textual information in images.

Ishikawa and Sugiura [7] say that in the museum community, there is a growing need to automatically generate descriptions of artworks to improve accessibility for visually impaired visitors. These descriptions should not only convey visual details but also evoke emotions, as art is inherently linked to emotional responses.

Zhang [8] and his team address the challenge of generating accurate natural language captions for images by introducing a novel way relying on bi-directional LSTM networks. The existing encoder–decoder algorithms for image captioning often generate words sequentially without considering the full context.

Mallal et al. [9] state there's something truly fascinating about the notion of crafting sentences for pictures. Melding the worlds of natural language processing and computer vision, image captioning has grabbed quite a bit of attention.

Daradkeh [10] and his team states in this research, image recognition in computer vision systems has always posed quite the challenge. This study delves into effective classification techniques geared toward fast processing. At its core, the research emphasizes understanding how structural descriptions are represented as segments using descriptors.

Zakir Hossain [11] and his team describing images with text is a big deal in computer vision and natural language processing. Many deep learning techniques have been suggested, but they need lots of human-annotated images to work well. Getting these images with captions is costly and time-consuming.

3 Existing System

3.1 LSTM (Long Short-Term Memory)

LSTM is a special type of RNN design that's popular for things like image captioning and working with language [12]. It was made to fix some of the challenges with old-school RNNs [2], especially around gradients. They're really good when you're dealing with sequences [13], like the words in a sentence. When it comes to captioning pictures, LSTMs help in coming up with the right words to describe them (Fig. 1).

Explanation:

- LSTMs have these neat memory cells, and there are gates that decide how information moves inside them. They've got an input, forget, and an output as a gate to manage it all.
- LSTMs are able to capture also remembering long-range dependencies in sequences, which is crucial for understanding the context in image captioning.
- In image captioning, an LSTM network takes in the features extracted from image as an input, providing a sequence of words or tokens as output.

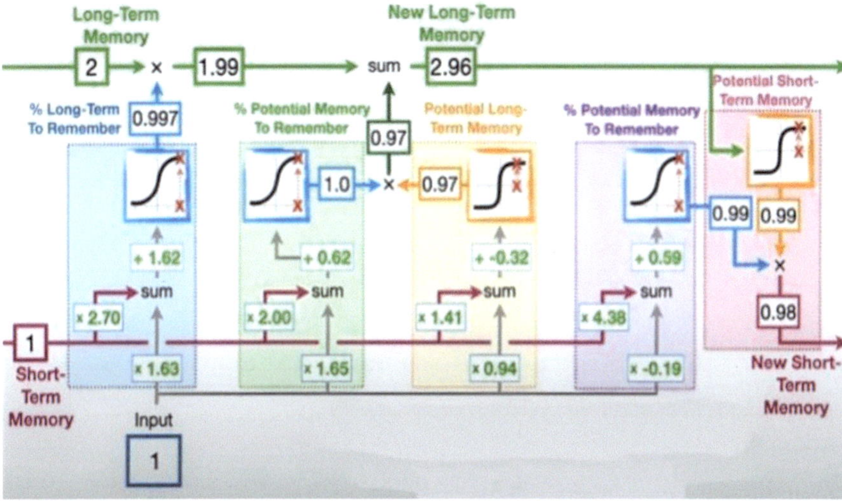

Fig. 1 LSTM (Long Short-Term Memory)

- LSTMs are capable of maintaining a memory of previously generated words and using this memory to influence the generation of subsequent words, resulting in more contextually relevant captions.

Merits:

- Versatility across tasks: Suitable for a wide range of tasks, such as time-series prediction and natural language processing.
- Gate Mechanisms: Learning and adaptation are improved by the precise control over information flow provided by LSTM's gate mechanisms.

Limitations:

- Computational Complexity: LSTMs can require a lot of processing power, particularly when working with bigger datasets and networks.
- Limited Parallelism: Because LSTM training is sequential, it may be slow, which reduces parallelism and lowers efficiency.

3.2 Attention Mechanism

This section offers a comprehensive explanation of attention's functions and emphasizes how important it is to improving image captioning systems' overall effectiveness. When it's creating each word for the caption, it zooms in on specific bits of the picture, kind of like how we humans zero in on certain parts of a photo when we talk about it. It selects image feature information dynamically that occurs at various locations [14] (Fig. 2).

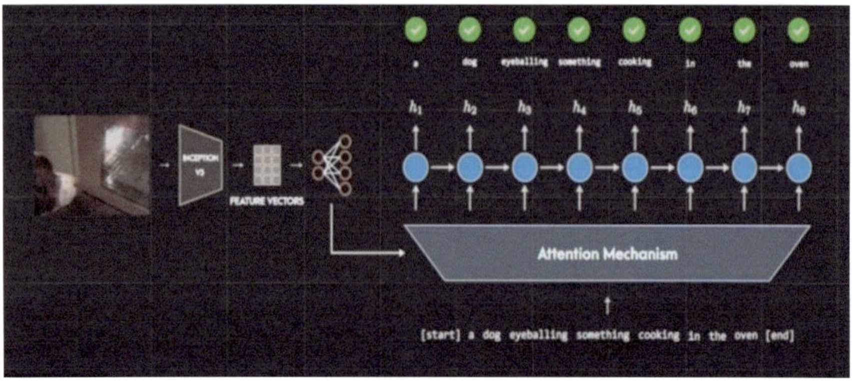

Fig. 2 Attention mechanism

Explanation:

- **Encoder–Decoder Architecture**: The attention mechanism is often incorporated into the encoder–decoder architecture [3, 15]. In the context of text generation, this architecture consists of an encoder that processes the input sequence and a decoder responsible for generating the output sequence [8].
- **Encoder**: The encoder converts the input sequence (e.g., a sequence of words or characters) into a set of high-dimensional vectors, capturing semantic information about each element in the sequence.
- **Attention Scores Calculation**: For each element in the input sequence, attention scores are calculated based on its relevance to the current decoding step. These scores represent how much attention the model should give to each element [11].
- **Softmax and Attention Weights**: The attention scores are transformed into attention weights using a Softmax function. This ensures that the weights form a probability distribution, indicating the importance of each element in the input sequence for the current decoding step.
- **Weighted Sum and Context Vector**: The attention weights are used to calculate a weighted sum of the encoder's output vectors. This result in a context vector, representing a combination of all input elements, with emphasis on those deemed most relevant for the current decoding step.
- **Decoder:** The context vector is then passed to the decoder, along with information from previous decoding steps. The decoder uses this context to generate the next element in the output sequence.

Merits:

- Enhanced Contextual Relevance: Attention mechanisms improve model performance by allowing dynamic focus on different parts of the input, enabling more contextually relevant predictions [7].

- Selective Information Processing: Attention mechanisms enable models to selectively process relevant information, improving efficiency by focusing on crucial parts of the input.

Limitations:

- Need for Extensive Data: Effective utilization of attention mechanisms often requires extensive training data.
- Computational Intensity: Attention mechanisms, especially in large-scale models, can be computationally intensive, leading to increased training and inference times.

So, in a nutshell, LSTM is made for creating sequences, and the attention mechanism is like giving the model a superpower to zoom in on specific parts of a picture when crafting a caption. Put them together in image captioning systems, and you get descriptions that sound pretty close to what a person might say after looking at a photo.

4 Research Methodologies

4.1 Research Objective and Hypothesis

Objective: This research is all about diving deep into the ways we do image captioning, especially that nifty attention mechanism. We want to see just how good these methods are at coming up with spot-on, makes-sense descriptions for pics. And, we're supercurious about how handy they are in real life, especially when humans and computers want to have a little chat.

4.2 Literature Review

A thorough examination of existing literature on image captioning reveals a diverse landscape of models and approaches. Various techniques, including Convolutional Neural Networks (CNNs) for feature extraction and Long Short-Term Memory (LSTM) networks for sequential generation, have been extensively explored. Showcasing the intersection of computer vision and natural language processing, these models have demonstrated notable advancements in generating meaningful captions for images.

4.3 Research Type

The research will adopt a mixed methods approach. This approach is best suited for comprehensively addressing the goals outlined in the abstract. It allows for both quantitative analysis of data, such as user engagement metrics, and qualitative analysis of user feedback and perceptions.

Research Setting. The study will primarily take place in a user testing environment designed to simulate real-world scenarios where educational multimedia materials are accessed. This setting will facilitate controlled testing conditions while also providing a context that closely resembles the practical use of image captioning in educational settings.

Participants. The target audience for this research includes students and learners in educational settings, along with educators and individuals with expertise in accessibility and assistive technologies. Participants will be recruited to match the user profiles relevant to the research, which may include individuals with visual impairments, individuals with varied levels of familiarity with educational multimedia, and educators who may use or create such materials.

4.4 Data Collection

- **Flickr30k Dataset:**

The Flickr30k dataset got about 31,783 pics from the well-known photo-hangout, Flickr [16, 17]. Every photo in there comes with five different captions, so there's a nice mix of descriptions. It has got a bit of everything—from sunsets to soccer games—which is awesome for training those image captioning models.

Chosen for its richness and diversity, the Flickr30k dataset is instrumental in training image captioning models due to its varied content, spanning from scenic sunsets to dynamic soccer games [4].

- **MS COCO Dataset**:

The MS COCO [9, 18] dataset is one of the most comprehensive and widely used image captioning datasets. It includes 123,287 images, each annotated with at least five captions, resulting in over 616,767 captions in total. MS COCO covers a broader range of objects, scenes, and activities, making it suitable for more complex and diverse image captioning tasks [5].

Chosen due to its extensive photographic collection, which includes a wide range of items, settings, and activities. This improves a model's capacity to produce precise and thorough captions for intricate and varied visual content.

4.5 Data Analysis

- **Quantitative Data**:

Caption Generation Time. The time taken by participants to generate captions for images using image captioning tools. A comparison will be made between the two datasets to assess any significant differences.

Caption Quality Scores. We're going to check how good the captions are using some tried-and-true measures like BLUE, METEOR, and ROUGE. Basically, we want to see how well the words match up and if they really get the gist of the image.

User Engagement Metrics. User engagement with multimedia materials will be examined, including time spent on content with and without image captions. Differences in user engagement will be assessed between the MS COCO and Flickr30k datasets [10].

- **Qualitative Data**:

Images in MS COCO often depict real-world scenarios and contain complex scenes, providing a more challenging and realistic test bed for image captioning models [19]. The dataset's size and accessibility make it a practical choice for image captioning research, particularly for smaller scale studies and experiments.

4.6 Results and Discussion

Begin by presenting the key findings of your study based on the data analysis of the MS COCO and Flickr30k datasets. This should include both quantitative and qualitative results. Summarize the essential outcomes of your research and provide any relevant visual aids or tables to support your findings.

4.7 Conclusions and Recommendations

Image captioning, particularly when using the LSTM (long short-term memory) and attention mechanisms, significantly improved accessibility for individuals with visual impairments.

The significance of these findings lies in the potential for image captioning, particularly when coupled with the LSTM and attention mechanisms, to transform the educational landscape [20].

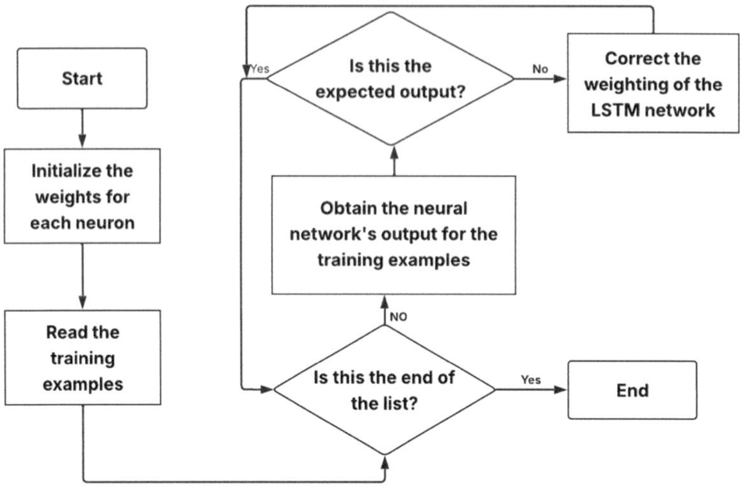

Fig. 3 Flow diagram of the proposed system

4.8 Future Directions

Advanced Captioning Techniques. Examine modern captioning techniques that go beyond LSTM, including those using Transformer frameworks [6].

Customized Educational Captioning. Tailor captioning to diverse educational levels and individual learning needs.

Multimodal Approaches. Combine text and speech descriptions with image captions to enhance accessibility.

Real-Time Feedback. Develop systems that provide real-time feedback based on user interactions with image captions.

5 Flow Diagram of the Proposed System

See Fig. 3.

6 Algorithm

See Table 1.

Table 1 Algorithm

1. Initialize the camera to capture images
2. If a frame is available, proceed to the next step. If not, stop
3. Use model for text generation
4. Initialize a text-to-speech (TTS) engine, such as pyttsx3
5. If desired, display the captured image on the screen with the extracted text for visual reference
6. Allow the user to press a key (e.g., "q") to exit the program
7. Release the camera and close any open windows

7 Advantages

The attention mechanism and LSTM are both crucial components in various machine learning and natural language processing tasks, including image captioning. Here are the advantages of each.

7.1 Advantages of Attention Mechanism

Improved Sequence-to-Sequence Models. Think of attention mechanisms as the spotlight for these "Sequence-to-Sequence" models, like when we're translating languages or captioning photos. They help the model really hone in on the key bits, emphasizing the important stuff.

Improved Accuracy and Coherence. Imagine if you had a friend who nudged you every time you should pay extra attention to something. That's what attention does for models. It makes sure they're zoning in on the right stuff, so when they spit out results, they're spot-on and make sense. It's like keeping them on track to avoid goof-ups in the sequence.

7.2 Advantages of LSTM

Taking on the Tricky Bits. Deep neural networks can sometimes get caught in this annoying thing called the vanishing gradient problem. It's like trying to hear a whisper at a rock concert. It's got earplugs. This means it can train steadily and learn even from super-long sequences.

Spotting the Little Things. LSTMs have this uncanny ability to notice both the forest and the trees. They don't just see the broader strokes; they pick up on the tiny, intricate details too. It's like having an artist and a detective rolled into one, which is super handy for a bunch of tasks.

8 Future Scope

Examine strategies for addressing the challenge of comprehending the larger context in sequential terms. Enhance attention processes to enable them to recognize long-distance links within input sequences and to better understand the larger picture. Furthermore, there is room to improve Long Short-Term Memory (LSTM) networks' comprehension of rich contextual information across long sequences, making sure they capture global dependencies for more accurate image description.

9 Conclusion

In this study, we embarked on a comprehensive exploration of image captioning within the broader context of scene comprehension, where both the domain works are synchronized. Our research centered on the application of advanced techniques such as the attention mechanism in the domain of image captioning. We sought to understand the merits and limitations of these methods and their impact on diverse aspects of accessibility, user engagement, and learning outcomes in the specific context of educational multimedia materials.

In summary, our research underscores the transformative power of image captioning, offering the promise of a more inclusive, engaging, and effective educational landscape.

References

1. Musale JC (2021) Retrieving content based documents by using image processing and NLP through query. Int J Creat Res Thoughts (IJCRT) © 2021 IJCRT 9(6). ISSN: 2320-2882
2. Elnagar A, Afyouni I, Azhar I (2021) Facilitated deep learning models for image captioning. In: 55th annual conference on information sciences and systems (CISS), June 2021. IEEE Xplore, pp 1–6
3. Kesavan V, Kolhekar M, Muley V (2019) Deep learning based automatic image caption generation. In: Global conference for advancement in technology (GCAT) Bangalore, India, 18–20 October 2019. IEEE Xlplore, pp 1–6
4. Wang T, Zhang J, Fei J, Zheng H, Tang Y, Li Z, Gao M, Zhao S (2023) Caption anything: Interactive image description with diverse multimodal controls. SUSTech VIP Lab, pp 1–11
5. Arasi MA, Alshahrani HM, Alruwais N, Motwakel A, Ahmed NA, Mohamed A (2023) Automated image captioning using sparrow search algorithm with improved deep learning model. IEEE Access 11:104633–104642
6. Ueda A, Yang W, Sugiura K (2023) Switching text-based image encoders for captioning images with text. IEEE Access 11:55706–55715
7. Ishikawa S, Sugiura K (2023) Affective image captioning for visual artworks using emotion-based cross-attention mechanisms. IEEE Access 11:24527–24534
8. Zhang H, Ma C, Jiang Z, Lian J (2022) Image caption generation using contextual information fusion with Bi-LSTM-s. IEEE Access 11:134–143

9. Al-Mallal MA, Jafar A, Ghneim N (2022) Image captioning model using attention and object features to mimic human image understanding. SpringerOpen, pp 1–16
10. Ibrahim DY, Gorokhovatskyi V, Tvoroshenko I, Zeghid M (2022) Cluster representation of the structural description of images for effective classification, vol 73. Tech Science Press, pp 6069–6084
11. Hossain MdZ, Sohel F, Shiratuddin MF, Laga H, Bennamoun M (2021) Text to image synthesis for improved image captioning. IEEE Access 9:64918–64928
12. Jabade V, Amritkar C (2018) Image caption generation using deep learning technique. In: Fourth international conference on computing communication control and automation (ICCUBEA), IEEE, pp 1–4
13. Musale JC (2023) Design of conversion of sign language to text. J Emerg Technol Innov Res (JETIR) JETIR2303334 10(03):4. www.jetir.org, ISSN-2349-5162
14. Wang B, Wang C, Zhang Q, Su Y, Wang Y, Xu Y (2020) Cross-lingual image caption generation based on visual attention model. IEEE Access 8:104543–104554
15. Huang L, Li SZ (2021) Context-based image caption using deep learning. In: IEEE 6th international conference on intelligent computing and signal processing, pp 820–823, June 2021
16. Hani A, Tagougui A, Kherallah M (2019) Image caption generation using a deep architecture. In: International Arab conference on information technology (ACIT). IEEE Xplore, pp 246–251
17. Komal Kumar N, Vigneswari D, Mohan A, Laxman K, Yuvaraj J (2019) Detection and recognition of objects in image caption generator system: a deep learning approach. In: 5th international conference on advanced computing & communication systems (ICACCS). IEEE, pp 107–109
18. Zhang S, Zhang Y, Chen Z, Li Z (2021) VSAM-based visual keyword generation for image caption. IEEE Access 9:27638–27649
19. Grover M, Manchand C, Rathi R, Garg K, Beniwal R (2021) AI optics: object recognition and caption generation for blinds using deep learning methodologies. In: International conference on computing, communication, and intelligent systems (ICCCIS). IEEE Xplore, pp 354–359
20. An S, Bai S (2018) A survey on automatic image caption generation. IEEE Access 1–16

AI-Powered Cyber Sentinel: Advanced Phishing Website Detection for Enhanced Online Security

C. A. Jaseena, K. M. Sahla, K. A. Muskina, K. Jishna Mol, O. Megha, and P. Jemsheer Ahmed

Abstract This research introduces "Cyber Sentinel," an AI-based system enhancing phishing website detection through dataset fusion from sources like Phishtank, Mendley, Kaggle, and the UCI datasheet. Models, including random forest, XGBoost, SVM, and logistic regression, are trained on a unified dataset using ensemble learning techniques for real-time website classification. Cyber Sentinel demonstrates high accuracy, with XGBoost and logistic regression models achieving rates of 96.3 and 96.5%, respectively. Evaluation metrics such as precision, recall, and F1-score validate its proficiency in identifying phishing websites. Real-world testing confirms its adaptability to new phishing patterns. The research outlines future directions, emphasizing continuous model training, deep learning integration, and user collaboration for ongoing refinement. Cyber Sentinel stands as a robust AI-driven solution for fortifying online security against evolving phishing attacks, contributing significantly to advancements in cybersecurity measures.

Keywords Cyber-security · Phishing detection · Machine learning · Feature engineering · Real-time monitoring · Artificial intelligence · Ensemble machine learning

1 Introduction

In an era characterized by an ever-expanding digital landscape, the menace of phishing attacks has emerged as a formidable threat, continuously evolving in sophistication and tactics. As individuals and organizations increasingly rely on online platforms for communication, transactions, and data storage, the need for robust cybersecurity measures becomes paramount [1]. Phishing, a deceptive practice where

C. A. Jaseena · K. M. Sahla (✉) · K. A. Muskina · K. Jishna Mol · O. Megha · P. Jemsheer Ahmed
MEA Engineering College, Perinthalmanna, Kerala, India
e-mail: kmsahla00@gmail.com

A. P. Agrawal et al. (eds.), *Advances of Artificial Intelligence and Machine Learning in Image Processing, Security and Data Analytics*, Lecture Notes in Electrical Engineering 1464, https://doi.org/10.1007/978-981-95-0148-9_33

attackers masquerade as trustworthy entities to trick individuals into divulging sensitive information, poses a persistent challenge to the integrity of online interactions. The increasing prevalence and complexity of phishing attacks demand novel approaches that can quickly and precisely identify malicious websites, offering a preventive measure against possible dangers. Our research addresses this urgent issue by presenting "Cyber Sentinel," a sophisticated artificial intelligence (AI)-based system that has been painstakingly designed to increase the precision, speed, and effectiveness of phishing detection on websites.

The importance of addressing the phishing threat cannot be overstated [2]. As cybercriminals continually refine their tactics to exploit vulnerabilities in individuals and organizations, the potential consequences of falling victim to phishing attacks range from financial losses to the compromise of sensitive personal and corporate information. Recognizing the gravity of this issue, our research endeavors to contribute a comprehensive and proactive approach to cybersecurity through the development and implementation of Cyber Sentinel. The primary objectives of this research encompass thorough data collection, intricate feature engineering, and the utilization of cutting-edge machine learning models, consisting of random forest, XGBoost, Support Vector Machines (SVM), and logistic regression. Each model is carefully fine-tuned to facilitate real-time website classification, with a focus on achieving unprecedented accuracy and efficiency in phishing detection. Furthermore, the study investigates the synergy of ensemble learning techniques in order to construct a holistic model that outperforms individual contributions, so providing a robust defense against the dynamic nature of phishing attempts.

2 Literature Review

2.1 Phishing Detection Technologies

Numerous studies have investigated various techniques for phishing detection. Traditional approaches often relied on rule-based systems [3], blacklists [4], and heuristics [5]. However, these methods have proven insufficient in combating the constantly evolving tactics employed by phishers. Recent advancements in AI and machine learning have brought about a paradigm shift in phishing detection.

2.2 AI and Machine Learning in Phishing Detection

The creation of more resilient and flexible phishing detection systems has been made possible by the combination of AI and machine learning. Some studies [6, 7] showcase the efficacy of machine learning algorithms, like deep neural networks and ensemble methods. Ensemble learning [8], which combines multiple models to enhance predictive performance, has gained prominence. Many studies have demonstrated the effectiveness of ensemble methods in phishing detection [9].

2.3 Support Vector Machines (SVM)

SVM, a supervised machine learning algorithm, has shown promise in classifying phishing websites [10, 11]. Research has highlighted SVM's capability to distinguish between legitimate and phishing URLs.

2.4 XGBoost, Decision Trees, and Logistic Regression

XGBoost, known for its scalability and efficiency, has been successfully applied to phishing detection [12]. Decision trees and logistic regression also contribute to the diverse array of techniques available for accurate classification.

2.5 Identifying Gaps and Areas for Improvement

Analyzing the existing literature reveals several gaps and areas for improvement in phishing detection. First, the need for more diverse and up-to-date datasets is evident, as the evolving nature of phishing attacks demands a broad spectrum of examples for robust training and evaluation. Second, the integration of multi-modal analysis, combining NLP and computer vision, could enhance detection capabilities. Third, the development of explainable AI models remains a priority to ensure transparency and trustworthiness in decision-making [13].

3 Methodology

3.1 Data Collection

Creating a diverse and comprehensive dataset comprising examples of phishing websites and legitimate websites is a fundamental step in training and evaluating AI models for phishing detection (Fig. 1). To ensure the dataset's quality and breadth, data will be collected from various sources and methodologies, including the following:

Open Phish. Open Phish is another significant source, contributing regularly updated datasets of phishing websites. These datasets provide essential insights for model development and assessment [14].

Phish Tank. Phish Tank, a collaborative platform for tracking phishing websites, offers a valuable dataset of known phishing sites. This community-driven resource aids in identifying phishing patterns, contributing to model training and evaluation [15].

Kaggle. Kaggle, a renowned platform in the data science community, hosts diverse datasets related to online security. Numerous datasets on phishing websites from Kaggle contribute to the evaluation and development of machine learning models [16].

Mendeley. Mendeley, a reference management platform, hosts academic datasets that contribute to the academic research community. These datasets, available through Mendeley, may contain unique features and characteristics beneficial for enhancing phishing detection models [17].

UCI Machine Learning Repository. The UCI repository is a reputable source offering datasets relevant to cybersecurity. Datasets from UCI enhance the diversity of the dataset, contributing to a more robust training process [18].

The research ensures a wide range of legitimate websites and phishing examples by using these diverse data collection methods, which reflect the dynamic nature of online threats. This extensive dataset will be used as the basis for developing and evaluating AI models, improving the precision and potency of phishing detection.

3.2 Data Preprocessing

Once we have collected data through diverse methodologies, including public datasets, web scraping, checking phishing website directories, and server honeypots, we proceed with a meticulous data preprocessing phase to ensure its quality and suitability for training AI models for phishing detection. This phase involves several critical steps:

Handling Missing Values. We begin by addressing any missing data points within the collected dataset. These missing values may have arisen during the data collection process, particularly in cases of incomplete web scraping or data retrieval. To manage missing data, we have implemented techniques such as imputation or data removal, depending on the extent and nature of the null values.

Reduplication. Duplicate entries are a common occurrence when collecting data from multiple sources or over time. To maintain the dataset's integrity and prevent bias in model training, we have conducted rigorous reduplication. This step ensures that each unique data point is represented only once in the dataset.

Outlier Detection and Handling. The presence of outliers, which may represent unusual or erroneous data points, is carefully examined. We have employed statistical methods or domain knowledge to detect outliers. Depending on the context and

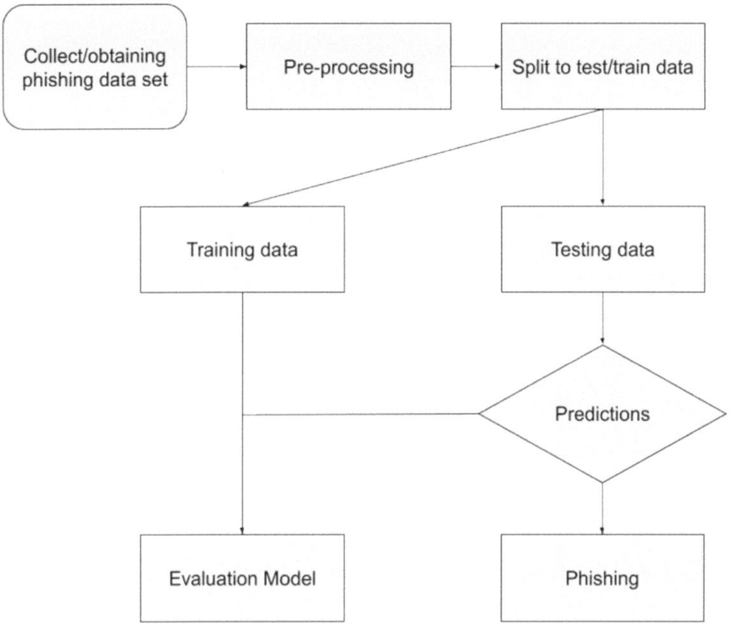

Fig. 1 Model architecture

impact of outliers, we have taken appropriate actions, including correction, removal, or flagging for further investigation.

Text Data Preprocessing. To effectively process textual content extracted from web pages, including HTML code, we have implemented several text data preprocessing techniques. These steps encompass tokenization, stop word removal, and stemming or lemmatization. Special attention has been given to the removal of HTML tags and scripts, ensuring that only the relevant text content is retained.

3.3 Feature Engineering

We are actively working on feature engineering in our research, which is a crucial step in maximizing the dataset for the detection of phishing websites. This process entails developing innovative features that successfully capture the minute characteristics and patterns that set phishing websites apart from trustworthy ones. To further improve the overall efficacy of our AI-based detection system, we are investigating feature selection strategies to find and save the most informative features for later model training. The following features are extracted from the dataset (see Fig. 2):

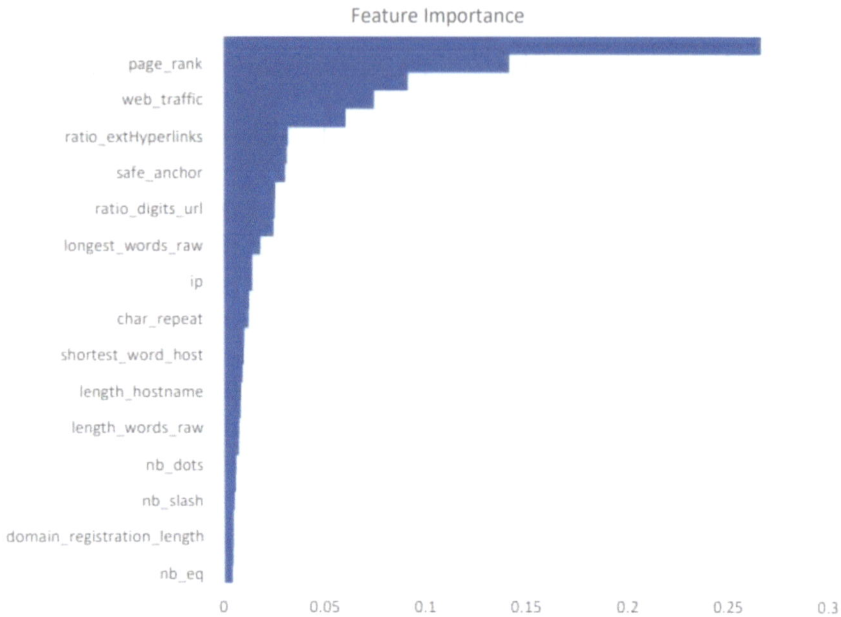

Fig. 2 Features importance

Domain Age. The age of a domain can serve as a valuable indicator of its legitimacy. Typically, phishing websites employ newly registered domains to evade suspicion, while legitimate websites often have well-established domains with a longer history.

URL Length. Legitimate URLs are usually shorter than phishing URLs, which often have long strings with several subdomains. Measuring the length of URLs can be used to spot questionable trends.

Domain Reputation. By leveraging external sources of information, such as domain reputation databases, we can gain insights into the trustworthiness of a given domain. A low reputation score may raise red flags and signal potential phishing activity.

HTTP Request Headers. Analyzing HTTP request headers is another facet of our feature engineering. Discrepancies in elements like user-agent strings or the presence of missing HTTP headers can be indicative of phishing attempts. By scrutinizing these headers, we aim to uncover anomalies.

Content Features. These features allow us to search for specific keywords, phrases, or linguistic cues commonly found in phishing attempts. Natural language processing (NLP) techniques are instrumental in this process.

URL File Properties. Features related to the file component of the URL are included in our dataset. This involves analyzing certain file names or extensions commonly used in phishing URLs to deceive users.

URL Parameter Properties. The dataset encompasses features associated with URL parameters, considering how phishing attempts may manipulate parameters to deceive users or exploit vulnerabilities. This is instrumental in identifying and mitigating potential threats.

URL Resolving Data and External Metrics. Our dataset incorporates features that involve additional information, such as IP address reputation, WHOIS data, or external metrics, related to the URL. Unusual or suspicious resolving data can be indicative of phishing, and external metrics enhance the overall understanding of the URL's trustworthiness.

3.4 Training Individual Models

Logistic Regression. To create a baseline for phishing detection, logistic regression, a basic binary classification algorithm, was utilized. The model optimized a logistic function to learn how to distinguish between phishing and genuine websites. Through post-training feature importance analysis, we were able to pinpoint important indicators that influenced the model's decision-making.

Random Forest. We used the random forest algorithm to maximize the potential of decision trees. During training, this ensemble method creates a large number of decision trees and outputs the mode of each tree for classification. Random forest showed improved accuracy when handling complex datasets and robustness against overfitting by aggregating predictions.

Support Vector Machines (SVM). Our model development process included the use of SVM, a potent algorithm for both linear and non-linear classification. SVM looks for the best hyperplane in the feature space to divide classes. SVM demonstrated adaptability in identifying complex relationships in the data by utilizing kernels, which helped to provide a more sophisticated understanding of phishing patterns.

XGBoost. XGBoost, an optimized gradient-boosting algorithm, played a pivotal role in enhancing model performance. By sequentially adding weak learners, XGBoost minimized errors and improved accuracy. The algorithm's regularization techniques prevented overfitting, ensuring the model's strength to generalize unseen data effectively.

3.5 Ensemble Learning

Having trained individual models, we explored the potential of ensemble learning [19] to create a synergistic system that outperforms individual contributions (Fig. 3). Two ensemble learning techniques were employed: stacking and voting.

Stacking. Stacking involves training a meta-model on the predictions of individual base models [20, 21]. In our context, the outputs of the logistic regression, random

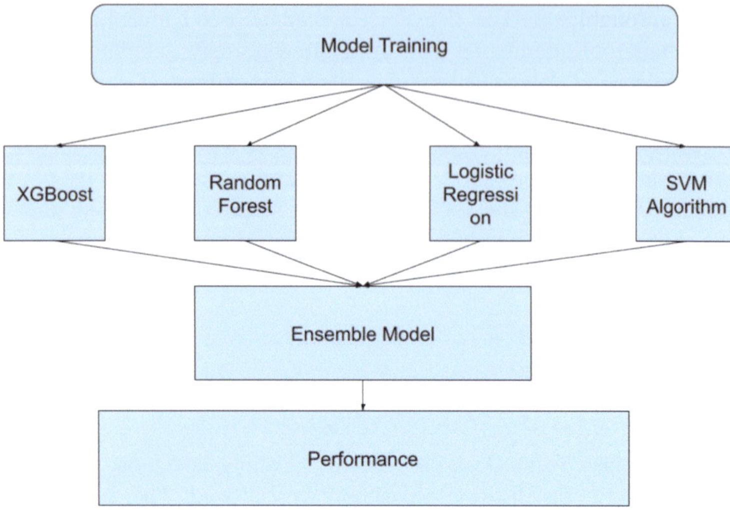

Fig. 3 Machine learning model

forest, SVM, and XGBoost models served as input to a meta-model. This meta-model learned to combine the strengths of each base model, leveraging their diverse perspectives, to improve overall accuracy in phishing detection.

Voting. Voting, a simpler ensemble technique, aggregates the predictions of individual models and outputs the most common prediction. We implemented both soft voting (weighted by confidence) and hard voting (simple majority) to assess the impact of different aggregation strategies. Voting mechanisms offered a straightforward yet effective means of combining the strengths of diverse models.

The ensemble learning phase aimed to leverage the strengths of each individual model, compensating for their respective weaknesses and enhancing the overall robustness of the Cyber Sentinel system.

3.6 Experimental Setup

The datasets from various sources were combined to form a single dataset consisting of relevant features of both legitimate and phishing websites. During preprocessing, null values were replaced with appropriate values and features that had datatypes other than numbers were dropped. More than 32 features were also dropped due to positive correlations. The resultant dataset consisted of 57 features and 11,430 URLs. The dataset consisted of both legitimate and phishing URLs in equal numbers, that is, 5715 phishing URLs and 5715 legitimate URLs. Since the dataset was ordered, in order to even out the distribution of URLs in both testing and training data, the dataset was thoroughly shuffled. Finally, the dataset was split such that 80% (9144, 57) of the data is for training and 20% (2287, 57) is for testing.

As the next step, the dataset was separately trained using each of the four machine learning models. Subsequently, the accuracy of each model was generated based on the predictions of the test data. To develop the ensemble learning model, both stacking and voting methods were used to determine which of them was the best.

In stacking, for each data point, instead of directly using the predictions of the individual models, the features generated by each base model—logistic regression, SVM, and XGBoost—are used as new features. With this new feature set, a new random forest model is trained, which will be the meta-model. This meta-model is used to make the final predictions. During voting, we first assign weights to each of the models based on their individual predictions. For each data point in the test data, the final prediction is made based on the weighted votes from each of the models. The label, legitimate or phishing, with the majority of the votes becomes the final prediction of the ensemble model.

4 Evaluation Results

Several key metrics were employed to comprehensively evaluate the performance of Cyber Sentinel. The models achieved commendable accuracy on the test set, ranging from 82.3 to 96.5%. XGBoost and logistic regression outperformed the other models with accuracy rates of 96.3 and 96.5%, respectively.

For phishing instances, the F1-score, recall, and precision ranged from 0.90 to 0.99, indicating a strong ability to correctly identify phishing websites. Legitimate instances exhibited high precision, recall, and F1-score, showcasing the model's proficiency in accurately classifying legitimate websites. The macro-average also reflects balanced performance across both classes (Table 1).

Each individual model—logistic regression, random forest, SVM, and XGBoost—underwent a detailed performance analysis (Table 2).

1. Logistic Regression: Demonstrated a solid baseline performance, particularly suitable for scenarios where interpretability is crucial. However, its simplicity might result in lower accuracy compared to more complex models.

Table 1 Performance metrics

	Precision	Recall	F1-score	Support
Phishing	0.96	0.97	0.96	1149
Legitimate	0.97		0.96	1137
Accuracy			0.96	2286
Macro-average	0.96	0.96	0.96	2286
Weighted average	0.96	0.96	0.96	2286

Table 2 Individual model analysis

Index	ML model	Train accuracy	Test accuracy
1	Random forest	0.952	0.939
2	XGBoost	1.0	0.963
3	Support vector machine	0.941	0.923
4	Logistic regression	0.979	0.965

2. Random Forest: Excelled in handling complex datasets, showcasing robustness against overfitting. Its ensemble nature allowed it to capture nuanced relationships within the data.
3. SVM: Showed versatility in capturing both linear and non-linear patterns, making it adept at discerning intricate features in phishing instances.
4. XGBoost: Leveraged gradient-boosting techniques to sequentially improve model accuracy, demonstrating resilience against overfitting and enhancing predictive capabilities.

As for the ensemble learning models, the stacked model performed better than the voting model as the performance of some of the individual models was not commendable. But in terms of computational intensity, the voting method was more effective since a large dataset and four machine learning models were used for this.

5 Conclusion and Future Works

In conclusion, positive outcomes have been obtained from the creation and assessment of Cyber Sentinel, a sophisticated artificial intelligence-based system for phishing website identification. By combining several datasets, carefully preprocessing them, and applying machine learning models like random forest, SVM, XGBoost, and logistic regression, Cyber Sentinel was able to identify possible phishing attacks with increased precision, speed, and effectiveness. Utilizing the individual model strengths through voting and stacking, two ensemble learning techniques produced a synergistic system that outperformed its component parts. A detailed assessment of Cyber Sentinel's performance was provided by the comprehensive evaluation metrics, which included precision, F1-score, recall, and accuracy. These metrics confirmed Cyber Sentinel's efficacy in traversing the complex world of phishing attempts.

While Cyber Sentinel has proven effective in its current form, there are several avenues for future research and enhancement. The model is also not yet effective enough to make accurate real-time predictions. The model needs to be refined more so that the predictions can be made more accurately in less time. While creating a phishing website detector, the most important feature should be being up to date as phishing attack methods are constantly evolving. So, we need to address this issue and come up with a solution so that the model can detect even recently created phishing URLs, like constantly updating the dataset and training the model. We are also planning to make this into a Chrome extension so that it could assist users in real-time detection of phishing links. The extension will be able to generate a warning alert as soon as there is a risk of phishing. In pursuing these future directions, Cyber Sentinel has the potential to evolve into a more robust, adaptive, and comprehensive solution for addressing the multifaceted challenges posed by phishing attacks in the digital era.

References

1. Alkhalil Z et al (2021) Phishing attacks: a recent comprehensive study and a new anatomy. Front Comput Sci 3:563060
2. Alabdan R (2020) Phishing attacks survey: types, vectors, and technical approaches. Future Internet 12(10):168
3. Moghimi M, Varjani AY (2016) New rule-based phishing detection method. Expert Syst Appl 53:231–242
4. Prakash P, Kumar M, Kompella RR, Gupta M (2010) Phishnet: predictive blacklisting to detect phishing attacks. In: 2010 proceedings IEEE INFOCOM, March 2010
5. Gupta BB, Jain AK (2020) Phishing attack detection using a search engine and heuristics-based technique. J Inf Technol Res 13:94–109
6. Rashid J, Mahmood T, Nisar MW, Nazir T (2020) Phishing detection using machine learning technique
7. Tang L, Mahmoud QH (2021) A survey of machine learning-based solutions for phishing website detection. Mach Learn Knowl Extr 3:672–694
8. Kalabarige LR, Rao RS, Abraham A, Gabralla LA (2022) Multilayer stacked ensemble learning model to detect phishing websites. IEEE Access 10:79543–79552
9. Gupta DK, Gandotra E, Mohan Y, Thakur SS (2023) Analysis of ensemble methods for phishing detection. Springer eBooks, pp 85–100. https://doi.org/10.1007/978-3-031-34873-0_4
10. Aksu D, Abdulwakil A, Aydin MA (2017) Detecting phishing websites using support vector machine algorithm. Pressacademia 5(1):139–142. https://doi.org/10.17261/pressacademia.2017.582
11. Niu W, Zhang X, Yang G, Ma Z, Zhuo Z (2017) Phishing emails detection using CS-SVM. IEEE Xplore. https://doi.org/10.1109/ISPA/IUCC.2017.00160, https://ieeexplore.ieee.org/document/8367390
12. Sadaf K (2023) Phishing website detection using XGBoost and Catboost classifiers, February 2023. https://doi.org/10.1109/icsca57840.2023.10087829
13. Do NQ, Selamat A, Krejcar O, Herrera-Viedma E, Fujita H (2022) Deep learning for phishing detection: Taxonomy, current challenges and future directions. IEEE Access 1
14. OpenPhish - OpenPhish Database, openphish.com. https://openphish.com/phishing_database.html
15. PhishTank > Developer Information, phishtank.org. https://phishtank.org/developer_info.php
16. "Phishing Dataset for Machine Learning," www.kaggle.com. Available: https://www.kaggle.com/datasets/shashwatwork/phishing-dataset-for-machine-learning
17. Ariyadasa S, Fernando S, Fernando S (2021) Phishing websites dataset, vol 1. data.mendeley.com, November 2021. https://doi.org/10.17632/n96ncsr5g4.1, https://data.mendeley.com/datasets/n96ncsr5g4#:~:text=The%20dataset%20consists%20of%20a
18. UCI Machine Learning Repository, archive.ics.uci.edu. https://archive.ics.uci.edu/dataset/327/phishing+websites
19. Ardabili S, Mosavi A, Várkonyi-Kóczy AR (2019) Advances in machine learning modeling reviewing hybrid and ensemble methods. Springer International Publishing, International conference on global research and education. Cham

20. Kalabarige LR et al (2022) Multilayer stacked ensemble learning model to detect phishing websites. IEEE Access 10:79543–79552
21. Ghaleb Al-Mekhlafi Z et al (2022) Phishing websites detection by using optimized stacking ensemble model. Comput Syst Sci Eng 41(1):109–125

Network Session Intrusion Detection by Random Forest and Modified TLBO

Yashwant Kumar Kori, Rajesh Kumar Boghey, and Ram Kumar Sahu

Abstract Communication Network of devices resolves different major minor problems. Secure communication increases the reliability and authenticity of the network. Many of researchers proposed different models for network security. This paper has proposed a model for intrusion detection in network. Network nodes' behavior was used for the training of Random Forest (multiple decision trees). Proposed trained model identifies the session class, hence 6 decision trees were developed. In order to reduce the training feature vector modified teacher learning based optimization algorithm was proposed. In this modified TLBO more than one teacher trains and students will learn individually. Experiment was done on real network dataset NSL200. Results shows that proposed model has improved various evaluation parameters.

Keywords Clustering · Genetic algorithm · Intrusion detection · Random forest

1 Introduction

The usage of technology has increased dramatically in the digital era, as has the interdependent nature of different networks and computer systems. On the other hand, as our dependency on technology has grown, so too have malware and breaches of security. Intrusion Detection Systems, or IDSs for short, have become essential tools for stopping illegal access to networks and computer systems as a result of the ubiquity of these threats.

Y. K. Kori (✉) · R. K. Boghey · R. K. Sahu
Department of CSE, TIT Excellence, Bhopal, MP, India
e-mail: yashwantkori123@gmail.com

R. K. Boghey
e-mail: rajeshboghey@gmail.com

R. K. Sahu
e-mail: ramkrsahu@gmail.com

A. P. Agrawal et al. (eds.), *Advances of Artificial Intelligence and Machine Learning in Image Processing, Security and Data Analytics*, Lecture Notes in Electrical Engineering 1464, https://doi.org/10.1007/978-981-95-0148-9_34

Computer networks and systems are now an essential part of our everyday life, having an influence on a variety of industries such as business, healthcare, education, and the armed forces. The increasing reliance on these systems has led to an increase in the possible threats to their security. IDSs are essential for spotting security lapses and bolstering computer system security. They keep a close eye on system logs, network traffic, and other activities and swiftly notify users when they see any unusual activity. Conventional intrusion detection systems (IDSs) use preset rules and signatures to detect malicious activity; however, this approach might overlook real attacks and result in a high rate of false alarms. Moreover, hostile actors are always changing their strategies to avoid detection, which makes the work of intrusion detection systems more difficult [1–4].

A Network Intrusion Detection System (NIDS) holds a crucial role in safeguarding network security by identifying and signaling potential intrusions to the relevant authorities. In terms of detection techniques, NIDS can be categorized into two main types, as outlined in [3]:

Anomaly Detection: This method entails the creation of a reference point for regular network behavior. Any deviations from this established norm are identified as potential intrusions, signaling a potential breach in the network's security.

Misuse Detection: In this method, predefined attack patterns are stored in a database. When network activity matches these attack patterns, it is identified as an attack.

As cybercrime continues to evolve and diversify, the anomaly detection system has proven to be more effective than the misuse detection system for constructing a robust Network Intrusion Detection System. Anomaly detection is better suited to identifying unknown attacks, offering enhanced protection against emerging threats and unauthorized access attempts.

2 Related Work

An innovative Deep Learning (DL)-based intrusion detection solution for Internet of Things devices is presented by Awajan et al. [5]. In order to identify malicious traffic that may start an assault on linked Internet of Things devices, this intelligent system employs a four-layer deep Fully linked (FC) network architecture. To simplify implementation, the suggested system has been designed to be independent of communication protocols. Through the experimental performance analysis, the suggested system shows dependable performance for both simulated and actual invasions. With an average accuracy of 93.74%, it can identify Black hole, Distributed Denial of Service, Opportunistic Service, Sinkhole, and Work hole attacks.

The goal of NishikaGulia et al.'s study in [6] was to manage the attacks by integrating machine learning (ML) technology with the Group-ABC (G-ABC) artificial bee colony (ABC). The G-ABC has been used to determine further simulation findings once the IDS detector has been put into practice. The assessment has been conducted with metrics like F-measure, accuracy, precision, and recall. Numerous attacks have been identified, including viruses, backdoors, denial-of-service (DoS) assaults, user to root (U2R), probing, and root to local (R2L).

A stacking-based ensemble technique that surpasses the general state of the art for NIDS was presented by Ali et al. [7]. A variety of attack scenarios were applied to the network topology built with the help of the Graphical Network Simulator-3 (GNS-3), together with benign user traffic. Using cic flow meter, key flow characteristics are retrieved for every attack and their behavior is assessed. The performance of various machine learning techniques is evaluated after being applied to the characteristics that were taken out of the traffic data.

An ideal framework for an image-based network intrusion detection system was presented by Siddiqi et al. [8]. The framework combines an image modification and enhancement technique with an enhanced feature selection flow. To attain overall efficiency, the suggested structure first minimizes the number of characteristics. The non-image data is then converted to images. The altered photos are further improved using a deep-learning classifier to provide efficient anomaly detection.

An intrusion detection system model that combines many detection methods into a single system is presented by Mora-Mora et al. [9] with the intention of simulating the overall behavior of the applications. The study also suggests a new, updated system named graph that combines and represents the data from many approaches into a unified data structure. The system combines the outcomes of the many detection methods applied in the global model using a deep neural network. The study's findings demonstrate the improvement in detection outcomes over the usage of individual strategies; the suggested model produces reduced false positive rates and greater detection rates.

3 Methodology

This section provides brief of proposed Modified Genetic Algorithm (Teacher Learning Based optimization) and Random Forest (MGARF). Figure 1 illustrates a block diagram of the proposed model, encompassing dataset processing, dimension reduction, and training blocks. A detailed explanation of each block is provided in this section under distinct headings.

Data Cleansing: The initial input data comprises various features, each bearing its own significance. This stage involves the cleaning of data by eliminating extraneous or irrelevant information from the dataset. For instance, in the dataset used for this study, certain fields, such as session ID, connection type, and transferring protocol, were removed to streamline the dataset [10, 11].

$$PND - Dataset\_Cleaning(ND) \tag{1}$$

In Eq. 1, ND represents the NSL dataset, and PND corresponds to the clean dataset matrix. The processed dataset is organized into a matrix with rows representing sessions and columns representing the feature set of each session.

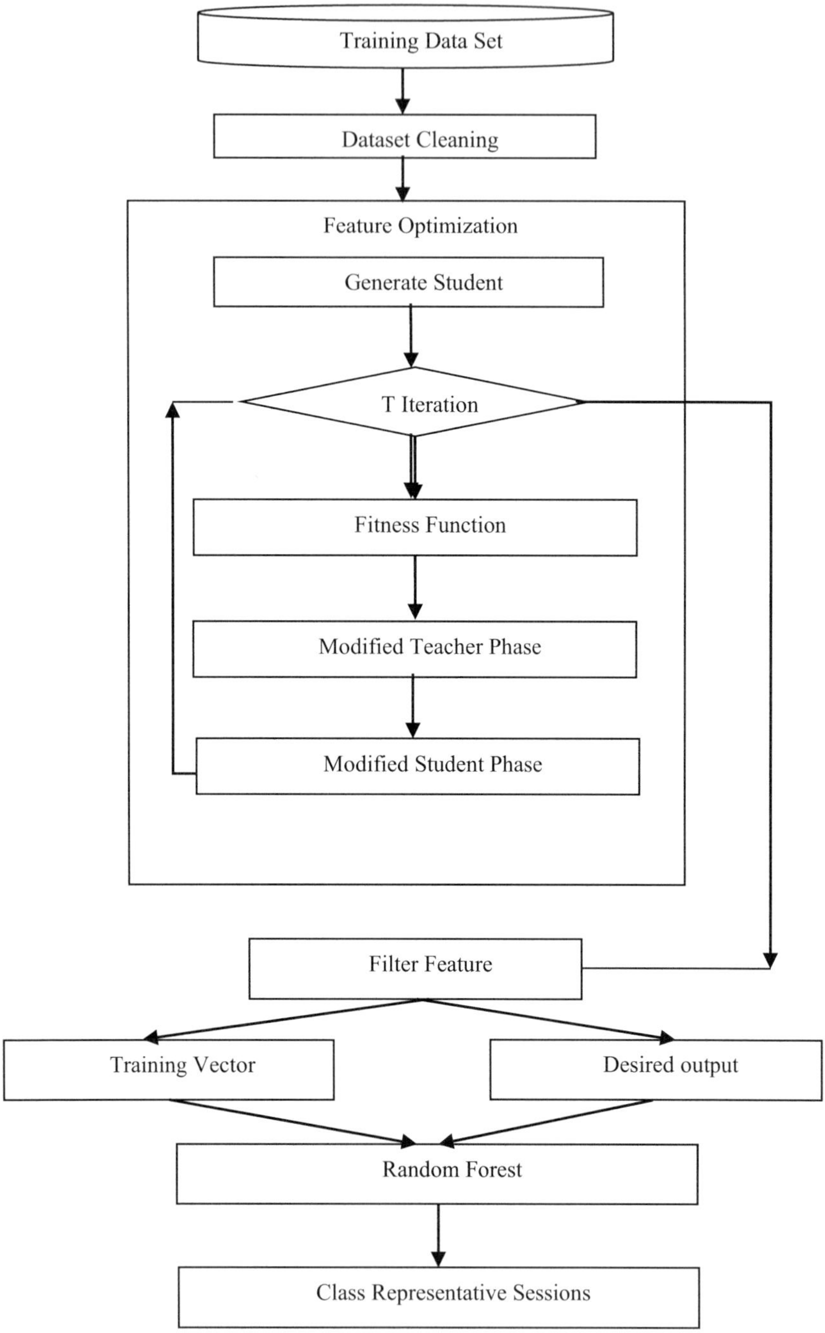

Fig. 1 Block diagram of MGARF network intrusion detection

Optimizing Features: The input CD matrix undergoes further processing through a Modified Genetic Algorithm, based on Teacher Learning, to diminish the values of the training vector and enhance learning accuracy.

Teacher Learning Optimization Algorithm: In this algorithm, each chromosome is regarded as a student. The primary aim of this algorithm is to increase the leaning of class or improving the student intelligence [12, 13].

3.1 Generate Students

In this context, 'Students' represent a group of chromosomes, with each chromosome being a potential solution for an optimized feature set. A Student is essentially a vector composed of 'n' elements, where 'n' signifies the number of columns in CD. Each element in the student vector is a binary value, where '1' indicates that a feature is considered for training, and '0' signifies that the feature is not included in the population. If 'p' denotes the number of Students generated, then the Student population matrix 'S' has dimensions of 'pxn'. The selection of 'f' features in a vector is accomplished by a random value generator function, Gaussian.

$$\text{SP - Student\_population}(p,\ n,\ f) \tag{2}$$

3.2 Fitness Function

To assess the performance of each Student, they are ranked based on a distance metric. The evaluation of this distance is computed as the fitness value. Each Student's feature vector is provided to a Random Forest for cluster representative identification, and the detection accuracy of the work is determined [12]. This detection accuracy value serves as the distance parameter within the algorithm.

3.3 Update Student Position

After obtaining the fitness value 'F' via the fitness function, the feature vectors are sorted in descending order. The best Student is then selected from the available chromosomes within the population.

3.4 Modified Teacher Phase

The success of a Genetic Algorithm relies on the modification of chromosomes. Therefore, based on the parameter 'X', the number of randomly altered positions in the Student feature vectors is adjusted. It's important to note that this operation is not applied to the best local Student [13, 14]. In this step, for each Student, 'X' positions are randomly modified, transitioning from zero to one or from one to zero, in alignment with the feature sets of the best three local Students.

3.5 Modified Student Phase

In this modified step instead of making groups of student for further learning, this model modifies it by mutation operation. In this phase, a student learn things on its own without the help of group best student.

3.6 Update Population

These Students are subsequently subjected to a fitness assessment, comparing them with the parent Students. If a child Student displays superior values, the parent Student is replaced; otherwise, the parent continues. After this stage, if the maximum number of iteration steps is reached, the process proceeds to the feature filtering stage. Otherwise, the fitness value of each Student is evaluated.

3.7 Filter Feature

Once the iterations are completed, the best Student is identified from the last updated population. Features with a value of 'one' in the chromosome are considered selected features for the training vector, while those with 'zero' are regarded as unselected. This stage also involves the creation of the desired output matrix.

3.8 Random Forest Based Cluster Representative

The feature set obtained from the previous algorithm is utilized to identify cluster representatives using a Random Forest (RF) model [15]. This helps in the identification of session classes based on the distance vector from the representative feature set.

3.9 Proposed Algorithm

Input: ND//NSL Dataset.

Output TRF: Trained Random Forest.

- PND-Dataset_ Cleaning (ND)
- SP-Student_ Population (m, n)
- SP-Student_ population (p, n, f)
- Loop 1: iteration
- F-Fitness (SP, PND)
- SP-Teacher (SP, F)
- SP-Student (SP)
- End Loop
- F-Fitness (SP, PND)
- Best-max (F)
- TV-Training_ Vector (SP [Best], PND)
- DO-Desired_ Output (PND)
- End Loop
- TRF-Random_ Forest (TV, DO)

4 Experiment and Results

MATLAB software was used to develop the comparison model and the MGARF model. The experimental system has a sixth-generation Intel Core i3 CPU with 4 GB of RAM. The source of the input–output (IO) dataset was reference [16]. A cloud malicious session detection model developed in previous work [12] was compared with MGARF.

5 Results

Precision value of intrusion detection models shown in Table 1 against different testing dataset. It was found that MGARF has used modified teacher optimization algorithm and improved the training feature set that directly enhanced precious values. Precision values was improved by 0.041% as compared to existing model.

Table 2 shows intrusion detection recall values with different testing dataset sizes. It was found that proposed model MGARF has increases the recall value by 1.866% as compared to existing methods.

Table 3 is inverse average of f-measure and recall. It was found the f-measure got improved by the use of random forest for the testing of intrusion detection. Measure values was improved by 3.05% as compared to existing method.

Table 1 Network security models comparison based on precision for different dataset size

Dataset size	Previous work	MGARF
6000	0.9561	1
8000	0.956	1
12,000	0.9579	0.9992
14,000	0.9575	0.9977
16,000	0.9566	0.997

Table 2 Network security models comparison based on recall for different dataset sizes

Dataset size	Previous work	MGARF
6000	0.977	0.9969
8000	0.977	0.9972
12,000	0.976	0.995
14,000	0.9767	0.9945
16,000	0.9773	0.9933

Table 3 Network security models comparison based on f-measure for different dataset sizes

Dataset size	Previous work	MGARF
6000	0.966	0.9984
8000	0.966	0.9986
12,000	0.967	0.9971
14,000	0.967	0.996
16,000	0.9668	0.9952

Table 4 Network security models comparison based on accuracy for different dataset size

Dataset size	Previous work	MGARF
6000	96.47	99.83
8000	96.49	99.85
12,000	96.55	99.6917
14,000	96.53	99.5857
16,000	96.51	99.4938

Accuracy of intrusion detection models in two class was shown in table 4. It was found that in case of two class where first is normal and other is attack detection accuracy is above 96% in all models. Further it was found that proposed MGARF has improved the feature learning models by modified teacher optimization algorithm. It was found that MGARF has improved the detection by 3.18% as compared to previous model.

MGARF model uses random forest training model has improved the work detection accuracy. This was achieved by the use of optimized feature set as well. Table 5

Table 5 No attack class detection accuracy of intrusion detection comparing models

Dataset size	Previous work	MGARF
6000	95.614	100
8000	95.6134	100
12,000	95.78	99.9214
14,000	95.7544	99.7716
16,000	95.66	99.7177

Table 6 DOS class detection accuracy of intrusion detection comparing models

Dataset size	Previous work	MGARF
6000	97.785	99.954
8000	97.7058	99.9663
12,000	97.798	99.8427
14,000	97.76	99.8456
16,000	97.8	99.797

Table 7 Probe class detection accuracy of intrusion detection comparing models

Dataset size	Previous work	MGARF
6000	81.985	99.4485
8000	82.2946	99.2918
12,000	82.35	98.6213
14,000	82.23	98.486
16,000	82.66	98.259

shows that MGARF model has improved the no attack class detection accuracy by 4.19% as compared to previous model.

Denial of Service attack different set of dataset results were shown in Table 6. Table 6 shows that MGARF model has improved the DOS class detection accuracy by 2.11% as compared to previous model.

MGARF model uses random forest training model has improved the work probe detection accuracy. This was achieved by the use of optimized feature set as well. Table 7 shows that MGARF model has improved the probe attack class detection accuracy by 16.51% as compared to previous model.

R2L attack different set of dataset results were shown in Table 8. Table 8 shows that MGARF model has improved the R2Lclass detection accuracy by 56.5% as compared to previous model.

Table 9 shows intrusion detection class detection accuracy of comparing models on different testing dataset sizes.

Table 8 R2L class detection accuracy of intrusion detection comparing models

Dataset size	Previous work	MGARF
6000	36.7347	89.79
8000	36.9231	92.3077
12,000	33.33	91.6667
14,000	29.914	89.7436
16,000	32.28	88.189

Table 9 U2R class detection accuracy of intrusion detection comparing models

Dataset size	Previous work	MGARF
6000	50	50
8000	50	50
12,000	40	25
14,000	40	25
16,000	42.857	14.2857

6 Conclusion

This work proposed Modified Genetic Algorithm (Teacher Learning Based optimization) and Random Forest (MGARF). In order to reduce the training feature vector modified teacher learning based optimization algorithm was proposed. In this modified TLBO more than one teacher trains and students will learn individually. Using the RF model, the feature set that was acquired from the previously mentioned method was utilized to identify the cluster representative. Experiment was done on real dataset of different class of attacks. Results show that Measure values were improved by 3.05% as compared to existing method. DOS detection accuracy value by 2.11% as compared to existing methods. Similarly R2L increased by 6.25%. MGARF model has improved the probe attack class detection accuracy by 16.51% as compared to previous model. In future researcher can apply model in other sets of networks.

References

1. Rupa Devi T, Badugu S (2020) A review on network intrusion detection system using machine learning. Springer International Publishing
2. Thakkar A, Lohiya R (2021) A review on machine learning and deep learning perspectives of IDS for IoT: recent updates, security issues, and challenges, vol 28, no 4. Springer Netherlands
3. Parkar P, Bilimoria A (2021) A survey on cyber security IDS using ML methods. In: Proceedings of the 5th international conference on intelligent computing and control systems ICICCS 2021, pp. 352–360

4. Shukla P (2018) ML-IDS: a machine learning approach to detect wormhole attacks in Internet of Things. In: 2017 intelligent systems conference IntelliSys 2017, January 2018–September 2018, pp 234–240. https://doi.org/10.1109/IntelliSys.2017.8324298
5. Awajan A (2023) A novel deep learning-based intrusion detection system for IoT networks. Computers 12:34
6. Gulia N, Solanki K, Dalal S, Dhankhar A, Dahiya O, Salmaan NU (2023) Intrusion detection system using the G-ABC with deep neural network in cloud environment. Sci Program
7. Ali M, Haque M, Durad MH et al (2023) Effective network intrusion detection using stacking-based ensemble approach. Int J Inf Secur 22:1781–1798
8. Siddiqi MA, Pak W (2022) Tier-based optimization for synthesized network intrusion detection system. IEEE Access 10:108530–108544
9. Mora-Mora H, Mora-Gimeno FJ, Volckaert B, Atrey A (2021) Intrusion detection system based on integrated system calls graph and neural networks. IEEE Access 9:9822–9833
10. Mahdavinejad MS, Rezvan M, Barekatain M, Adibi P, Barnaghi P, Sheth AP (2018) Machine learning for Internet of Things data analysis: a survey. Digit Commun Netw Science Direct 4(3):161–175
11. Liu J, Yang D, Lian M, Li M (2021) Research on intrusion detection based on particle swarm optimization in IoT. IEEE Access 9:38254–38268
12. Ghanem WAHM et al (2022) Cyber intrusion detection system based on a multiobjective binary bat algorithm for feature selection and enhanced bat algorithm for parameter optimization in neural networks. IEEE Access 10:76318–76339
13. Puralachetty MM, Pamula VK, Gondela LM, Akula VNB (2016) Teaching-learning-based optimization with two-stage initialization. In: 2016 IEEE students' conference on electrical, electronics and computer science (SCEECS)
14. Venkata Rao R, Patel V (2013) An improved teaching-learning-based optimization algorithm for solving unconstrained optimization problems. Sci Iran 20(3)
15. Savargiv M, Masoumi B, Keyvanpour MR (2021) A new random forest algorithm based on learning automata. Comput Intell Neurosci 2021, Article ID 5572781, 19 pp
16. https://www.unb.ca/cic/datasets/nsl.html

Navigating Complex Texts with Semantic Extraction and Condensing

R V Rahul Krishna, Boppana Vegesh Sai, Puttamraju Sai Manaswi, and Arpita Gupta

Abstract In India, dealing with legal documents can be quite a chaotic affair, often filled with complexity and disorder. The existing records within the Indian judiciary are no exception to this rule (Parikh, V, Mathur, V, Mehta, P, Mittal, N, Majumder: Lawsum: A weakly-supervised approach for Indian legal document summarization, P. (2021)). Compounding the issue is the absence of a readily available dataset that provides concise summaries of these legal documents. To tackle these challenges head-on, a comprehensive method is proposed to navigate through these intricate texts. The approach involves using semantic extraction techniques coupled with a condensed summarization method specifically tailored for legal documents. The goal is simple: to bring order to the disorderly nature of Indian legal records and create a structured resource that can benefit legal professionals and researchers alike. The method doesn't aim to overhaul the system; instead, it seeks to extract and preserve the essential semantic elements from these legal documents. The idea is to analyze legal texts more efficiently by focusing on what truly matters. Ultimately, the strive is to provide a practical solution for those who grapple with the complexities of Indian legal documentation.

Keywords Natural language processing · Image processing · Legal document summarization · Document analysis · Machine learning

1 Introduction

In today's world, where information is overflowing, summarization has become a crucial tool. It's like the magic wand that transforms extensive and complex content into bite-sized, easy-to-digest forms, all while keeping the essence intact [2, 3].

R. V. Rahul Krishna (✉) · B. V. Sai · P. S. Manaswi · A. Gupta
Department of Computer Science and Engineering, Koneru Lakshmaiah Education Foundation, Hyderabad, Telangana, India
e-mail: rahul.rangavazzala@gmail.com

A. P. Agrawal et al. (eds.), *Advances of Artificial Intelligence and Machine Learning in Image Processing, Security and Data Analytics*, Lecture Notes in Electrical Engineering 1464, https://doi.org/10.1007/978-981-95-0148-9_35

Presently, the generation of legal document summaries entails a substantial allocation of human resources, alongside the abundance of online information within the legal domain, which underscores the significance of legal text processing as a pivotal area of research [4]. This process is characterized by labor-intensive procedures, prolonged time requirements, and significant associated costs [5]. This paper takes a deep dive into the world of unraveling complicated texts, shining a spotlight on the innovative technique of semantic extraction and condensing legal document summarization. In a landscape flooded with textual data from various domains, the ability to quickly pull out the relevant information has become essential. Summarization isn't just a fancy trick; it plays a crucial role in diverse fields like journalism, academia, business, and tech-driven realms like natural language processing and machine learning.

Navigating the intricate world of legal documents is no small feat, and that's where legal document summarization comes into play. It's like the superhero of information processing, transforming extensive and complex legal texts into simplified yet comprehensive formats while preserving the crucial data needed for making informed decisions [6]. Legal documents, often riddled with complex jargon, specialized terms, and references to intricate laws, pose a challenge even for seasoned legal professionals. The mission? To distill these documents into clear and understandable forms without compromising their substance or meaning—a significant undertaking.

The need for legal document summaries has become increasingly evident for several reasons. First and foremost, time is of the essence. Reading and dissecting legal documents thoroughly demands a substantial investment of time and effort. Legal practitioners are on the lookout for streamlined methods to quickly access crucial insights without having to navigate the entire document. Second, there's a noteworthy potential for cost-cutting. Reduced reading time translates into substantial savings in the expenditure associated with legal services, fostering competition among law firms and, ultimately, leading to cost savings for clients.

Legal document summarization isn't just about efficiency and cost; it holds the promise of enhancing accuracy in legal conclusions. By ensuring the inclusion of vital information, these summaries play a pivotal role in scenarios where deciphering legal language can be intricate and perplexing. Accessibility is another key dimension. Legal documents often employ intricate jargon that can be challenging for non-legal professionals to grasp. Summarization bridges this gap, making legal information more comprehensible and applicable to a wider audience. Moreover, the compliance facet underscores the importance of summarized legal documents in ensuring adherence to the intricate web of rules and regulations, especially in sectors like finance and healthcare, where regulatory alignment is imperative.

In this paper, a novel approach is introduced that combines semantic extraction techniques with a condensed legal document summarization framework. Through a comprehensive exploration of this approach, the aim is to showcase its effectiveness in addressing the inherent complexities of legal texts. The goal is to contribute to streamlined information retrieval, steadfast compliance, improved accuracy, and accessibility.

2 Related Work

Text summarization is gaining a lot of attention these days, and for good reason—it's like a shortcut for readers navigating through dense books and grappling with complex subjects. In the realm of legal scenarios, researchers have been exploring various methods to pull out crucial information and trim down lengthy documents. Back in the day, early information extraction systems were all about rules [7–10], and image processing techniques were also made use of by scanning the hardcopy document and generating its corresponding computer-readable softcopy version of the file [11]. They're still being studied and tweaked [12–16] to tackle the real-world challenges faced by extraction systems. Multi-task Deep Learning techniques were implemented for legal document translation, summarization, and multi-label classification, for use-case-specific objectives.

Below is a quick overview of some ongoing research works sharing the same goals as our approach of condensing legal document summaries and extracting semantics.

2.1 Legal Document Summarization Techniques

In the pursuit of creating concise summaries for legal materials, previous studies have explored various strategies. For instance, a methodology that relied on linguistic patterns and subject-matter expertise to generate coherent summaries of legal documents [17] emphasized the crucial task of identifying key terms and relationships within the legal papers to simplify the summarization process. Building on a similar thread, there was a potential of automated methods, employing natural language processing to craft succinct summaries of court case materials [18]. Generic techniques using neural networks for the summarization task for Indian legal judgment documents were applied, wherein two neural network architectures for the corresponding task were implemented by utilizing the word and the sentence embeddings for capturing the semantics [19]. These endeavors underscore the significance of both language analysis and subject-matter knowledge in pinpointing crucial information within the complex realm of legal texts [20].

2.2 Extractive Summarization Approaches

Our approach zeroes in on extractive summarization, and it's been quite the talk of the town in research circles. Back in 2004, Erkan and Radev crafted an approach that as-signs significance ratings to phrases within the text, strategically picking out the most instructive statements for summarization [21]. Taking cues from this, Mihalcea and Tarau (2004) delved into the world of graph-based algorithms, aiming to spotlight key phrases in legal texts for effective summarization [22]. These extractive methods lay the foundation for our methodology, which is all about extracting those vital semantic components from the extensive legal documents. Bayesian Optimization-based Score Fusion of Linguistic Elements was employed for scoring essential sentences in legal document summarization, involving a fusion of various statistical and semantic approaches that focus on extractive approaches [23].

2.3 Abstractive Summarization and Paraphrasing

Even though extractive approaches have been quite popular, abstractive summarization has also caught the attention of researchers. Take Zajic and the team in 2007, for instance—they delved into the challenge of paraphrasing legal writings to make them more condensed and understandable [24]. Their work puts a spotlight on the importance of maintaining the meaning while reorganizing text, aligning with our goal of streamlining legal documents without compromising their integrity. It's like a delicate dance of condensing without losing the essence.

2.4 Semantic Analysis in Legal Summarization

Summarizing legal documents has evolved, and now it often involves diving into semantic analysis. In 2015, Chung and colleagues brought forth a semantic-based approach for summarizing court cases [25]. Their method included ontologies and semantic linkages to uncover essential concepts within legal documents. This focus on semantic comprehension resonates with the core principle of our method—to utilize semantic extraction for enhancing the precision and coherence of summarization. It's all about digging into the semantics to bring out the essence of legal content. The sentence scoring approach, termed DCESumm, encompasses both supervised sentence-level summary relevance prediction and unsupervised clustering-based document-level score enhancement, as well as a few other methodologies that were implemented to precisely delineate the scoring of sentences [26].

2.5 Domain-Specific Summarization

Legal texts are a breed of their own, demanding specialized methods for effective summarization. Giannopoulou and team (2014) recognized this and devised a system tailored for summarizing patents—a domain with its share of nuanced language and intricate material [27]. Their approach, employing domain-specific patterns and linguistic analysis, aligns with our focus on semantic extraction from legal documents. It underscores the importance of developing procedures that navigate the subtleties of legal language. While enormous models in natural language processing showcase impressive performance across various tasks, some researchers argue that these models might lack a true understanding of language and may not be robust enough for critical domains [28–32]. Current research in text summarization, especially in the legal realm, provides valuable insights into approaches and strategies for navigating complex texts. Efficiency, accuracy, and accessibility in understanding intricate legal papers are at the core of our proposed approach, which combines semantic extraction with condensed summaries of legal materials. It's about finding that sweet spot where technology meets the unique challenges of legal language.

3 Proposed Model

Introducing DELSumm (Delete and Retrieve Summarization), an algorithm designed for extractive summarization of legal case documents. Leveraging an optimization framework, DELSumm integrates domain knowledge into an objective function with constraints, employing Integer Linear Programming (ILP) for maximization. The algorithm takes a case document with labeled rhetorical segments and the desired summary length L as inputs, producing a summary of at most L words that represent each segment.

The model is an unsupervised technique finely tuned to navigate the unique complexities found in legal documents. By seamlessly blending synergistic semantic extraction and condensing procedures, DELSumm takes on the challenge of deciphering intricate texts in the legal arena. It's an innovative project leveraging cutting-edge methodology to craft comprehensive and enlightening summaries of legal case studies (Table 1).

3.1 Input and Rhetorical Function Labeling

Here, a case document becomes the starting point for the summarization process. Here, every sentence in the case file gets tagged with its appropriate rhetorical function. Categories like Facts, Issues, Final Judgment, and Arguments are applied to

Table 1 Raw legal text versus normalized text

S. no.	Raw legal text	Normalized text
1	"...petition filed by the appellant under Section 13 of the Hindu Marriage Act, 1955 on grounds of separation."	"...petition filed by the appellant under Section 13 (for divorce) of the Hindu Marriage Act, 1955 on grounds of separation."
2	"...they approached the High Court of Kerala by way of W.P. (C) No. 2329 of 2014."	"...they approached the High Court of Kerala by way of Writ Petition under 226 and 227 of the Constitution No. 2329 of 2014."

each sentence, providing a detailed categorization of rhetorical functions that lays down the groundwork for a summary that's concise and focused.

3.2 Guidelines for Representation and Statute Frequency

The proposed mode incorporates specific criteria that outline the ideal representation for each of these rhetorical elements within the summary. These criteria are carefully crafted based on domain-specific knowledge, ensuring that the final summary encapsulates the essential elements of the case. The summarization process is about condensing and distilling the core aspects that matter in the legal context. It also takes into consideration the expected frequency of statute terms relative to content words. This keen awareness of the unique language patterns in legal writings ensures that the summarization process is finely tuned to the intricacies of legal language.

3.3 Internal Optimization Through Integer Linear Programming (ILP)

Integer linear programming (ILP) based optimization is the core strategy used in DELSumm. It aids the model in creating a coherent and condensed summary that aligns with the specified requirements for each rhetorical function. The method meticulously selects and merges sentences from the input case document. Through the power of ILP, it not only boosts the informativeness of the summary but does so while staying true to the defined constraints.

ILP formulation is designed to enhance the informativeness of sentences and leverage domain-specific vocabulary, such as terms from a legal dictionary. The objective function within the model aims to maximize the presence of both informative sentences and crucial content words. The formulation includes constraints addressing summary length, content word inclusion, sentence, and content word relationships, along with ensuring a minimum sentence inclusion for each segment.

To solve the ILP, any suitable solver, like the GUROBI optimizer, can be applied. Notably, the model incorporates a mechanism to handle redundancy. In cases where two sentences exhibit similarity, this method avoids including both in the summary, effectively eliminating redundancy and producing a more refined and concise summary (Fig. 1).

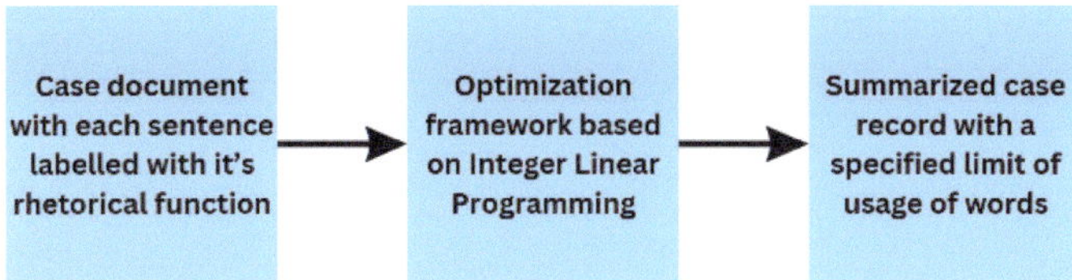

Fig. 1 A holistic approach to the DELSumm model

3.4 Delete and Retrieve Strategy

The "Delete and Retrieve" tactic forms the foundation of DELSumm's distinctive methodology. It works by eliminating particular words or phrases from the input text, which reduces the text's substance to its essential elements. The remaining words are then used to construct a summary using a language model. With the help of this method, DELSumm may draw attention to the most important details, producing summaries that effectively capture the key points of the case document.

3.5 Preservation of Meaning and Versatility

The delete and retrieve strategy allows DELSumm to delicately trim down the content while keeping a tight grip on the essence, ensuring that the summary is both succinct and true to the original meaning. It is a linguistic balancing act, maintaining clarity without losing the nuance.

3.6 Broad Applicability and Beyond Summarization

Beyond the usual summarizing tasks, DELSumm also showcases remarkable performance in a range of natural language processing applications, including sentiment transfer and style transfer, highlighting its adaptability and versatility. It incorporates the Delete and Retrieve strategy and is guided by rhetorical function labeling and ILP optimization. The model produces insightful summaries while preserving the essential elements of the original text, making it a valuable asset in the field of legal documentation by performing the steps as shown in Fig. 2. Distilling legal content, the model enhances understanding and application.

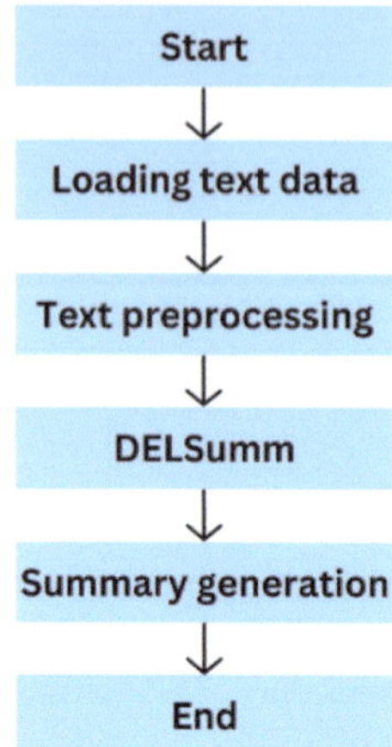

Fig. 2 Steps involved in the proposed model

4 Results and Discussion

During the methodical evaluation, a systematic comparison was conducted between assessments from legal experts, Law Expert A1 and Law Expert A2, with outcomes generated by the DELSumm model. This rigorous assessment focused on the in-depth analysis of three legal documents—Sample_1 (1953_L_1), Sample_2 (1953_S_23), and Sample_3 (1954_M_25) as shown in Table 2.

The meticulous examination revealed compelling evidence of DELSumm's consistent and highly accurate performance in generating summaries aligned with evaluations by subject-matter legal experts. The precision exhibited underscores the model's potential as a reliable tool for distilling complex legal documents. DELSumm emerges as an invaluable companion in legal analysis, presenting substantial advancements in the field. This research not only contributes meaningful insights but also introduces a practical, reliable solution for enhancing the summarization of intricate legal content within a formalized methodology. Figure 3 depicts a visual representation of the comparative performance of the three sample documents.

Table 2 Predicted result analysis using term frequency—inverse document frequency

Sample file	Law expert A1	Law expert A2	DelSumm model
Sample_1 (1953_L_1)	0.9810029223645412 ≈ **98.10%**	0.9807641110449942 ≈ **98.07%**	0.9649209828877827 ≈ **96.49%**
Sample_2 (1953_S_23)	0.9598236304044414 ≈ **95.98%**	0.9642714020590469 ≈ **96.42%**	0.9830001708578402 ≈ **98.30%**
Sample_3 (1954_M_25)	0.9771921118343009 ≈ **97.71%**	0.977304069518567 ≈ **97.73%**	0.9717560854639198 ≈ **97.17%**

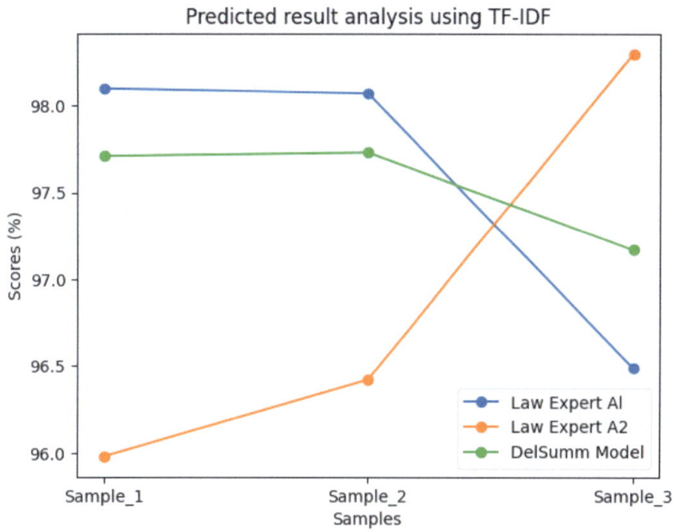

Fig. 3 Line graph depicting the result analysis performed using TF-IDF

5 Conclusion

In this study, DELSumm is proposed as an unsupervised algorithm that systematically incorporates domain knowledge for the extractive summarization of legal case documents. The strengths of this approach are multifaceted. It systematically encodes the necessary domain knowledge for legal document summarization into a computational framework. Despite being an unsupervised approach, it demonstrates performance comparable to supervised learning models trained on extensive datasets. The model achieves a balanced representation from all rhetorical segments in the generated summary, addressing a significant gap in previous approaches. Therefore, the proposed model is characterized by its flexibility and generalizability, enabling the summarization of documents from various jurisdictions. The observed capabilities of DELSumm suggest a potential revolution in the analysis of legal documents, serving as a robust tool for practitioners and legal experts to navigate the intricate landscape of legal texts with efficiency. Further, there are plans to apply DELSumm to documents from different jurisdictions and generate diverse types of summaries for various stakeholders to analyze its performance.

References

1. Parikh V, Mathur V, Mehta P, Mittal N, Majumder P (2021) Lawsum: a weakly-supervised approach for indian legal document summarization
2. Popescu MC, Grama L, Rusu C (2020) On the use of positive definite symmetric kernels for summary extraction. In: 13th international conference on communications (COMM), June 2020. IEEE, pp 335–340
3. Ghosh S, Dutta M, Das T (2022) Indian legal text summarization: a text normalization-based approach. In: 19th India council international conference (INDICON), November 2022. IEEE, pp 1–4
4. Kanapala A, Pal S, Pamula R (2019) Text summarization from legal documents: a survey. Artif Intell Rev 51:371–402
5. Jain D, Borah MD, Biswas A (2021) Summarization of legal documents: where are we now and the way forward. Comput Sci Rev 40:100388
6. Bhattacharya P, Hiware K, Rajgaria S, Pochhi N, Ghosh K, Ghosh S (2019) A comparative study of summarization algorithms applied to legal case judgments. In: Advances in information retrieval: 41st European conference on ir research (ECIR), proceedings, Part I 41, April 14–18. Springer International Publishing, Cologne, Germany, pp 413–428
7. Appelt DE, Hobbs JR, Bear J, Israel D, Tyson M (1993) FASTUS: a finite-state processor for information extraction from real-world text. IJCAI 93:1172–1178
8. Cunningham H (2002) GATE: a framework and graphical development environment for robust NLP tools and applications. In: Proceedings of 40th annual meeting of the association for computational linguistics (ACL 2002), pp 168–175
9. Shaalan K, Raza H (2008) Arabic named entity recognition from diverse text types. In: Advances in natural language processing: 6th international conference, GoTAL 2008 Gothenburg, Sweden, 25–27 August 2008, proceedings. Springer, Berlin Heidelberg, pp 440–451
10. Riloff E (1993) Automatically constructing a dictionary for information extraction tasks. AAAI 1(1):2-1
11. Sil R, Roy A, Dasmahapatra M, Dhali D (2021) An intelligent approach for automated argument based legal text recognition and summarization using machine learning. J Intell & Fuzzy Syst 41(5):5457–5466
12. Ciravegna D.F (2001) Adaptive information extraction from text by rule induction and generalization. In: Proceedings of the 17th international joint conference on artificial intelligence (IJCAI2001)
13. Jayram TS, Krishnamurthy R, Raghavan S, Vaithyanathan S, Zhu H (2006) Avatar information extraction system. IEEE Data Eng Bull 29(1):40–48
14. Muslea I (1999) Extraction patterns for information extraction tasks: a survey. In: The AAAI-99 workshop on machine learning for information extraction, vol 2(2), July 1999
15. Shen W, Doan A, Naughton JF, Ramakrishnan R (2007) Declarative information extraction using datalog with embedded extraction predicates. VLDB 7:1033–1044
16. Turmo J, Ageno A, Catala N (2006) Adaptive information extraction. ACM Comput Surv (CSUR) 38(2):4-es
17. Moens MF (2001) Innovative techniques for legal text retrieval. Artif Intell Law 9:29–57. Springer
18. Duan X, Zhang Y, Yuan L, Zhou X, Liu X, Wang T, Wang R, Zhang Q, Sun C, Wu F (2019) Legal summarization for multi-role debate dialogue via controversy focus mining and multi-task learning. In: 28th ACM international conference on information and knowledge management, November 2019, pp 1361–1370
19. Anand D, Wagh R (2022) Effective deep learning approaches for summarization of legal texts. J King Saud Univ-Comput Inf Sci 34(5):2141–2150
20. Elnaggar A, Gebendorfer C, Glaser I, Matthes F (2018) Multi-task deep learning for legal document translation, summarization and multi-label classification. In: Proceedings of the 2018 artificial intelligence and cloud computing conference, December 2018, pp 9–15

21. Erkan G, Radev DR (2004) Lexrank: Graph-based lexical centrality as salience in text summarization. J Artif Intell Res (JAIR) 22:457–479
22. Mihalcea R, Tarau P (2004) Textrank: bringing order into text. In: Proceedings of the 2004 conference on empirical methods in natural language processing, July 2004, pp 404–411
23. Jain D, Borah MD, Biswas A (2023) Bayesian optimization based score fusion of linguistic approaches for improving legal document summarization. Knowl-Based Syst 264:110336
24. Liang P, Potts C (2015) Bringing machine learning and compositional semantics together. Annu Rev Linguist 1(1):355–376
25. Ma Y, Zhang P, Ma J (2018) An ontology driven knowledge block summarization approach for Chinese judgment document classification. IEEE Access 6:71327–71338
26. Jain D, Borah MD, Biswas A (2023) A sentence is known by the company it keeps: improving legal document summarization using deep clustering. Artif Intell Law 1–36
27. Liu Z, Zhang J, Qin T, Qu Y, Li Y (2022) One-to-many comparative summarization for patents. Scientometrics 127(4):1969–1993. Springer
28. Marcus G, Davis E (2019) Rebooting AI: building artificial intelligence we can trust. Vintage
29. Marcus G (2020) The next decade in AI: four steps towards robust artificial intelligence. arXiv: 2002.06177
30. Nie Y, Williams A, Dinan E, Bansal M, Weston J, Kiela D (2019) Adversarial NLI: a new benchmark for natural language understanding. arXiv:1910.14599
31. Jin D, Jin Z, Zhou JT, Szolovits P (2020) Is bert really robust? a strong baseline for natural language attack on text classification and entailment. In: Proceedings of the AAAI conference on artificial intelligence, vol 34(05), pp 8018–8025, April 2020
32. Liu X, Cheng H, He P, Chen W, Wang Y, Poon H, Gao J (2020) Adversarial training for large neural language models. arXiv:2004.08994

MIX
Papier aus verantwortungsvollen Quellen
Paper from responsible sources
FSC® C105338
FSC
www.fsc.org

If you have any concerns about our products,
you can contact us on
ProductSafety@springernature.com

In case Publisher is established outside the EU,
the EU authorized representative is:
Springer Nature Customer Service Center GmbH
Europaplatz 3, 69115 Heidelberg, Germany

Printed by Libri Plureos GmbH
in Hamburg, Germany